SEISMIC RETROFITTING: LEARNING FROM VERNACULAR ARCHITECTURE

SEISMIC-V: VERNACULAR SEISMIC CULTURE IN PORTUGAL
RESEARCH PROJECT FUNDED UNDER THE NATIONAL RESEARCH AGENCY FCT

Seismic Retrofitting: Learning from Vernacular Architecture

Editors

Mariana R. Correia
CI-ESG, Escola Superior Gallaecia, Vila Nova de Cerveira, Portugal

Paulo B. Lourenço
ISISE, University of Minho, Faculty of Engineering, Guimarães, Portugal

Humberto Varum
CONSTRUCT-LESE, Faculty of Engineering, University of Porto, Porto, Portugal

CRC Press
Taylor & Francis Group
Boca Raton London New York Leiden

CRC Press is an imprint of the
Taylor & Francis Group, an **informa** business
A BALKEMA BOOK

Cover drawing designed at CI-ESG
Research Centre at Escola Superior Gallaecia
[Centro de Investigação da Escola Superior Gallaecia]

CRC Press/Balkema is an imprint of the Taylor & Francis Group, an informa business

Typeset by MPS Limited, Chennai, India
Printed and bound in Great Britain by CPI Group (UK) Ltd, Croydon, CR0 4YY

Published by: CRC Press/Balkema
P.O. Box 11320, 2301 EH Leiden, The Netherlands
e-mail: Pub.NL@taylorandfrancis.com
www.crcpress.com – www.taylorandfrancis.com

ISBN: 978-1-138-02892-0 (Hbk + CD-ROM)
ISBN: 978-1-315-64739-5 (eBook PDF)

Seismic Retrofitting: Learning from Vernacular Architecture – Correia, Lourenço & Varum (Eds)
© 2015 Taylor & Francis Group, London, ISBN 978-1-138-02892-0

Table of contents

Seismic Retrofitting: Learning from Vernacular Architecture – Correia, Lourenço & Varum (Eds)
© 2015 Taylor & Francis Group, London, ISBN 978-1-138-02892-0

Preface

Local communities have adapted for centuries to challenging surroundings, resulting from unforeseen natural hazards. Vernacular architecture reveals often very intelligent responses when adjusting to the environment. So, the questions were: How did local populations prepare their dwellings to face frequent earthquakes? How could seismic retrofitting perseverance be identified in vernacular architecture? It was to respond to this gap in knowledge, that the research project 'Seismic-V: Vernacular Seismic Culture in Portugal' was submitted for approval by the Portuguese National Research Agency FCT. The Foundation for Science and Technology validated the project with an excellent evaluation and funding. The jury panel enhanced the project's outcomes, as an important contribution for the population's safety.

The research project was coordinated by Escola Superior Gallaecia, as project leader, and by the Departments of Civil Engineering at the University of Minho and the University of Aveiro, as partners. Relevant findings and project results were accomplished thanks to a consistent cross-collaboration between the three institutions, which addressed a complementary expertise within the research project.

The fundamental contribution and aims of this publication were to enhance the disciplinary interest in vernacular architecture and its contribution to risk mitigation in responding to Natural Hazards; to encourage academic and scientific research collaboration among different disciplines, while contributing to the improvement of the vernacular architecture, which more than half of the world's population, still inhabits nowadays.

This publication is structured in 6 parts: the first is dedicated to the framework of the research; the second part concerns Local Seismic Culture (LSC) around the world; the third part focusses on the identification of LSC in Portugal; the fourth part is devoted to the LSC assessment by regions, the fifth part concerns the typology performance study related to 3 identified housing typologies; and finally the sixth part, closing the publication, concerns the conclusions of the project and its recommendations.

The emerged findings brought consistent and systematic outcomes, reaching different publics, through different publications and the project's website. The entailed research methodology also emerged as a result of the project, as it could be extrapolated and applied to other contexts, creating further findings. The research revealed the existence of a local seismic culture, in terms of reactive or preventive seismic resistant measures, able to survive, in areas with frequent earthquakes, if properly maintained.

'Seismic retrofitting: learning from vernacular architecture' brings together 43 chapters with new perspectives on seismic retrofitting techniques and relevant data addressing vernacular architecture, an amazing source of knowledge still relevant, in the present world. The publication gathered the contributions of international researchers and experts, invited as key-references in the disciplinary field. 50 authors presented case studies from Latin America, the Mediterranean, eastern Asia and the Himalayas region. There are references to examples from at least 18 countries, on 4 continents. This is the case of Algeria, Bolivia, Bhutan, Chile, China, Egypt, El Salvador, Greece, Haiti, Italy, Japan, Mexico, Morocco, Nepal, Nicaragua, Peru, Romania, Taiwan, and a closer detailed analysis of Portugal.

The research project and this publication were possible thanks to the funding granted by FCT – Foundation for Science and Technology, in the framework of the Portuguese research project Seismic-V (PTDC/ATP-AQI/3934/2012), Scientific Research Projects and Technological Development Program. The research project received the Aegis of the Chair UNESCO – Earthen Architecture | ICOMOS – CIAV | ICOMOS-ISCEAH | PROTERRA Iberian Network and the Institutional support provided by UNIVEUR-Ravello, Italy, and the DRCN – Northern Portugal Regional Directorate for Culture.

To all the authors, collaborators, and consultants that contributed to the research project and to this publication, with quality, consistency and high standards, thank you.

Mariana R. Correia, Paulo B. Lourenço, Humberto Varum
Editors of the publication, July 2015

 ISBN 978-1-138-02892-0

Opening remarks

The Escola Superior Gallaecia (ESG) is a university school in architecture and urban design, art and multimedia, located in the north of Portugal. The school is a member of the *UNESCO Chair-Earthen Architecture and Sustainable Development*, since 2005 – one of five European institutions, and of 44 institutions worldwide. Due to achieved results, in 2012, UNESCO Chair agreement with ESG was renewed. This brought a new reflection on the school's scientific research and cooperation, with R&DT having a significant contribution for Gallaecia's strategy and scientific impact.

CI-ESG, Research Centre at ESG (http://www.esg.pt/ciesg/) was then created with 3 main broad scientific research areas: Architecture & Heritage; Urban Design, Territory & Landscape; and Arts & Design. Each includes different fields of study and expertise. In Architecture & Heritage, four specific research areas were identified: Earthen Architecture, Sustainable Architecture, Vernacular Heritage and Military Heritage. Scientific research is addressed through formal projects integrated in financed programmes, but also through consultancy to regional Portuguese authorities, as well as Galician entities, in Spain.

Regarding research projects, the school has submitted and won several projects, funded by national and international research programmes: *One National project*: CATPAP – Architectural and Landscape Heritage Catalogue of Alto Minho region (2004–2006) (http://esgallaecia.inwebonline.net); *Two Iberian project*: CADIVAFOR – Cataloguing, Digitalization and Return of Value to the Defensive Fortresses of the frontier Galiza-North Portugal (2006–2007) (www.cieform.org); Natura Minho-Miño: digital database of the region's landscape (2008–2013); *Three European projects*: "Houses and Cities Built with Earth: Conservation, Significance and Contribution to Urban Quality" (2005–2006); "Terra Incognita – Conservation of European Earthen Architecture" (2006–2007); "Terra Europae – Earthen Architecture in Europe" (2009–2011) (http://culture-terra-incognita.org).

In 2012, two projects were approved with ESG, as project leader: The European project "VerSus – Lessons from Vernacular Heritage to Sustainable Architecture" (2012–2014) (www.esg.pt/versus) and the National FCT scientific project "SEISMIC-V – Vernacular seismic culture in Portugal" (www.esg.pt/ciesg). This was of major importance as it acknowledged the school's expertise to lead projects and to establish new front lines of research. Both research projects had a wide scientific dissemination and relevant data collection for an integrated literature review, through the international conference CIAV 2013 | 7°ATP | VerSus (www.esg.pt/ciav2013), organized in October 2013, by ESG and ICOMOS-CIAV, and held in Vila Nova de Cerveira, Portugal.

Furthermore, since 2005, the university school made an important effort to co-publish and support the edition of twelve books, concerning CI-ESG research (http://www.esg.pt/index.php/en/publicacoes). The increase of financial approval of ESG research projects at national and international level, proved a firm progress, in terms of rigor and quality of results. The strengthening of inter-institutional cooperation and the accomplishment of R&DT findings resulted in an interdisciplinary cooperation with excellent results. The potential of current developments, predict a solid growth of Gallaecia, as a higher education institution at national and international levels.

We renew our commitment for quality and high-standards, towards a scientific, cultural and educational high-level institution.

Mariana Correia
President of the Board of Directors and Director of CI-ESG Research Centre
Escola Superior Gallaecia, Vila Nova de Cerveira, Portugal

Seismic Retrofitting: Learning from Vernacular Architecture – Correia, Lourenço & Varum (Eds)

Research project framework

PROJECT IDENTIFICATION

SEISMIC-V: Vernacular Seismic Culture in Portugal

FUNDING ENTITY

FCT – Portuguese Foundation for Science and Technology

Programme: Scientific Research Projects and Technological Development
Main Scientific Area: Environment, Territory and Population – Architecture
Project Reference: PTDC/ATP-AQI/3934/2012
Prime Investigator (IP): Mariana Rita Alberto Rosado Correia

PROJECT TEAM

Project Coordinator

Escola Superior Gallaecia – Foundation Convento da Orada, Vila Nova de Cerveira, Portugal
Team: Mariana R. Correia (project coord.), Gilberto D. Carlos, David L. Viana, Goreti Sousa, Ana Lima, Filipa Gomes, Jacob Merten, Sandra Rocha & Sousa.

Project Partner

University of Minho, Guimarães, Portugal
Team: Paulo B. Lourenço (partner coord.), Graça Vasconcelos, Javier Ortega.

Project Partner

University of Aveiro, Aveiro, Portugal
Team: Humberto Varum (partner coord.), Aníbal Costa, Hugo Rodrigues, Ricardo Barros, Alice Tavares Ruano, António Figueiredo, Dora Silveira.

INTERNATIONAL CONSULTANTS

Ferruccio Ferrigni | Julio Vargas Neumann

AEGIS

Chair UNESCO – Earthen Architecture, Building Cultures & Sustainable Development
ICOMOS-CIAV – International Council on Monuments and Sites |
International Scientific Committee for Vernacular Architecture
ICOMOS-ISCEAH – International Scientific Committee on Earthen Architectural Heritage
PROTERRA – Iberian-American Network on Earthen Architecture and Construction

INSTITUTIONAL SUPPORT

DRCN – Northern Portugal Regional Directorate for Culture, Portugal
UNIVEUR – European University Centre for Cultural Heritage, Ravello, Italy

Institutional support and Acknowledgements

PROJECT LEADER

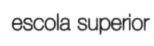

escola superior gallaecia

Fundação Convento da Orada

PARTNERS

Universidade do Minho

universidade de aveiro

AEGIS

UNESCO – Organisation des Nations Unies pour l'éducation, la science et la culture | uniTwin – Chaire UNESCO Architecture de terre

ICOMOS - CIAV International Committee of Vernacular Architecture – International Council on Monuments and Sites

ICOMOS - ISCEAH International Scientific Committee on Earthen Architectural Heritage – International Council on Monuments and Sites

PROTERRA

INSTITUTIONAL SUPPORT

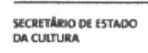

GOVERNO DE PORTUGAL – SECRETÁRIO DE ESTADO DA CULTURA

CENTRO UNIVERSITARIO EUROPEO PER I BENI CULTURALI

FUNDING

FCT – Fundação para a Ciência e a Tecnologia – MINISTÉRIO DA EDUCAÇÃO E CIÊNCIA

ACKNOWLEDGEMENTS

This publication is supported by FEDER Funding through the Operational Programme Competitivity Factors – COMPETE and by National Funding through the FCT – Foundation for Science and Technology, within the framework of the Research Project Seismic V – Vernacular Seismic Culture in Portugal (PTDC/ATP-AQI/3934/2012).

Part 1: Framework

Seismic Retrofitting: Learning from Vernacular Architecture – Correia, Lourenço & Varum (Eds)
© 2015 Taylor & Francis Group, London, ISBN 978-1-138-02892-0

Vernacular architecture: A paradigm of the local seismic culture

F. Ferrigni
European University Centre for Cultural Heritage, Ravello, Italy

ABSTRACT: After the Irpinia 1980 Italian earthquake, it turned out that the usual tools of the seismic engineers were not useful enough to analyse the historical retrofitting, following the seismic shocks. In these buildings, damages were mainly caused by the lack of or bad maintenance.
In 1985, the European University Centre for Cultural Heritage launched a research, to assess the traditional seismic-proof techniques existing around the world. This made it possible, to define the mastery of these retrofitting techniques, and the consistent behaviour of 'Local Seismic Culture' (LSC). This paper deals with the different kinds of LSC, depending on the intensity/ recurrence of earthquakes, which shows the effectiveness of traditional techniques against all components of seismic shocks. It also proposes an introductory tutorial, to recognise LSC elements in vernacular architecture.

1 INTRODUCTION

On the 23rd of November 1980, a 6.5 Maw earthquake hit the Irpinia region, in Italy, heavily damaging more than 120 historical centres. During the following years, the protection of historical built-up areas began to focus on scientific and political debates. At the time, the methods to calculate/ check masonry structures were rough: POR Method, based on a very simplified model of isolated buildings, was unusable for the connected and irregular historical built-up areas. The evolution to FEM (Finite Elements Method) offered, to seismic engineers, more precision in structural calculation of buildings having irregular geometry; but the difficulty in knowing the exact kind of materials in each point of structures, made it difficult to apply the FEMs to the historical built-up.

Nevertheless, there was evidence of a fact: despite all numerical simulations show that the majority of historical buildings had collapsed, they were, in fact, standing. On the other hand, it is easy to recognize, in regions where the seism is recurrent, that local communities necessarily had to develop "seismic-proof" construction techniques. So, when trying to answer the question "what is it necessary to do, to protect the historical built-up areas?" a new question emerges: "what have builders/users of historical built-up environment done, to protect it against the seismic shocks in seismic prone regions?"

Moving from this question, in the frame of EUR-OPA (Eur-Opa Major Hazards Agreement, a Program of the Council of Europe), in 1983 the European University Centre for Cultural Heritage (EUCCH) of Ravello launched a research line on traditional seismic-proof technologies around the world. The desk analysis showed many ancient methods used to dissipate the energy generated by seismic tremors in monuments: Greek temples (Touliatos, 1994), ancient pagodas in China (Shiping, 1991), or Japan (Tanabashi, 1960). Field researches carried out in Italy, Greece and France added more examples of traditional seismic-proof techniques in vernacular architecture (Helly, 2005). At the same time, researches on Naples' historical built-up area, carried out after the 1980 Irpinia EQ, showed that the extent of the observed damage was more closely correlated with the population density, than with the age of the buildings, and with their materials (De Meo, 1983). In other words, the correct use, and the appropriate maintenance of buildings resulted, at least as important, as the performance of the structures.

At the end of the research, all traditional seismic-proof techniques both common in monuments, and in vernacular architecture –, as well as the use of the buildings have been considered the result of a "Local Seismic Culture" (LSC), defined as the "combination of knowledge on seismic impacts on buildings and behaviours in their use, and retrofitting consistent with such knowledge" (Ferrigni, 1985).

During the following years the LSC research line has been focused on retrofitting of vernacular architecture in seismic regions, by means of a series of intensive courses organised on "Reducing vulnerability of historical built-up areas by recovering the Local Seismic Culture". The aim of the courses was to make aware architects and engineers that "reinforcing" historical buildings, using materials and techniques different than the original ones may be dangerous, as well as to supply them with sample criteria to recognise the seismic-proof elements in vernacular architecture.

The history of the research on LSC around the world; the complete analysis of seismic-proof features of monuments and vernacular architecture in seismic prone regions; the systematisation of different kinds of LSC, depending on recurrence/intensity of earthquakes; and the usefulness of a LSC approach in retrofitting historical built-up, have been published in "Ancient buildings and Earthquakes" (Ferrigni et al., 2005).

2 THE DIFFERENT KINDS OF LSC

In ancient societies, knowledge was handed down by word of mouth, from master to apprentice. The assumption for a technique to be validated and the relevant know-how to become firmly established was that one, and the same generation should be in a position to analyse the damage caused by an earthquake, define new construction techniques, or carry out repairs, test them during the next earthquake and hand down its own conclusions to the next generation.

However, if the time lag is very long (one century or more), empirical knowledge is most unlikely to take root. In practice, there is evidence that a community will be able to develop empirical know-how concerning the most effective construction techniques, only if the interval between two successive earthquakes does not exceed 40–50 years. Although specific construction techniques may be well-established locally, a generation, which has never experienced an earthquake, will not really develop an awareness of the seismic-proof function of given features, comparable to that, which induced the previous generations to adopt them. It can be assumed that the observation of the different impact of a new earthquake on different buildings will generate new awareness and rekindle knowledge, which had been forgotten. This will determine, after an earthquake (EQ), a sudden increase in the level of knowledge, and a corresponding expansion of the LSC. Subsequently, the physiological tendency towards repressing the memory of that event will result in the loss of the knowledge acquired, and the resulting culture may even be more fragmentary than it was before the tremor occurred (Fig. 1a).

On the contrary, if the earthquakes follow upon one another, at intervals of time, which enable a generation to hand, their experience down to the next generation by word of mouth (40–60 years), the local seismic culture will be embedded.

Provided these assumptions are met, the effectiveness of building techniques can be tested in a sufficiently frequent and reliable way, and the criteria adopted in building and repairing buildings can become the common heritage of a community (Fig. 1b)

But the degree to which empirical knowledge takes root does not only depend on the frequency of seismic events. The damage caused by small-scale tremors, albeit frequent, will not be severe enough to allow people to select the most effective techniques. Nor will the impact of exceptionally severe earthquakes. If all the buildings are destroyed, it will be difficult to assess

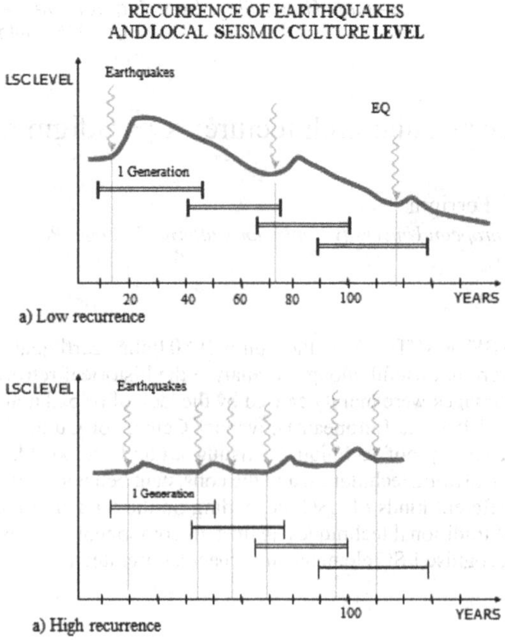

Figure 1. If the recurrence of EQ is low (a), the experience of a generation can't be transmitted to the next one. In regions having a high recurrence (b) the "seismic-proof" knowhow stay, an LSC can root (credits: Ferruccio Ferrigni).

the comparative effectiveness of different technical devices. Not to mention the fact that a catastrophic earthquake wipes out the very memory of techniques previously in use, because it destroys existing documents, and causes people's death, who were conversant with their contents.

According to the MCS scale for classifying earthquakes, the intensity of an earthquake is generally based on the type of damage observed in the buildings that are affected. Major cracks and fractures which, though sizable, do not jeopardise the stability of the buildings, are classified between intensity II and V. An earthquake of intensity VI or VII produces more serious cracks, but only a small section of the buildings collapse in part, or suffers from damage to the point where they become unstable. An earthquake, which shakes the outer walls of a large number of buildings (causing the walls to come apart at the corners) is classified as intensity VIII. Situations where buildings actually collapse are typical of intensity IX. An earthquake of intensity X causes extreme damage to all buildings, besides numerous building failures. Intensify XI results in the near total destruction of the built-up area, including constructions such as bridges and dams, as well as cracks in the ground.

With close approximation it is possible to assume, therefore, that a LSC will only emerge as a result of severe, but not catastrophic earthquakes. In practice, earthquakes between intensity VII and X, on the MCS scale.

In conclusion, only earthquakes, which follow upon one another at suitable time lags, and with an 'appropriate' recurrence/intensity combination, are likely to result in a sound LSC. This observation implies two highly interesting consequences for the LSC approach. If the intensity/recurrence rates of earthquakes experienced in different parts of a country or region in a diagram is reported, it is possible to identify the areas where it is most likely to find traces of an LSC. In addition to this, it is also possible to tell, in advance, if the LSC of a given place is a set of construction 'techniques' in use, regardless of the occurrence of any earthquakes; or if it has established itself in the wake of an earthquake, based on the carried out 'retrofitting' actions (though a combination of these is also possible).

The Fig. 2 shows the position of the case studies analysed by EUCCH's field research, in a EQs recurrence/intensity diagram. In the diagram the 'optimal' range of recurrence and intensity, facilitating the rooting of an LSC is marked. Starting from this, it is possible to identify different 'LSC domains':

- Places with a high recurrence of major earthquakes: memory of the effects of the earthquakes remains alive. Buildings invariably show a number of built-in features, which make them more earthquake-resistant (and the 'seismic-proof techniques' that are observed there will point to what it may be called a 'prevention LSC');
- Places exposed to medium/high intensity, but not frequent, earthquakes: memory of their effects fades as time passes. Another earthquake causes damage; the buildings must be reinforced haphazardly, by adding new structures or mending the existing ones (in this case the LSC concerned is said to be 'retrofitting-oriented', and it is revealed by the 'anomalies' recognised in the buildings).
- Places comprised within the part of the diagram, in which earthquakes are less frequent and/or of lesser intensity: here any traces of a LSC will be hardly found.
- In addition, there are places where earthquakes, though rare, occur with catastrophic intensities. Here, too, it is unlikely to find a well-rooted or generally recognised LSC, while it is highly probable that entire cities had to be relocated, and the new ones reconstructed according to imported and/or imposed technical criteria (in these cases it might even be spoken of an 'imposed by decree LSC').

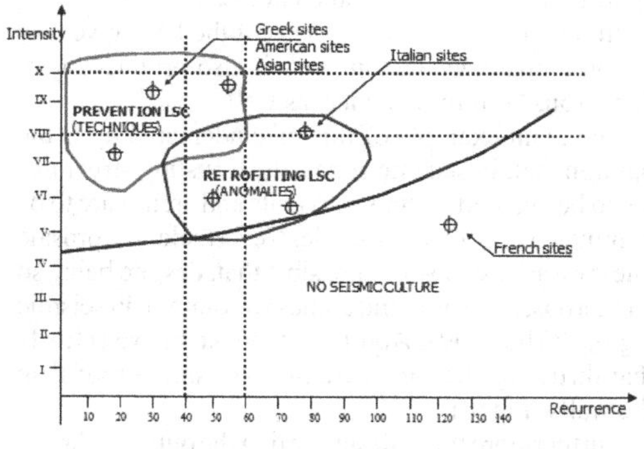

Figure 2. By locating a specific site in a Recurrence/Intensity diagram one can find out in advance if there is a reasonable probability of finding evidence of LSC and, if so, what type is it (prevention/retrofitting LSC) (credits: Ferruccio Ferrigni).

3 THE LSC: A COMPLETE RESPONSE TO THE SEISMIC SHOCK

Leaving aside the specific parameters of each earthquake (magnitude, frequency, duration, ground motion, etc.), and generally speaking, it might be said that the forces that act on buildings during an earthquake are of three kinds: vertical, horizontal and torsional. In mechanical terms this means that a seismic shock suddenly increases the vertical load on bearing and horizontal structures, while generating shearing stress at the base of buildings, as well as a twisting motion of their vertical edges.

It is true to say that man-made artefacts are designed to resist gravity, but these are seldom capable of withstanding horizontal forces. The sudden increase in vertical load caused by vertical components of seismic shock is thus absorbed fairly easily by the structures (the only elements that are likely to suffer damage are the "accessory" ones: cantilevered structures, such as stairways, balconies and the like); while in the absence of specific precautionary measures, neither horizontal movements (which generate shearing stress), nor those, which cause the walls to come away at the corners (i.e. torsional stress), will be absorbed.

In general terms it is right to say that only techniques, which make a building significantly more resistant to shearing and torsional stress, can be termed 'seismic-proof'.

Nonetheless, in VA are easily recognised specific features and/or elements for resisting all the components of seismic shock. In Algiers' medina, the cantilevered structures are reinforced by sub-vertical rafters (Fig. 3, left). In the same medina, the technique

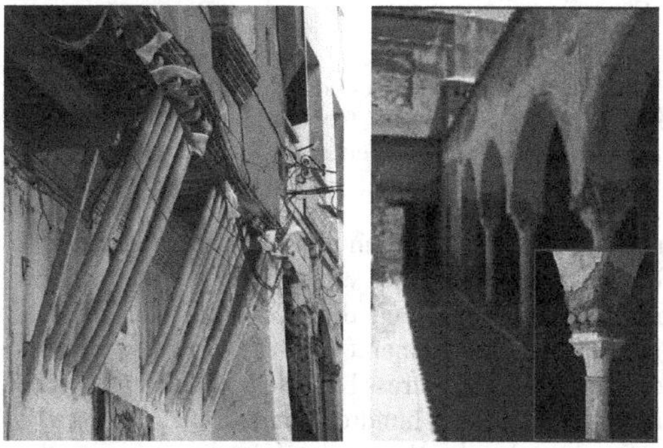

Figure 3. Historical (XVIII) seismic-proof techniques in Algiers' Medina: left, reinforced cantilever; right, cypress wood rollers protecting the column against the horizontal forces transmitted by the overhead structure (credits: Ferruccio Ferrigni).

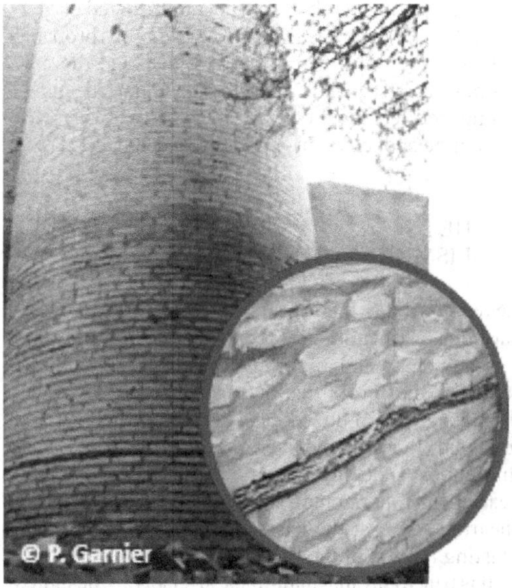

Figure 4. Historical seismic isolator in Uzbekistan (XI) (credits: Philippe Garnier).

Figure 5. The inserted tables don't resist to the traction. So they can't carry out a "belt" function, but they increase the friction (credits: Ferruccio Ferrigni).

used to protect the columns against the horizontal forces is even more sophisticated: rollers are inserted between the capital and the overhead wall (Fig. 3, right). The Uzbek minaret (Fig. 4) is very well protected against shear stress by a 'seismic isolator', very up-to-date: just nine hundred years old.

On the other hand, in seismic prone regions, the insertion of wood tables in masonry is common in historical buildings. This feature is very interesting, because the aim of the tables is not to "belt" the walls, as usually presented, but to increase the friction and/or to cut the diagonal cracks. In Mytilini Greek Island

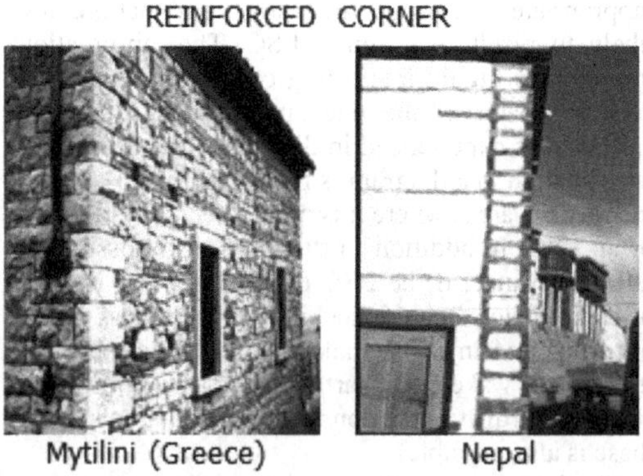

Figure 6. Different techniques but same target: to increase the resistance of the corner to the torsional components of EQs (credits: CRAterre).

all houses, rich or poor, are built in listed masonry (Fig. 5). The detail of table's junctions shows clearly that the table has no resistance to the traction, so it can't be a "belt".

Techniques reinforcing corners are typical features aimed to increase the resistance to the torsional components. Greek and Nepalese houses show a clearly more expensive technique in reinforcing the corner (Fig. 6). An over cost acceptable, just if the reinforcement of the corners is considered important, because of its seismic-proof effectiveness.

4 THE RISKS ARISING FROM THE LOSS OF THE LSC

In many countries, the traditional seismic-proof techniques have survived until today. The Chinese kiosk and the Nepalese house (Fig. 7) prove how the LSC is rooted in these populations. In any case, re-evaluating the traditional seismic-proof techniques is not as important for new constructions, as it is to appropriately retrofit the ancient ones.

In fact, not only does the loss of the LSC give way to some misunderstandings, as it also produces very dangerous "reinforcements" as well.

The main example of this misunderstanding is the opinion that, in seismic zones, the thrusting structures are to be avoided. Outcome: vaults and arches are to be eliminated as soon as possible. Nevertheless, a prosaic question arises: how is it possible that, despite being so dangerous, are vaults and arches so common in seismic regions? The 2009 L'Aquila earthquake showed clearly that thrusting structures are more seismic-proof than the walls (Fig. 8).

Furthermore the wall supporting the outward thrusting structures has to be designed and built in such a way, to absorb the horizontal forces produced by vaults and arches, so they will easily absorb the increase in charges originated by the earthquake. On the other hand, vaults and arches can warp under the alternate

Figure 7. Where the EQs are frequent and strong the seismic-proof techniques persist until today. The Chinese kiosk uses the same structural scheme codified since the XVII for the temples. The corner of the Nepalese house is reinforced with the same technique than the ancient one (see Fig. 5) (credits: CRAterre).

seismic motion, so they collaborate in "metabolising" the energy transmitted by the EQ.

Another false view concerns wooden floors, considered too weak and, above all, unable to offer the 'rigid diaphragm' required by seismic codes. So, usually, it is suggested to replace wooden floors with concrete ones. The 1997 Umbria Marche earthquake, showed the effects of this retrofitting. After the 1979 Umbria EQ many buildings have been reinforced by inserting, complying with the Italian seismic code, a rigid diaphragm. But, due the Umbria Marche EQ in 1997, the "reinforced" buildings suffered from heavy damages (Fig. 9). Why? In fact, wooden floors are deformable and follow the walls (Fig. 10a), while concrete ones are crushproof, do not follow the walls and generate outward thrusts (Fig. 10b). Moreover, during the deformation, the friction between tables and beams of the wooden floor, contributes to absorb the energy

Figure 8. L'Aquila EQ 2012. The vaults are almost intact, the walls collapsed (credits: Ferruccio Ferrigni).

transmitted by the earthquake, whereas the rigidity of the concrete floor makes very low its collaboration in energy absorbing. Result: The wall 'explodes' (Fig. 9).

5 SOME SUGGESTIONS TO ANALYSE LSC CONSTRUCTION OR RETROFITTING TECHNIQUES

Manuals on local seismic-proof construction techniques should normally be compiled by technical personnel, of at least university level. However, this research activity can also be carried out by a local team of professional consultants and students (including at least one architect, one civil engineer acquainted with seismic-proof structures, one geologist with knowledge of seismology and one historian), using guidelines that have been successfully tested in all case studies performed to date.

In compiling a manual on a given town (or towns in an area), the following questions should be answered, although other questions may be required, according to the local context.

1. Preliminary research

   1.1. Is the town, on which the LSC revival action focuses, located in an area said homogeneous in terms of seismic history, construction techniques and resources available at the time of the most severe earthquakes? If not, which other towns have to be exposed to further research, in order to obtain useful information?

Figure 9. This building had been "retrofitted" by changing the original wood floors with the concrete ones, to obtain, according to the Italian seismic code, a rigid diaphragm. But a new EQ "peeled" the perimeter walls (credits: Ferruccio Ferrigni).

1.2. Which earthquakes, of intensity ranging between VIII and X on the MMS scale, have struck the town? Were other towns in the area affected? Have there been other earthquakes of similar intensity, which struck other towns, but not the one in question?
1.3. How large was the built-up area[1] of the town (and other towns in the area) at the time of major earthquake occurrences? Which part of the built up area was affected?
1.4. What resources (construction material, capital, know-how) did the town have access to, at the time of these earthquakes? How have these

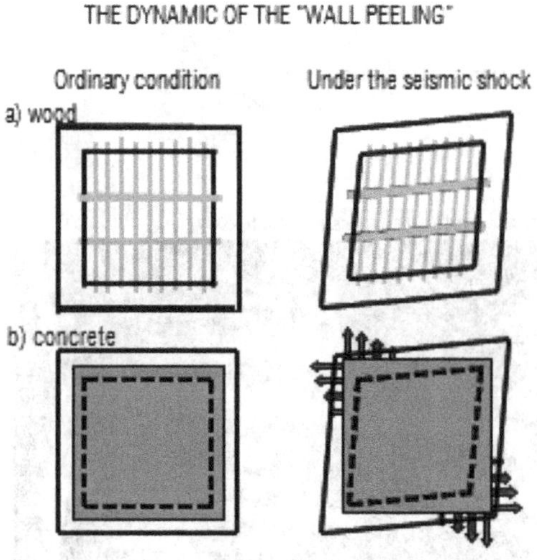

Figure 10. In traditional wood floors beams and tables follow the deformation of the perimeter walls, collaborating in energy dissipation too. The concrete floor is very rigid, can't follow the walls and produces outward thrust, causing the 'peeling' (credits: Ferruccio Ferrigni).

factors changed since then? What political relations existed between the various towns in the area? Which administrative organisation was in charge? What were trade relations like within the area? What were trade relations like between the area in question and other areas?
1.5. Have official earthquake-protection regulations ever been drafted for compulsory application in the area at question? Is there any knowledge of technical recommendations or traditional techniques used in the past?

2. How to identify anti-seismic techniques
2.1. What historical construction techniques (and building types) are found in the town and the area at issue? Can they be dated to the time of major earthquakes? Or to economic, political and/or administrative events? Do any of these techniques stem from anti-seismic regulations?
2.2. Are there any construction techniques/building types found only in some of the towns of the area? Or only in some parts of the town at question? Or only in some parts of other towns?
2.3. If so, do these construction techniques/building types provide effective protection in the event of an earthquake? Do they present any analogies with traditional techniques found in other areas, or other countries, which have been defined as anti-seismic by the scientific community?
2.4. Do the construction techniques/building types in the town differ from those in other towns with access to similar resources, but which have not been struck by earthquakes of similar intensity or frequency? Are they similar to those of towns with an analogous seismic history? Were there

[1] In order to obtain information on the town's built-up surface area at the time of major earthquakes, it is normally necessary to consult archive sources integrated with other documentation from the time. This research is often difficult and not always feasible. However, useful information can be collected through cross-reference analysis of the built-up area's morphology (i.e. architectural styles, road network), recurring construction features (internal organisation of the individual homes, how they are interrelated) and construction techniques (materials used, level of craftsmanship, size of constructions, particular technologies or construction features).

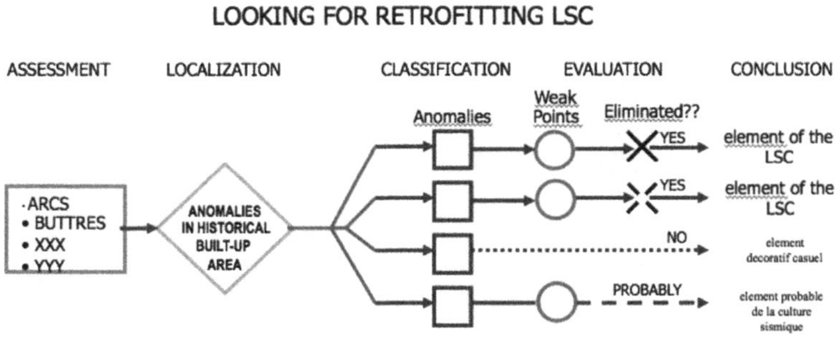

Figure 11. Looking for retrofitting LSC we have: first, to compare the distribution of anomalies with the historical built-up existing at time of major EQs, then to classify them (reducing/no reducing the comfort, close/no close to a weak point), then to evaluate their effect (eliminated/no/probably eliminated the weak situation). Finally we can surely consider as an element of the retrofitting LSC the "anomalies" reducing the comfort and eliminating the weak points (credits: Ferruccio Ferrigni).

any significant variations following major earthquakes? Are they present in systems with access to the same resources, and which have had a similar political and social history, but a different seismic history?

2.5. Could the choice of construction techniques/building types have been determined by different needs, objectives or circumstances? In particular, could they have been conditioned by the lack of other materials? By the decision to imitate techniques adopted in traditionally more important towns? By requirements dictated by health, military or religious considerations?

2.6. To confirm the anti-seismic worth of the construction techniques/building types in question, are they ever (or only rarely) found in towns/areas with similar resources, but a different seismic history?

2.7. Have the construction techniques/building types been modified in recent years? Are they compatible with current requirements? If not, how can they be made compatible?

3. How to identify the historical retrofitting

3.1. What construction 'anomalies' can be detected in the town's built-up areas? Are they the same as those referred to as anti-seismic reinforcement works in other areas or in towns built with similar construction techniques?

3.3. What is the main role of the anomalies identified? Do they make the dwelling more comfortable, or increase the strength of the building; or both? Has any increased structural strength been obtained at the expense of the dwelling's comfort? Or to the detriment of public areas?

3.3. With reference to the size of the built-up area at the time of major earthquakes, can a particular distribution of certain anomalies be noted in any one part of the town, as compared to others? Or in one town, rather than in others? Are these anomalies the same as those referred to as post-earthquake reinforcement or repair works?

3.4. Are there any parts or features of the built-up area, which are vulnerable and/or were presumably vulnerable at the time of major earthquakes?

3.5. Have the identified anomalies eliminated/reduced the vulnerable features? Have they strengthened the building?

Some findings resulting from the experience accumulated to date suggest using these tutorials in assessment schemes. The flow-chart in Figure 11 synthetises the tutorials.

REFERENCES

De Meo, P. (ed.) (1983). *Indagine sui danni del terremoto nell'edificato storico di Napoli*, Research carried out by the University of Naples "Federico II".

Ferrigni, F. (1985). La Cultura Sismica Locale: che cos'è, come recuperarla, perché. *Opening conference of the Intensive Course on reducing vulnerability built-up areas by recovering the Local Seismic Culture.* Centro Universitario Europeo per i Beni Culturali, Ravello.

Ferrigni, F., Helly, B., Mauro, A., Mendes Victor, L., Pierotti, P., Rideaud, A. & Teves Costa, P. (2005). Ancient Buildings and Earthquakes. The Local Seismic Culture approach: principles, methods, potentialities. Ravello: Centro Universitario Europeo per i Beni Culturali, Edipuglia srl.

Helly, B. (ed.) (2005). *Case studies* in *Ancient buildings and earthquakes.* Edipuglia, Bari: Conseil de l'Europe Strasbourg.

Shiping, H. (1991). The Earthquake-Resistant Properties of Chinese Traditional Architecture. *Earthquake Spectra,* 7 (3), pp. 355–389.

Tanabashi, R. (1960). *Earthquake resistance of traditional Japanese wooden structures.* Proceeding of the Second World Conference on Earthquake Engineering at Tokyo and Kyoto, Tokyo.

Touliatos, P. (1994). Traditional seismic-resistant techniques in Greek monasteries. Conference on *X Intensive Course on reducing vulnerability built-up areas by recovering the Local Seismic Culture.* Centro Universitario Europeo per i Beni Culturali, Ravello. Accord Partiel OUvert, Conseil de L'Europe, Strasbourg.

Seismic Retrofitting: Learning from Vernacular Architecture – Correia, Lourenço & Varum (Eds)
© 2015 Taylor & Francis Group, London, ISBN 978-1-138-02892-0

Vernacular architecture?

G.D. Carlos, M.R. Correia & S. Rocha
CI-ESG, Escola Superior Gallaecia, Vila Nova de Cerveira, Portugal

P. Frey
Ecole Polytechnique Fédérale de Lausanne, Lausanne, Switzerland

ABSTRACT: The first challenge to encounter, when addressing a comprehensive understanding of the vernacular architecture problematic, is to try to define the etymological sense and the epistemological mean of the term. This paper intends to approach the evolution of the 'vernacular architecture' designation, and to reflect about its current conceptual application, considering its most substantial ideological divergences over time. The study was based on the recognition of two fundamental phases: the autonomy and the consolidation of the field of study, where the development of the object's differentiation was determinant. Consequently, it will address a comprehensive rehearsal for a terminology specification, according to the most disseminated descriptions and their corresponding authors. Finally it will open the discussion for the concept revision, due to the present socio-cultural context, focusing on the influence of the industrialization and globalization processes, the changes on the notion of local production, and the suitability to use natural materials from the local environment.

1 INTRODUCTION

90 to 98 percent of the world's total building stock is considered to be vernacular architecture, according to Oliver (2003) and to Rapoport (2006). From the billion of buildings existent worldwide, Vellinga, Oliver and Bridge (2007, p.3) refer to 80% or even a higher proportion, to be vernacular architecture. Therefore, it is clear that more than a half of the actual population lives in vernacular dwellings.

Considering that 7 billion people live in the planet and 4 billion people leave in informal houses, from which 1 billion of urban poor live in slums (UN-HABITAT, 2006), it is relevant to discuss if informal housing is vernacular architecture? What is vernacular architecture? Worldwide, if the majority of people live in vernacular buildings, why has this architecture been less acknowledged? Is it due to its significance, then, what does it mean and what is the value of 'vernacular architecture'? Does the actual changing world is also altering the perception of what has been considered to be vernacular architecture?

2 CONCEPTUAL DEVELOPMENTS

2.1 *Autonomy of the Field of Study*

"Architecture without Architects attempts to break down our narrow concepts of the art of building by introducing the unfamiliar world of non-pedigreed architecture. It is so little known that we don't even have a name for it. For want of a generic label, we shall call it vernacular, anonymous, spontaneous, indigenous, rural, as the case may be."

Rudofsky (1990, p.2)

The attempt to implement concrete descriptions/labels to the nature of informal architecture, as opposed to the classical nature architecture has caused numerous ideological differences. If conceptually, its definition is generally recognised, the main problem occurs, when it is intended to set limits or transition areas to certain etymological domains. According to Papanek (1995), this is due curiously to the key stakeholders (architects, historians, authors or critics) that, in addition to systematically promoting fallacies, also tend to address vernacular architecture under reducing prospects. A result of individual interpretations, whose relevance can only be considered mono-disciplinary targets.

"...it is precisely because so many architectural critics tried to fit theories of vernacular architecture in a single category that few explanations maintained consistency"

Papanek (1995, p.151).

Thereafter, Papanek defends that in order to understand this architecture, resulting of multiple causes, or more specifically, the result of a dynamic system of those same causes; the perspective must be directed through the process and not just be supported by its product. This condition was identified very earlier, by some of the pioneers of this field of study, such as Rudofsky or Rapoport. The interest increasingly develop especially since 1950, and generated two marked tendencies: 1) No conceptual spin-off, where there was clearly an attempt to group similar character architectures without specific discrimination; 2) The determination of the particular essence of differentiated study objects that could bring greater scientific objectivity to the study field.

The first trend encompasses the concept expressed by Sybil Moholy-Nagy (cited by Oliver, 2003, pp.12–13), which already in 1957, recognized that "*all classifications applied to anonymous architecture must remain arbitrary and unsatisfatory*". It was a premonitory expression regarding its theoretical differences that have invariably marked the evolution of studies in this scientific area. However, Moholy-Nagy very skilfully put it into perspective by focusing instead on his study object and its significance. Bernard Rudofsky, one of the most accountable for the international dimension given to this field of study, demonstrates, with some irony, a taste for this orientation expressed in the notorious exhibition '*Architecture without Architects: A Short Introduction to Non-Pedigreed Architecture*'.

According to some authors, it was actually Bernard Rudofsky, the responsible for definitely launching the interest for this heritage in the recent and current architectural culture, even regardless the quantity and quality of work that preceded it. Even if '*Architecture without Architects*' does not involve a large scientific consistency of study or of methodological rigor, in terms of selection criteria, classification or representation, the way in which Rudofski depicted the exhibited examples, truly conditioned the perspective from which the architects were to address the area in the future (Duarte Carlos, 2014).

The first approach on this sort of architecture emerged naturally through the first Ethnographic reports, initially on the form of small monographs or articles published in specialised journals, dedicated exclusively to small rural settlements, tangible, with which the authors shared some relation or affinity (Ordem dos Arquitectos, 2004, p. xxii).

Its origin derives from the increase of social, cultural and anthropological sciences, in the early twentieth century. With the consequent development of Ethnography and Human Geography, the building, especially that of traditional nature, constitutes now a valuable testimony for understanding the communities and their cultural evolution. At an early stage the fascination will lay in the most exotic and more culturally contrasting civilizations, subsequently also the inner reality of European countries started to raise attention. Albert Demageon (1920) stands out within the study of rural territory and its special relation with the human habitat, becoming therefore an unavoidable personality in the social sciences. This author will reveal a new dimension from the '20s on, in what could be understood as a required study for understanding the phenomenon matrix of the cultural identity of each nation.

With further analysis, some features, such as the characterisations of building techniques, required the participation of technical expertise. If the simple enunciation process was possible through popular stories and local artisans, its systematisation, registration and classification called for a more scholarly knowledge and competence of the building and its representation methods. That led to the first real involvement of the Architect with this reality, even though as an auxiliary staff.

To refer is also the alienation of the European nations, which shortly after the physical and economic post-war reconstruction imposed political resistance and ideological mistrust to all that had originated from foreign sources. The acknowledgment and consent of the political forces, even if for the wrong reasons (Ordem dos Arquitectos, 2004 p.xi), would eventually be critical to the involvement of the Architects as professionals in the study of Traditional Heritage.

This involvement soon would claim to the political authorities the practical and exclusive implementation of field surveys to the informal architecture, as a way to assimilate the values of national tradition. In a restricted professional cycle and addressing the necessary change and growth of academic training, these professionals would be responsible for the dawn of the theoretical awareness of intellectual circles, which, at a later stage, stimulated rebound and adherence in leading academic centres through the restructuring of its educational curricula, particularly with the insertion in architecture courses of subjects from the Geographic, Anthropology and Ethnography extents (Diez-Pastor, 2012).

Another decisive factor for the introduction of the subject in the professional field of architecture was the consolidation of the anti-historical intellectual gap in the architecture process. This was expressed by the rationalist movement that dominated almost all instances of the international panorama. This field of study had already known relevant precursors in a first reaction, of Ruskin and Morris, as an alternative to the dominant neo-classicism of the nineteenth century. Despite never having imposed to its ideological opponent, the picturesque perspective of Arts and Crafts, continued to captivate and to stimulate Architects over time for a closer observation of the vernacular heritage, even without a massive influence (Toussaint, 2009 p. 59).

This same legacy would, ironically, be constituted as a valid alternative at the announcement of the exhaustion of modernism within its anti-historicist logic. It began to take shape as a critical reaction to what they considered the dehumanisation of the International style, especially from the mid-twentieth century on. It found references in those who, like Alvar Aalto, never abstracted from the physical and cultural reality where their works were, and later, it found compelling themes, in the premises of Norberg-Schulz, to experience a theoretical reformulation (Cerqueira, 2005, p. 46).

From the general literature review, it is possible to infer that this scientific field of study has gone through a regular evolution that came from the Ethnographic reports of national comparison before achieving the Morphological Surveys of Regional character. This perception is furthermore supported by the identification of a specific Bibliography set. Despite the years that separate the publications and the methodological differences that outlined them, these ethnographic works, with all their insufficiencies and imperfections, launched the foundations and scientific parameters of the classification of Vernacular Architecture. At this time, it was nominated as Regional, Indigenous, Rural or Popular, being the last designation the most used due to its lower ideological discrimination (Sabatino, 2011)[1].

Unlike *Vernacular*, the application of the term *Popular* does not appear to derive from a systematic evolution of clearly defined ideological frameworks. Its use may be consciously unifying, as it seems to happen in the authors of greater antiquity, or drastically circumscribed, as it was in a more recent tendency of some authors, when referring to informal contexts with unsorted models, or in the process of systematization.

The first application of the term *Popular* architecture, as a unifying terminology, arises as opposed to the Monumental classification. It designated the whole architecture accomplished through empirical knowledge and the exclusive use of local resources – material and human –, which consciously brought distinction from real traditional examples. This first discrepancy between Popular and traditional, although with recognition of some conceptual overlap, can be found in the early twentieth century studies, particularly during the 30s. This option is expressed, for example, in the pioneering researches of Torres Balbás or Fernando Garcia Mercadal (Calatrava, 2007), whose terminology, at that time, was far from considering the implications of industrial means of production, the dissemination logics, and distribution of global markets. Given the description made by the authors, the original term application was virtually identical, or at least comprehensive of the vernacular expression, whose specification would take more than 40 years to be developed.

The perceived influence and impact from the authors, especially the Latin authors, would lead to the consolidation of this interpretation of the popular terminology in the academic environment, to the detriment of other designations. Thus, the lower appropriation of the term 'vernacular', according to these authors, lies precisely in the redundancy of the concept. This also explains the predilection of some countries to the designation of the term popular architecture, beyond the subsequent international recognition of their etymological differentiation. For Flores (1973) and Llano Cabado (1983), the essential principles of interpretation of the term Popular Architecture are: the relation with the environment; the binomial indoor-outdoor; and the preference for a basic spatial design. These are the reasons for its conceptual genesis. Formally this architecture should allude to the local building tradition, thus referring to the use of indigenous materials. Programmatically, the Popular Architecture presents itself as super-specialized, although conditioned to the flexibility of the materials available, establishing definitive functional archetypes, both in the general nature of the building, and in the specific characteristics of the elements (Duarte Carlos, 2014).

At the end of the 60s of the twentieth century, the Western world, much by the fault of the exhaustion of the modernist movement and the scepticism caused by the industrial production, returned to the study of traditional values that characterised the diversity of Popular Architecture. It was Amos Rapoport (1972), who brought consistency to the term 'vernacular architecture', in a set of reflections published in "House, Form & Culture". According to Vellinga (2013), Rapoport meant to establish the study field in the academic environment, joining the scholars that developed it, though not in a concerted manner, and thus allowing a greater disciplinary cooperation. Rapoport intended to contribute to the recognition of architecture as a cultural expression, understanding the culture and the local building tradition as a repository of the cultural identity of specific communities. Thus, understanding the architecture as a formal language, the ascription of the term 'vernacular', earns enough objectivity, in that it aims to stabilise the name that corresponds to "... architectural language of the people with their ethnic, regional and local 'dialects'" (Oliver, 2006, p.17). It also reflects a critical perspective on the institutionalization of the education and the acknowledgement of the relevance of self-learning processes, supported by common social relations, according to Ivan Illich conception (Frey, 2006). Though pragmatic and elementary, characteristics that erroneously decreased a historiography perspective of Vernacular Architecture before the twentieth century (Rudofsky, 1990), this phenomenon represents a conditioned reaction to the culture of the civilization that build it. According to Oliver (2006, p.17), it is also important to refer, the fact that the term 'vernacular' is a linguistic designation that portrays the articulation of speech of the non-erudite population. In addition, language is always adjusted to its cultural context, which as a communication system cannot forget its anthropological and social dimensions. Language, whatever its type or nature, will always be a cultural expression, and, in this order of terms, the architecture – regardless of its variants – may be considered as a formal result of cultural expression, both conceptual and interpretative, if these will ever be inseparable, as Christian Norberg-Schulz (1967) noted.

The impact of the theoretical ground of this term gains followers, mainly from the 80's on, and the systematic adoption of the term in the works of authors, such as Paul Oliver or Brunskill will be decisive for the

[1] Considering in particular the Italian Regime, the different terminology acquired strong political connotations.

recognition of this designation. According to Rapoport (2006), the disciplinary autonomy of the study field is definitely provided by the publication 'Encyclopaedia of Vernacular Architecture of the World', coordinated precisely by Paul Oliver (1997).

2.2 *Consolidation of the field of study: Specification of the object*

Despite the recent developments of this area of study and its thematic enlargement, it is common the use of the terms: Traditional, Vernacular and Popular Architecture as indiscriminately synonyms. This is far from the real potential of each term, despite of their close relation.

The excessive relation and affinity of the study objects, described later on this article, establishes a significant confusion in its actual designation. This is due not only to its conceptual overlap, as to the lightness with which, most of the texts, insufficiently documented, address these issues.

Common to all, is the Ethnographic contributions, whose evolution of the processes of archaeological analysis has boosted, from the study of primitive buildings as major anthropological evidence. It is certain that this interest will quickly transfuse to the domains of the Morphological study, applied from the methodological model of Auzelle, already integrated in the pedagogical concepts of the leading academies of Europe (Duarte Carlos, 2014).

Its descriptive and consensual basis is understood as opposed to the term of the Historical Architecture of scholar character that Baker (1999) characterised as Monumental. This architecture is intellectual and of symbolic aspirations, taking the previous designation due to its framing with streams, movements or conceptual affinities of theoretical ground, and presenting an aesthetic expression recognised and enclosed in what is frequently called the "architectural style", which invariably derives from the "artistic styles". Another deeply widespread expression that encompasses all these features as opposed to Classical Architecture is the Architecture of Anonymous, often mentioned by Pedro de Llano (1996), whose wide dissemination of the term, he associates with Bernard Rudofsky. This concept assumes a building activity without project, without technical representation, whose final shape is derived from the procedure established by a perfectly integrated empirical knowledge on local tradition. It does not mean however, that it does not hold a priori conceptual work, associated with the relationship between a particular need and a specific spatial response, with a practical interpretation in the typological development. It is understood the concept of anonymity devoted to the work as a metaphor of the full integration of it in the natural and cultural environments where it belongs, without pretensions of evidence in relation to the environment, referring to the initial ethnographic concept that establishes as priority the balance between man and the environment itself (Dias et al. 1969).

2.3 *Rehearsal of a terminology specification*

The comparative study of the reference authors aforementioned, their definitions and use of the terms seem to point to the following interpretive synthesis:

2.3.1 *'Traditional architecture'*

This is the broadest term; it derives from the actual application of the constructive tradition and empirical knowledge, based on oral transmission between generations. It explores mainly the regional peculiarities using local resources, such as environmental response and cultural event, disproving the technological sophistication of industrial essence and the materials associated to it. Contrary to other definitions, it may allow more significant investments, applied by higher social strata and, occasionally, integrate scientific knowledge, acquiring therefore a Monumental connotation. As a result, its formal process is more elaborate, resorting in more complex building systems, of less pragmatic execution, and integrating specialized actors, subjected to scholarly influences, even without scientific training. A significant portion of traditional architecture can be considered Vernacular and Popular.

2.3.2 *'Vernacular Architecture'*

Determined by Amos Rapoport, this term refers solely to the specific buildings in a certain geographical context, in response to the physical and cultural environments. It uses local techniques and local construction processes. It originates specific typological models, producing characteristic plastic elements of the area, that they are restricted. It is no longer used as a primitive building process, or it is not only reduced to self-construction by their owners, though they may recur to labour-skilled labour, usually consolidated in trades recognised by the community. It is a term close to the concepts of Regional architecture or Autochthonous Architecture (Rapoport, 1972). By definition, all Vernacular Architecture is always Traditional and it can also encompass a component of Popular architecture.

2.3.3 *'Popular Architecture'*

Popular Architecture is the extreme opposite to the application of scientific knowledge and to the monumental expression of architecture. Although sensitive to regional values, it is undoubtedly the one of most pragmatic implementation, overlapping economic optimisation to all the other parameters assigned to the Vernacular Architecture. This term is usually associated with poverty and modesty of its construction. Usually, its implementation is accomplished by the users themselves, or significant part of it is. Even when it gathers collective labour forces, there are no players in the thick of the process that are exclusively dedicated to the art of building. It also requires a minimum expense, often performed with precarious means and it may include all types of current buildings endow with the conditions mentioned. It does not provide a binding site affinity in terms of technique or materials used, the latter may be industrial or from outside the region where they are applied. It is in fact in these latest features that Popular Architecture distances from the fundamental principles of Vernacular Architecture and consequently from Traditional Architecture.

3 FINAL REMARKS

3.1 *Definition Evolution and Overlapping*

In abstract, the etymological distinction presented seems appropriate to an objective differentiation. It is however its application to certain objects (study), or when subjected to certain contexts, that can generate ambiguities or even conflicts of conceptual order.

The constructive culture cannot be understood as a sealed phenomenon. The necessarily development depends on a long and time consuming evolutionary process, from a certain point of origin to the general systematisation of a particular technique or process. After an accurate analysis, it is possible to perceive that there are no external sources to the region or outside the craft production related to Traditional Architecture. In turn, Popular Architecture does not discriminate adaptation to the outdoor or to industrial resources, provided that the local accessibility is ensured. The vernacular constructive culture, by definition, assumes that the techniques and the processes are assimilated and used intuitively by local communities. Thus, the exact definition of the origin of a particular feature or technique, an interpretation less 'conservative', should be considered more flexible.

Many vernacular solutions are today regarded as regional icons, aimed at long processes of appropriation, up to the point of becoming more characteristically representing their own region. As an example is the gradual replacement of roofs covered with thatch, typical of Atlantic regions in Europe, by the tile, introduced since the Romanization process in the territory. The introduction of lime, drastically revolutionised the properties and applications of mortars, constituting an essential element in many cultures of the Mediterranean region (Garcia Mercadal, 1926). Currently, this material is inseparable from several traditional construction techniques, although in many of its territories the extraction of the raw material base is not even possible. Did, in the appropriate time, this profound technological revolution sparked so many ideological differences? To what extent the use of Portland cement, or the application of cement panels, or zinc sheets, will not represent an analogue mixing process? It is precisely this condition that many authors advocate for emerging constructive cultures, or contexts in accelerated development process, arguing that certain communities may take ownership of technical and exotic materials, from a regionalist approach (AlSayyad, 2006). For many authors it is the consolidation of this process of 'acculturation' of external elements that distinguishes the Early Architecture from Vernacular Architecture, and enables the evolution from the first to the second (Dias et al. 1969 & Oliveira et al. 1998). In these situations, it is imperative to open a conceptual bracket, and to reshape the ontological reflection. In the current context, in any constructive culture in development, it is hard to alienate the phenomenon of industrial production and global cultural contamination (AlSayyad, 2006).

Nonetheless, the recognition of these new variables and its profound implications is far from determining the indiscriminate acceptance of models and techniques underlying it and that many want to impose by force of circumstances – or by fashion. The concept of vernacular architecture can and should be responsive to emerging issues. It is part of its nature. However, its unjustified flexibility should be carefully avoided and openly discussed.

ACKNOWLEDGMENTS

This paper is supported by FEDER Founding through the Operational Programme Competitivity Factors – COMPETE and by National Funding through the FCT – Foundation for Science and Technology within the framework of the Research Project 'Seismic V – Vernacular Seismic Culture in Portugal' (PTDC/ATP-AQI/3934/2012).

REFERENCES

AlSayyad, N. (2006). Foreword. In L. Asquith & M. Vellinga (eds). Vernacular Architecture in the 21st Century: Theory, Education and Practice. London: Taylor & Francis, p.xvii

Asquith, L. & Vellinga, M. (Eds.) (2006). Vernacular Architecture in the Twenty-First Century: Theory, education and practice. London: Taylor & Francis.

Baker, H. G. (1999). Análisis de la forma: Arquitectura e Urbanismo. Barcelona: Editorial Gustavo Gili.

Calatrava, J. (2007). "Leopoldo Torres Balbas: Architectural Restoration and the Idea of "Tradition" in Early Twentieth-Century Spain" in Otero-Pailos, J. (ed.) (2007). Future Anterior, Volume IV, Number 2. New York: University of Minnesota Press. p.40–49

Cerqueira, J. (2005). "O Estilo Internacional Vs. A Arquitectura Vernácula: O conceito de Genius Loci" Idearte – Revista de Teorias e Ciências da Arte. Ano 1, n°2, 2005. Porto. p.41–52

Demangeon, A. (1920). L'habitation rurale en France, essai de classification. Paris: Annales de Géographie.

Dias, J., Oliveira, E. V., Galhano, F. & Pereira, B. (1969). Construções Primitivas em Portugal, 1ª ed. Lisboa: Instituto da Alta Cultura.

Diez-Pastor, C. (2012). "Architectural Koinè: Architectural Culture and the Vernacular in 20th Century Spain" in ESAP-CEAA. Surveys on Vernacular Architecture, Their significance in 20th century architectural culture. Conference proceedings. Porto: ESAP-CEAA, p.182–201

Duarte Carlos, G. (2014). O Legado Morfológico da Arquitectura Vernácula. PhD Thesis. Coruña: ETSA|UdC

Flores, C. (1973). Arquitectura Popular Española. Vol. II. Madrid: Aguilar.

Frey, P. (2010). Learning from the Vernacular///. Towards a new vernacular architecture. Lausanne: Actes Sud

Gárcia Mercadal, F. (1926). Arquitectura mediterránea (I) [Mediterranean architecture]. Arquitectura, n°85, pp.192–197.
Llano Cabado, P. (1983). Arquitectura Popular en Galicia. Vol. I e II; Vigo: COAG.
Llano Cabado, P. (1996). Arquitectura Popular en Galicia: Razón e construción. Coruña: Edicións Xerais de Galicia.
Norberg-Schulz, C. (1967). Intentions in Architecture. Cambridge, MA: MIT Press.
Oliver, P. (ed.) (1997). Encyclopaedia of Vernacular Architecture of the World. Cambridge: Cambridge University Press.
Oliver, P. (2003). Dwellings: The Vernacular Houses World Wide. London: Phaidon Press.
Oliver, P. (2006). Built to meet needs: Cultural Issues in Vernacular Architecture. Oxford: Architectural Press.
Oliveira, E. V. & Galhano, F. (1998). Arquitectura Tradicional Portuguesa. Colecção Portugal de Perto. Lisboa: Publicações Dom Quixote.
Ordem dos Arquitectos (2004). Arquitectura Popular em Portugal. 4ª ed. Lisboa: Ordem dos Arquitectos. [1ª ed., SINDICATO NACIONAL DOS ARQUITECTOS, 1961. Lisboa: S.N.A.]
Papanek, V. (1995). Arquitectura e Design. Ecologia e Ética. Lisboa: Edições 70.
Rapoport, A. (2006). "Vernacular design as a model system" in Asquith, L. & Vellinga, M. (Eds.). Vernacular Architecture in the Twenty-First Century: Theory, education and practice. London: Taylor & Francis. pp.182–183
Rapoport, A. (1972). Vivienda y Cultura. Barcelona: Editorial Gustavo Gili. [1ª ed., 1969. EnglewoodCliffs, NJ: Prentice Hall]
Rudofsky, B. (1990). Architecture Without Architects: A Short Introduction to Non-Pedrigreed Architecture. Exhibition Catalogue, Museum of Modern Art (MoMA) New York, 9-11-1964 to 7-2-1965. 3ª ed. Albuquerque: University of New Mexico Press.
Sabatino, M. (2011). Pride in Modesty: Modernist Architecture and the Vernacular Tradition in Italy. Reprint edition. Toronto: University of Toronto Press, Scholarly Publishing Division.
Toussaint, M. (2009). Da arquitectura à teoria e o universo da teoria da arquitectura em Portugal na primeira metade do século XX. PhD Thesis. Lisbon: UL – FA.
UN-HABITAT: United Nations Human Settlements Programme (2006). The State of the World's Cities Report 2006/2007. 30 Years of Shaping the Habitat Agenda. London: Earthscan and UN-Habitat
Vellinga, M. (2013). 'The noble vernacular'. In The Journal of Architecture 18 (4), pp. 570–590
Vellinga, M., Oliver, P. & Bridge, A. (2007). Atlas of Vernacular Architecture of the World. London: Routledge.

Seismic Retrofitting: Learning from Vernacular Architecture – Correia, Lourenço & Varum (Eds)
© 2015 Taylor & Francis Group, London, ISBN 978-1-138-02892-0

Seismic-resistant building practices resulting from Local Seismic Culture

J. Ortega & G. Vasconcelos
ISISE, Faculty of Engineering, University of Minho, Guimarães, Portugal

M.R. Correia
CI-ESG, Escola Superior Gallaecia, Vila Nova de Cerveira, Portugal

ABSTRACT: Considering that vernacular architecture may bear important lessons on hazard mitigation, this chapter focuses on the European Mediterranean countries and studies traditional seismic-resistant architectural elements and techniques that local populations developed to prevent or repair earthquake damage. This area was selected as a case study because, as a highly seismic region, it has suffered the effect of many earthquakes along the history and, thus, regions within this area are prone to have developed a Local Seismic Culture. After reviewing seismic resistant construction concepts, a wide range of traditional construction solutions that, in many cases, have shown to improve the seismic performance of vernacular constructions of these regions is presented, as a contribution to the general overview of retrofitting building systems provided in this book. The main motivation is that most of these techniques can be successfully applied to preserve and to retrofit surviving examples without prejudice for their identity.

1 INTRODUCTION

The present chapter deals with vernacular architecture earthquake preparedness, and the methods adopted by local communities to repair and restore their dwellings in the Mediterranean region. Being an important seismic area within Europe, since the Mediterranean-Himalayan belt is responsible for 15% of the world seismic activity, Mediterranean communities have been exposed to long-term important recurrent earthquake hazard along the history and, subsequently, had to adjust to this risk, and had to make decisions, implementing plans and taking action for the protection of their built-up environment.

These efforts made by local populations led to the development of a Local Seismic Culture (Ferrigni, 1990). European Mediterranean regions, where local population have undertaken preventive measures aiming at minimising future losses in following earthquakes, gave rise to rather similar traditional seismic-resistant construction techniques. This is largely due to the traditional cultural connections between ancient and modern communities around the sea, and the fact that, because of the similar climate and geology, they share similar vernacular housing typologies, structural systems and materials.

Therefore, this chapter presents a comprehensive overview of the most common seismic-resistant provisions that can be traditionally identified in European Mediterranean vernacular architecture, particularly in Italy (Pierotti, 2001), Greece (Touliatos, 1992) and Turkey (Homan, 2004). These seismic resilient local building practices concern just some basic structural members of the building, or consist of an entire building structural system. In any case, the most successful ones have lasted for centuries, surviving numerous seismic events and proving their validity. As a sort of natural selection, if something has become traditional, it is because it has been effective in resisting past seismic events in the region and, more important, it can resist seismic events in the future.

2 CHARACTERISTICS OF VERNACULAR SEISMIC-RESISTANT CONSTRUCTIONS

Traditionally, the best and costliest materials, as well as the most advanced techniques, were traditionally reserved for temples and monumental buildings, as they were the buildings that were conceived to last over time. The sturdiest types of masonry were used to build bulky constructions, able to resist very large earthquakes, based solely on the strength, rigidity and good quality of the materials. However, the basic seismic-resistant concepts that eventually took root in the vernacular building culture of a seismic prone region had to make use of affordable and locally available materials. In addition, they also had to develop simple practices concerning construction aspects affecting the seismic vulnerability of their buildings other than the quality of the materials.

For example, regarding the geometry of their buildings, the main requirement is that buildings should be simpler, in order to have more seismic stability. The building should present symmetry in terms of mass and stiffness, both in plan and elevation, in order to reduce torsion; low height to base ratio, in order to minimise the tendency to overturn; and a compact plan with a low length to width ratio, in order to have similar resistance in every direction. Uniform elevation with equal floor heights and a low centre of gravity also reduce the building's vulnerability. The number of openings should be reduced, carefully and symmetrically distributed, with their frames properly reinforced.

With respect to construction solutions and materials, local populations acknowledged and accepted that it is not economically viable to construct every building to resist earthquakes without suffering deformation and damage, but the collapse of the structure must always be avoided. Thus, in order for the building to be able to deform while keeping the building standing, ductile materials, such as timber, are required, so that they can resist the tensile stresses. Enhancing the deformability of the structure requires that the load-carrying components are well coupled together, in order to form closed contours in vertical and horizontal planes. This way, the stress concentrations are avoided and forces are transmitted from one component to another even through large deformations. Another key aspect for a building to sustain damage without total collapse is the redundancy of the structural elements, so that failure of certain members is tolerated.

Lastly, the building seismic vulnerability will be highly reduced if it is in a good state of conservation, requiring proper maintenance and adequate post-earthquake repair and strengthening works. The potential resilience to earthquakes of vernacular constructions presenting earthquake resistant characteristics is considerable and worthy to be studied and recognised. However, due to the common lack of maintenance, vernacular constructions are extremely vulnerable to earthquake damage and need awareness and protection.

3 TRADITIONAL SEISMIC RESISTANT BUILDING PRACTICES

Although many of the following traditional practices may not have been originally conceived as earthquake resistant measures, they are actually efficient in enhancing the structural performance of buildings during earthquakes. Their use was spread out along the Mediterranean countries because, after an earthquake, reconstruction works tended to copy those designs that withstood the event and thus, these practices can be recognised as evidences of a Local Seismic Culture. However, that is also the reason why some of them can also be found in regions with low earthquake hazard.

3.1 *Elevation configuration*

As previously stated, a low centre of gravity reduces the building's vulnerability because it provides greater stability to the structure by concentrating more mass towards the ground. For that purpose, scarp walls have been used since the earliest civilizations, as they decrease the thickness of the upper floors' walls, and provide light timber floors. Another very common practice in regions, where stone and timber are available local materials, such as in the city of Xanthi, in Greece (Papadopoulos, 2013), was the combination of two different structural systems. The ground floor is built with heavy stone masonry walls, while a timber frame structure is used in the upper floors. The use of lighter stone masonry in upper floors is also a common technique in many masonry constructions in Italy (Ferrigni et al., 2005).

3.2 *Use of timber elements*

A very common vernacular practice that can be observed in many seismic prone areas of the Eastern Mediterranean countries, such as Greece (Vintzileou, 2011) and Turkey (Homan, 2004), consists of imparting ductility to the masonry wall by inserting timber elements within the wall, as reinforcement. The good seismic performance of this practice has been reported in many past earthquakes, such as in the 1999 Marmara earthquake in Turkey (Gülhan and Güney, 2000).

This technique dates as far back as the Minoan civilization in Bronze Age Crete. When applied within the rubble and ashlar masonry walls, these embedded timber reinforcements were, in many cases, sophistically arranged constituting a structural timber frame, extending from the foundations to the roof. In some other cases, just a few vertical or horizontal timber elements were inserted inside the walls, sometimes consisting of a rough timber grid of horizontal timber trunks or tree branches, lying longitudinally and transversally at different levels of the wall. Due to its continued use during the last 35 Centuries, this practice has nowadays become endemic of the vernacular way of building in these regions, as part of a Local Seismic Culture (Fig. 1).

The insertion of timber elements within the masonry is clearly a strengthening method, as their excellent tensile properties allow them to constitute successful slip planes and both, vertical and horizontal, shock absorbers, helping to dissipate relevant amounts of energy. In addition, by confining the masonry, they enhance its bearing capacity, its compressive strength, its shear strength, and its deformability properties

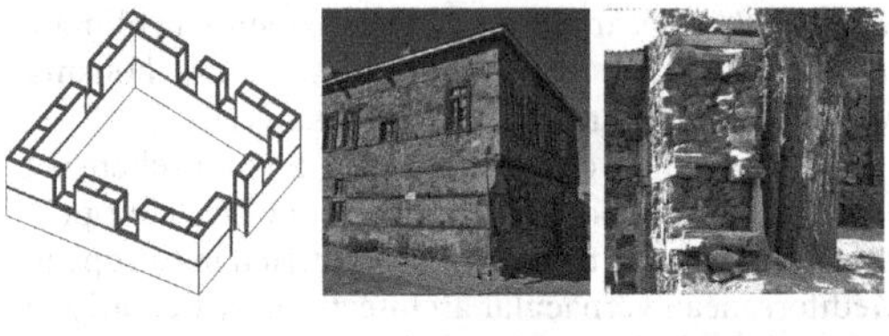

Figure 1. Typical timber reinforcements of traditional houses in: (left) Northern Greece (Touliatos, 2001); (right and middle) Erzurum, Turkey, known as *hatıl* (Inan, 2014).

(Vintzileou, 2008). Another main function of this technique is the fact that, by introducing timber elements at different levels within the height of the walls, longitudinally and transversally, it connects the different structural elements among themselves, tying the building and enhancing its box-behaviour. Moreover, when applied in multiple-leaf masonry walls, these longitudinal and transverse timber beams can help to increase the integrity of the entire wall, by tying the faces and preventing them from delamination.

3.2.1 *Structural timber frames: Historical earthquake protection regulations*

There are several particular cases throughout history, in which devastating earthquakes induced the development of official regulations for the reconstruction of the city through a post-earthquake concerted response that involved the government of that time. Some of these regulations included the design and introduction of new seismic-resistant construction systems, all of them based on the use of a structural timber frame. The most well-known example is the *Pombalino* building system, introduced in Portugal after the destructive 1755's Lisbon earthquake.

Nevertheless, this took place also in Calabria, in Italy, after the 1783's earthquake. A similar earthquake resistant system was developed after a scholarly commission was appointed by the government to study earthquakes, and to recommend reconstruction policies (Tobriner, 1983). A better seismic behaviour of timber structures during the earthquake was reported, forming the basis of the new system, known as *casa baraccata*. Timber elements included vertical, horizontal, diagonal bracing members, and transverse components, linking the two wall faces. Local communities embraced this system, acknowledging its good seismic-resistant characteristics, and they have continued to use them, becoming part of the traditional way of building of those regions and their Local Seismic Culture. Several ruined buildings testify the application of this system in reconstructed towns and cities in Calabria (Fig. 2).

Another example occurred in the island of Lefkas, in Greece, where the periodic recurrence of earthquakes led the inhabitants to improve the seismic resistance of their constructions, and to work out an indigenous structural system that effectively resisted earthquake loading. This system emerged from a long traditional practice, being a very illustrative example of the development of a Local Seismic Culture (Porphyrios, 1971). Additionally, after the 1825's destroying earthquake, its use was also imposed by the English government, who occupied the island at the time (Touliatos, 1992).

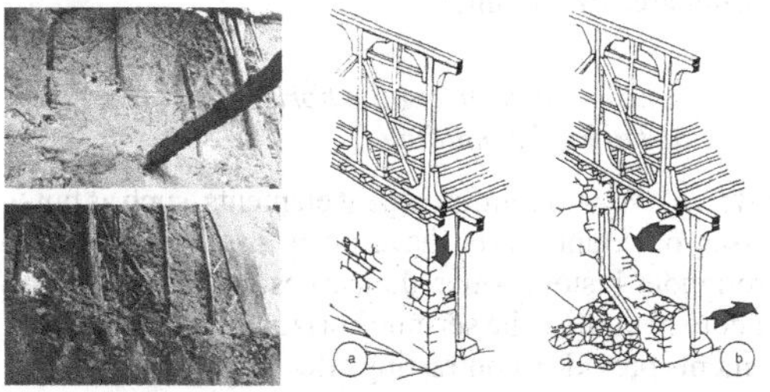

Figure 2. (left) *Casa baraccata* timber frame system in a vernacular construction in Calabria (credits: Tobriner, 1983); (right) Dual bearing structure in Lefkas Island, Greece (credits: Ferrigni et al., 2015).

Its most significant seismic resistant characteristic is the structural redundancy. On the ground floor, the buildings are constructed with load bearing thick masonry walls, but an independent timber frame is also present as a secondary structure. This way, the masonry walls can collapse in the event of an earthquake, which will tend to be thrown toward the exterior, because of the presence of the timber frame in the interior. The timber structure will not collapse, keeping the building standing and the roof intact, since it is supported by the timber frame, and thus, protecting the people inside the building. The masonry walls can be easily and rapidly repaired (Fig. 2).

Additionally, the upper floor is built with a highly perfected timber frame. So the weight is reduced, and the centre of gravity is lowered. The timber frame is composed by vertical, horizontal and diagonal members, forming different compartments filled with bricks held together with mortar. Timber elbows are also used to stiffen the connections between the vertical and horizontal elements, and to maintain the geometrical integrity of the structure. Partition walls are also built entirely in timber, and are of negligible weight. Today, this system is still common and widespread in the island, and has proven to behave well against earthquakes, such as in the 2003's earthquake, when none of them suffered total collapse, even though there were cases of three-story reinforced concrete buildings that did (Karakostas et al., 2005).

3.3 *Connection between structural elements*

Proper connections are essential for the vertical structural elements not to behave independently, ensuring the box behavior of the building so that the horizontal forces can be absorbed by walls in the same plane. This is one of the most effective measures against earthquakes, as the in-plane resistance of the masonry is significantly higher than its out-of-plane resistance (Lourenço et al. 2011). However, a full multi-connected box is often very far from reality in vernacular architecture. In many cases, single walls work separately, having to bear, by themselves, the portion of load that acts on them.

Traditionally, quoins were used to improve the connections between walls at the corners. The best quality, large and squared stone blocks were used at the corners to improve the adequate connection of the façades of the building, and to prevent their overturn, by creating efficient overlapping of the ashlars with the rest of the wall. They are a very common element in the stone masonry vernacular architecture in the Mediterranean countries.

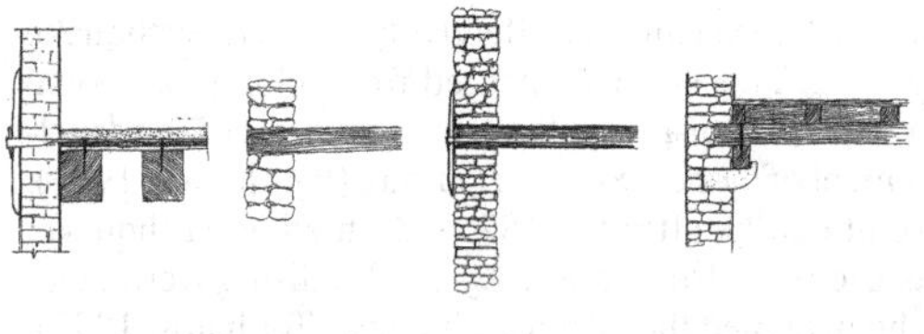

Figure 3. Examples of common reinforced floor-to-wall connections found in the Veneto region, Italy (credits: Barbisan & Laner, 1995).

Improving wall-to-floor and wall-to-roof connections has also been always a concern for builders in seismic areas. Many technical construction manuals arose during the nineteenth and early twentieth century in Italy, describing detailed methods on how to properly connect the floors to the vertical masonry walls (Barbisan and Laner, 1995), also acknowledging the importance of this aspect in seismic resistant building systems. As technical solutions, they were not commonly applied in vernacular architecture, but other traditional devices, which tried to copy them and to achieve the same effect, can be usually found.

Reinforced floor and roof-to-wall connections are traditionally achieved using wooden wedges to ensure a tight connection between the walls and the floor, or roof joists that pierce them. Also, transition elements, such as timber resting plates, or stone brackets, are applied to improve these connections. Metallic anchoring devices, such as metal brackets or steel straps, can also be found in vernacular architecture reinforcing the connections (Fig. 3).

3.3.1 *Ties*

The application of ties, making effective links to hold together the different parts of masonry structures, might be the most common ancient strengthening practice adopted to ensure the box behaviour of the building, and to improve its structural integrity. Given the fact that ties are relatively easy to implement in existing structures, before or after earthquake damage, they have been widely used for many centuries, and they can be systematically observed in highly seismic regions of Mediterranean Europe. They are introduced as a reinforcement measure, used to connect perpendicular load bearing walls, load bearing walls to interior walls, parallel load bearing walls, walls to floors and walls to roofs.

Ties connecting perpendicular walls provide lateral bracing. Ties connecting parallel walls are intended to avoid their out-of-plane collapse, but also to constrain the floors, facilitating the transfer of the load to the bracing orthogonal walls in the same plane, and improving the overall performance of the system. Actually, a common practice to vernacular architecture is the use of their own timber floor joists as ties between parallel walls. Ties have to be well restrained at the ends, commonly by steel anchor plates, in the case of steel tie rods, or, by wedges, in the case of wooden tie beams.

3.3.2 *Traditional jointing system*

An important feature about the connections is the type of joints used. In vernacular architecture, the jointing system for wooden structural elements has been traditionally made through flexible housed joints and wedges, which were actually an effective energy dissipation system in the event of an earthquake, because, while allowing the tightening of the joints, they effectively act as pin joints, also allowing some movement within the joints.

3.4 *Stabilisation of floors and roofs*

Concerning the stabilisation of roofs and floors, the traditional approach has consisted of improving their diaphragmatic behaviour by reducing their excessive deformability, and by adding in-plane and flexural stiffness. In this way, they are able to transfer the loads for a given direction of motion from the out-of-plane walls to the in-plane walls. This has been traditionally achieved through diagonal bracing and triangulation. A significant example, illustrating a Local Seismic Culture, can be found in Galaxidi, on the seismically hazardous Corinthian Bay in Greece, where the typical structural system applied consists of stiffening the ceiling through triangulation and proper coupling with the timber reinforcing components, located on top of the masonry walls (Touliatos, 2001).

3.5 *Reinforcement of the openings*

Seismic-resistant vernacular constructions usually present a reduced number of openings, and symmetry in their layout. Closed-up openings can be commonly identified in seismic prone areas, showing the inhabitants' awareness of the vulnerability of these elements. Several ways of reinforcing openings can be commonly observed, such as the use of relieving or discharging arches inserted within the wall, over the openings lintels. These are intentioned to lighten the load of the underlying element, and to better distribute the load path. Windows and doorframes are also traditionally reinforced with big stone or timber lintels, aimed at promoting enough resistance to bending stresses. Brackets are useful for reducing the free span of the lintel; and jambs are necessary because of the strong compression forces that concentrate in the bearing area of the lintel.

3.6 *Elements neutralizing the horizontal forces exerted by the building*

Different types of reinforcement elements, such as buttresses or counterforts, have been also widely used throughout history, since the earliest civilizations, in order to neutralise the seismic horizontal forces. These elements provide a contrasting effect against the buckling tendency of a wall, and are very common in most seismic prone regions in the Mediterranean (Pierotti, 2001).

Buttresses are the most common strengthening measure, aimed at counteracting the horizontal forces exerted by the building during earthquakes. These are also very commonly recognised in vernacular constructions. They consist of pier-like, massive local additions, generally built of masonry, whose working principle is to counter the rotation of the façade thanks to their sheer mass. They can be built at the same time as the building, as a deliberated feature, or they can be added on to older masonry as a reinforcement measure.

3.6.1 *Urban reinforcing measures*

In urban environments, other elements that perform a similar reinforcement task are the reinforcement arches, also known as buttressing arches. These are usually made of masonry and span the street, joining facing buildings. These alterations of the historical built-up areas effectively enhance the interaction between buildings, and lead to the collaborative action of neighbouring constructions and structural elements, enabling the horizontal movements to be redistributed among their vertical walls.

Urban reinforcement arches and buttresses can eventually transform into other urban elements that accommodate new uses, since their construction results in an increase in volume and new space available for the building. Therefore, these added structures can eventually become habitable and turn into loggias, vaulted passageways or arcades, fulfilling, simultaneously, a structural and a functional role, with the addition of new paths and rooms. Sometimes other urban structures, such as external stairs, can also fulfil a similar role, counteracting the rotation of the walls. These reinforcements are the characteristic historical solution to avoid the development of out-of-plane mechanisms, at an urban level, in villages built mainly of stone masonry, such as the Italian. These have become, indeed, part of their historical fabric (Fig. 4). They are a distinctly reinforcement measure because, when added to the buildings, they took space from the public use, narrowing the public space with a subsequent discomfort for the inhabitants and, thus, showing their seismic concern.

3.7 *Position within urban fabric*

Finally, the interaction between buildings has also a significant influence in their seismic performance. Different responses to the seismic action by neighbouring buildings can cause damage in the connecting borders, where stress concentrations are present. Different stiffness of the bodies like, for instance, reinforced concrete buildings adjacent to masonry house, introduce a severe risk of hammering actions to take place. However, an interaction between buildings can also have beneficial effects, and even prevent earthquake damage. Actually, the common vernacular tradition consists of making the neighbouring buildings to collaborate and to reinforce each other. Historical city centres are usually composed of many single buildings adjacent to one another, and structurally connected, thus achieving a structural continuity, and accordingly, reacting uniformly to seismic loading.

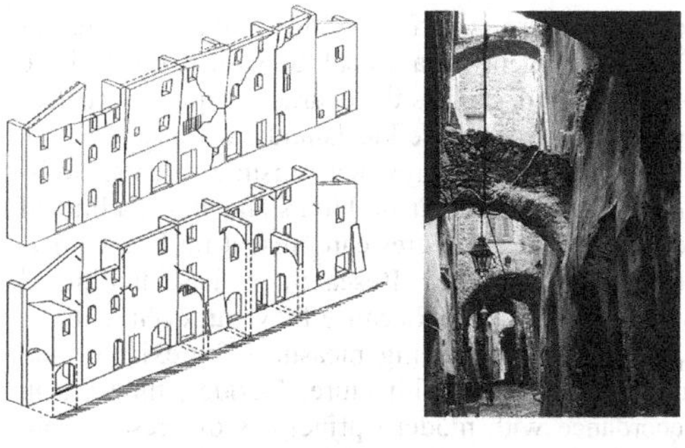

Figure 4. Building complexes in: (left) Anavatos village in Chios Island, Greece (Efesiou, 2001); (right) Mandraki, in Nysiros Island, Greece (credits: Ferrigni et al., 1995).

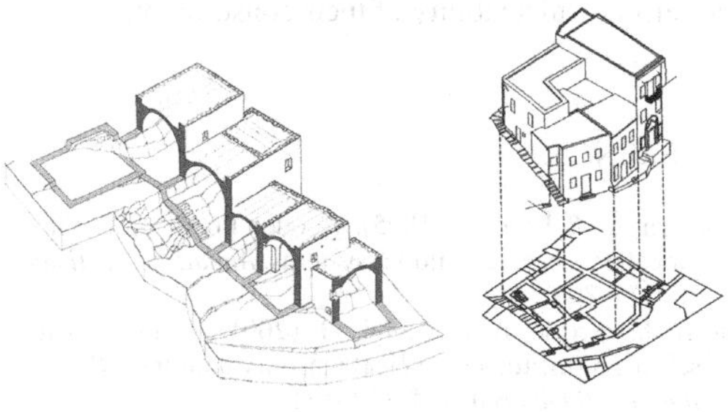

Figure 5. (left) Historical solution to out-of-plane mechanisms at an urban level (Borri et al., 2001); (right) Reinforcement arches in Dolce-Aqua, Italy (credits: Ferrigni et al., 1995).

There are several examples of strategies to resist earthquakes involving their position at the urban fabric (Fig. 5). In Chios Island, Greece, buildings were usually constructed in contact to one another, trying to make them to cooperate and to reinforce each other, by equilibrating the horizontal forces exerted by the domes, and providing more stable dynamic units (Efesiou, 2001). In Mandraki, main village on Nysiros Island, in Greece, the intricate composition of the historical building complexes also ensures a unified behaviour under horizontal loading (Touliatos, 2001).

4 CONCLUSIONS

This paper provides an overview of the most common seismic-resistant provisions, traditionally used in the vernacular architecture across the Mediterranean Sea, focusing on those construction characteristics that most influence their seismic behaviour: geometry, materials and construction solutions, openings characteristics, use of reinforcement elements and position within urban fabric. A wide range of traditional solutions, where each of these aspects can be observed, together with a cohesion in the use of specific seismic-resistant features, are present in some of the Mediterranean countries.

Even though associating changes or innovations in the construction techniques to the existence of a seismic culture is difficult, illustrative examples of the development of a Local Seismic Culture have been reviewed, such as the characteristic constructive system that arose in Lefkas Island, in Greece.

As reported in previous seismic events, well-constructed vernacular buildings showing traditional seismic-resistant features can present far less vulnerability than expected. Research in these traditional practices is justified, because they can eventually be applied as strengthening measures for existing and in-use vernacular architecture. Besides, they are in accordance with modern principles of preservation, regarding compatibility and authenticity, since they use similar materials and techniques than the original structures. Local communities should be encouraged to readopt some of these techniques, in order to reduce the seismic vulnerability of their constructions.

REFERENCES

Barbisan, U. & Laner, F. (1995). Wooden floors: part of historical antiseismic building systems. *Annali di Geofisica* 38, 775–784.

Borri, A., Avorio, A. & Cangi, G. (2001). Guidelines for seismic retrofitting of ancient masonry buildings. *Revista Italiana di Geotecnica* 4, 112–121.

Efesiou, I. (2001). Constructional analysis of the local structural system of the historic settlements of Anavatos in Chios island. In *International Seminar of Restoration of Historic Buildings in Seismic Areas: The Case of Settlements in the Aegean*. Lesvos Island, Greece.

Ferrigni, F. (1990). *À la recherché des anomalies qui protégent. Actes des Ateliers Européens de Ravello, 19–27 Novembre 1987.* Ravello: PACT Volcanologie et Archéologie & Conseil de L'Europe.

Ferrigni, F., Helly, B., Mauro, A., Mendes Victor, L., Pierotti, P., Rideaud, A. & Teves Costa, P. (2005). *Ancient Buildings and Earthquakes. The Local Seismic Culture approach: principles, methods, potentialities*. Ravello: Centro Universitario Europeo per i Beni Culturali, Edipuglia srl.

Gülhan, D. & Güney, I.Ö. (2000). The Behavior of Traditional Building Systems Against Earthquakes and its Comparison to Reinforced Concrete Frame Systems; Experiences of Marmara Earthquake Damage Assessment Studies in Kocaeli and Sakarya. *International Conference on the Seismic Performance of Traditional Buildings*. Istanbul, Turkey.

Homan, J. (2004). Seismic Cultures: Myth or Reality?. *Second International Conference on Post-Disaster Reconstruction: Planning for Reconstruction*. Coventry, UK.

Inan, Z. (2014). Runner beams as building element of masonry walls in Eastern Anatolia, Turkey. In M. Correia, S. Rocha & g. Carlos (Eds.), *Vernacular Heritage and Earthen Architecture: Contributions for Sustainable developments*: 721–726. London: Taylor & Francis Group.

Karakostas, C., Lekidis, V., Makarios, T., Salonikios, T., Sous, I. & Demosthenus, M. (2005). Seismic response of structures and infrastructures facilities during the Lefkada, Greece earthquake of 14/8/2003. *Engineering Structures* 27(2), 213–227.

Lourenço, P.B., Mendes, N., Ramos, L.F. & Oliveira, D.V. (2011). On the analysis of masonry structures without box behavior. *International Journal of Architectural Heritage: Conservation, Analysis, and Restoration* 5(4–5), 369–382.

Papadopoulos, M.L. (2013). Seismic Assessment of Traditional Houses in the Balkans – Case Studies in Xanthi. *Journal of Civil Engineering and Science* 2(3): 131–143.

Pierotti, P. (2001). *Culture sismiche locali*. Pisa: Edizioni Plus Università di Pisa.

Porphyrios, D.T.G. (1971). Traditional Earthquake-Resistant Construction on a Greek Island. *Journal of the Society of Architectural Historians* 30(1): 31–39.

Tobriner, S. (1983). La Casa Baraccata: Earthquake-Resistant Construction in 18th-Century Calabria. *Journal of the Society of Architectural Historians* 42(2): 131–138.

Touliatos, P.G. (1992). Traditional aseismic techniques in Greece. In L. Mendes Victor (ed.), *Proceedings of the International Workshop "Les systemes nationaux faces aux seismes majeurs"*. Lisbon: Centro de Geofisica, Universidade de Lisboa.

Touliatos, P.G. (2001). The box framed entity and function of the structures: The importance of wood's role. In *International Seminar of Restoration of Historic Buildings in Seismic Areas: The Case of Settlements in the Aegean.* Lesvos Island, Greece.

Vintzileou, E. (2008). Effect of Timber Ties on the Behavior of Historic Masonry. *Journal of Structural Engineering* 134(6): 961–972.

Vintzileou, E. (2011). Timber-reinforced structures in Greece: 2500 BC-1900 AD. In *Proceedings of the Institution of Civil Engineers (ICE), Structures and Buildings* 164(SB3): 167–180.

Seismic Retrofitting: Learning from Vernacular Architecture – Correia, Lourenço & Varum (Eds)
© 2015 Taylor & Francis Group, London, ISBN 978-1-138-02892-0

Practices resulting from seismic performance improvement on heritage intervention

R.F. Paula & V. Cóias
STAP – Reparação, Consolidação e Modificação de Estruturas, S.A., Lisbon, Portugal

ABSTRACT: Significant effort is being made to develop and install new solutions capable of enhancing the strength, ductility and energy dissipation capacity of ancient buildings, whilst respecting their original structural concept, and therefore their authenticity. Seismic performance improvement of historic constructions can be addressed by means of low intrusive interventions. A set of complementary low intrusive systems and techniques, designed for structural rehabilitation and seismic retrofitting, are presented, as well as a representative case study.

1 INTRODUCTION

The basic structural concept of ancient buildings relied on resistant masonry walls and on interior timber components such as the roof, floors and walls. In seismic prone areas, structural timber elements assemblage represented a very important feature in bracing the masonry walls, and enhancing the lateral stability of the constructions. This is particularly evident in the case of the Portuguese *Pombalino* buildings.

During a building lifetime, uncontrolled structural changes and lack of maintenance can negatively affect its strength and seismic capacity. Seismic vulnerability of ancient buildings is mostly due to: a) Significant modifications of the constructions, and of its original structural concept, like the adding of extra floors, the removal of walls and columns, notably on the ground floor, inadequate widening of openings in façades and the addition of steelwork and reinforced concrete elements (Cóias, 2007) (Lopes, 2010); b) Aging or deterioration of materials caused by harmful environments, affecting the structural capacity of the resistant elements; for example, fungal decay of timber exposed to water infiltrations; c) Deficiencies resulting from the original project and construction (low quality construction).

Significant effort is being made to develop and validate new seismic retrofit solutions, based on a low intrusive approach, with the aim of preserving and respecting historical buildings.

2 SOLUTIONS TO IMPROVE THE STRUCTURAL BEHAVIOUR

2.1 *Low intrusive approach*

As issued by ICOMOS Recommendations (ICOMOS, 2003), interventions in heritage buildings should be attained through minimum works necessary to guarantee safety and durability, and with the least damage to heritage values. Preference should be given to the materials and techniques that are the least invasive and the most compatible with the worthy features of the historic buildings.

When defining the strategy for the intervention, the basic principle to be considered is the conservation of the original construction concept and of the resistant components. Repair or strengthening of the original elements should be limited to the strictly necessary to preserve their structural function. Repair techniques ought to make use of compatible materials and lead to the restoring of the initial conditions of the elements, eliminating the anomalies. Strengthening should be achieved through low intrusive methods, resorting to advanced products or systems if required.

Seismic performance improvement solutions, capable of enhancing the strength, ductility and energy dissipation capacity of the buildings, while respecting their original structural concept, have been developed and installed. New products and techniques can play an important role in the seismic rehabilitation of constructions, by improving the structural behaviour of main masonry and timber framed walls, timber floors and connections between structural elements.

In order to ensure a satisfactory intervention, the works should be done by experienced and qualified operatives, and appropriate quality control procedures should be carried out.

2.2 *Reinforcement of masonry walls*

The most common structural rehabilitation techniques comprehend reconstruction of sections, consolidation injections, rendering of surfaces with a reinforcement material, and installation of transverse tie bars.

The first two techniques are used to recover and consolidate less resistant sections, often applied before strengthening. The injection of masonry walls, under controlled pressure and using a grout that contains an inorganic binder, improves its strength properties, as a result of its increased cohesion and density. This procedure is also used to seal cracks or fill voids.

Insertion of transverse tie bars through the thickness of a masonry wall is used to confine masonry or connect two leaves of a wall, increasing the compressive resistance of the element. Another strengthening method consists in the application of a reinforced render. However, if inappropriate or non-compatible materials are overlaid, the masonry will not have an adequate physical and mechanical behaviour.

To comply with the basic principles of conservation of historical buildings, a new system named UHPR – Ultra High Performance Render was developed under RehabToolBox R&D Project (Guerreiro, 2014). This system is based on the use of an advanced composite that is applied to the surface of the masonry. The composite is formed by a carbon fibre mesh, embedded in an inorganic matrix, that is compatible with the old masonry substrate, to which the composite is applied (Fig. 1). Basic conception of the system also comprises a spaced grid of connectors that are put through the thickness of the wall, fastening the composite render and working as confinement devices. The application of the UHPR system to the faces of a wall aims to improve its flexural capacity regarding both in-plane and out-of-plane actions. The technique allows merging two different, although complimentary, materials: very high resistance/weight ratio fibres with an inorganic lime-based render that is compatible with the materials that compose the old resistant masonry units, and that is also adequate to these substrate units' behaviour. Moreover, the success of the new system is also related to the distinctive way the matrix is applied, through a high velocity spraying process, as a better adhesion is achieved by using this special application method.

2.3 *Rehabilitation of timber*

Maintenance and conservation of old buildings frequently involve rehabilitation of certain members or parts of timber structures. While in some cases total substitution of the existing timber is inevitable, in other situations, in order to preserve the original materials and structure or even for economic reasons, it may be wiser to adopt a partial replacement of the decayed parts, as well as the reinforcement of the existing structural elements.

Figure 1. Masonry walls reinforced with UHPR. Detail of the spraying works (credits: STAP).

In combination with traditional carpentry methods, versatile solutions for timber restoration can be deployed to bring about low intrusive interventions, with less waste of original good material, low mass and reduced visual impact (Cruz, 2004) (Paula et al., 2006a).

The primary restoration material is the original timber component that is identified, whenever possible, during the inspection with regard to species, age, current condition, i.e., moisture and visual grading. The restoration components, that are introduced to site, forming an integral part of the rehabilitated structure, may be one or a combination of the following:

- Prefabricated timber elements, usually to replace a damaged portion or the whole member (Fig. 2), but also to reinforce and increase the density of actually functioning cross-sections;
- Single or multiple configurations of bars or plates, as overall strengthening components or shear connectors.

The prefabricated timber components should be of the same species of the timber to be rehabilitated, or compatible, in terms of its mechanical properties, moisture, durability and colour. However, if the durability of the original timber is rather insufficient, regarding the particular hazard class, timber with adequate natural durability or with a selected preservative treatment may be used. Timber is lighter, compared with most alternative materials, it has a high strength/weight ratio, and it is easy to handle. These are particularly useful properties at construction sites, where access and transportation are major problems.

Figure 2. Restoration of previous conditions. Preservation of the structural concept. Reconstruction of a timber framed wall. Reinforcement of joints (credits: STAP).

Strengthening components such as bars and plates can be metallic or of FRP materials. The metallic reinforcement elements can be of stainless steel, or steel adequately protected against corrosion. FRP bars or laminates may have different compositions, for example, unidirectional glass or carbon fibres agglutinated on a matrix of epoxy resin; unidirectional glass fibres on a polyurethane thermoplastic matrix (Rotafix, 2015).

In some cases, it might be necessary to use adhesives. Adhesives are usually a two or three part epoxy systems, which are used as the interface between timber and the strengthening components. The adhesives should be specifically formulated for timber engineering, to fix anchorages in timber and to fill drilled holes and slots, to embed metallic or FRP reinforcement elements. The adhesive should have low surface tension in order to obtain a good spreading of the material.

Many different repair and strengthening configurations are possible. New timber and strengthening components are used in several ways to reinforce existing, or added timber members and/or to join two or more sections (Cóias, 2007) (Paula et al., 2006a) (Ross, 2002) (Rotafix, 2015).

Metal plates and fasteners enable joints of higher strength and stiffness (Fig. 2). These materials are also commonly used in structural repair and reinforcement operations, either being inserted in critical cross-sections, or used for load transmission, when damaged beam ends are cut off, and replaced with new timber.

To provide 'hidden' strengthening solutions, steel bars or FRP composites can be embedded in existing wooden beams and joints. The system can be designed to increase the strength of beams deficient in flexure and/or shear. End connection details can also be incorporated to assist in load transfer between elements (Paula et al., 2006b).

If it is required to increase the bending capacity of a timber beam, then an epoxy may be used to bond steel or FRP rods into small slots along the line of the beam. The reinforcement can either be in the form of bars or laminates (Paula et al., 2006b).

A large number of timber anomalies are related to damage of one end of a member. The repair of this kind of deterioration can be done by replacing the wound part with a new section, aligned with the remaining member and replicating the outline of it (Fig. 3). In order to form a continuous member, it is necessary to make an inline joint between the two sections. For example, in the case shown in Figure 3, the connection was done with FRP rods, inserted in holes filled with epoxy adhesive. Beam-ends were repaired by drilling holes in the sound material, after the decayed end was removed. Then, resin was injected into the holes, and the rods were inserted into the holes. The decayed parts were replaced by prefabricated timber components that had a top slot to receive the pre-installed rods.

Instead of being installed *in situ* in the sound timber, the reinforcement elements can be pre-inserted in the prefabricated components at the workshop. The option for the best configuration depends, for example, on the constraints of accessibility, and the available area to place the prefabricated components.

As for the example presented in Figure 3, the repair technology was chosen for both on-site accessibility requirements and cost analysis. The intervention was completely carried out from below the beams, and it was not necessary to remove the wooden lining of the floor. The selected technique permitted the rehabilitation of the timber structural elements without extra weight; and total removal of the sound timber, facilitating significant cost savings, and with little disruption to the space below. On the other hand, if it is necessary to repair or upgrade timber beams above a decorative ceiling, the works can be done totally from top.

2.4 *Connections' enhancement*

Suitable structural connections play an important role in the enhancement of seismic response. Improvement of wall-to-floor and wall-to-wall connections

Figure 3. Preservation of timber beams of a floor. Repair of decayed ends (credits: STAP).

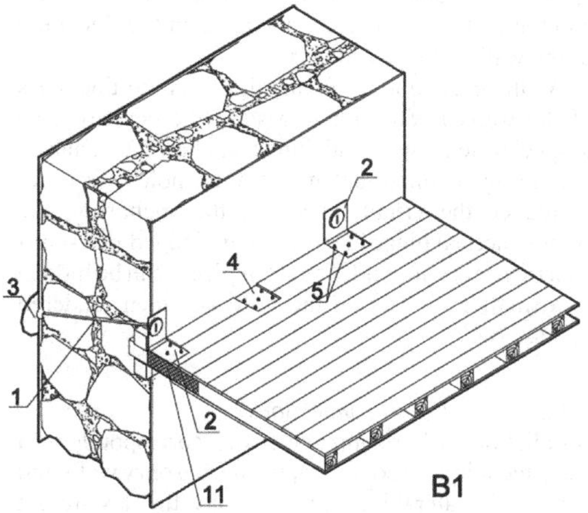

Figure 4. Wall-to-floor connecting devices (credits: STAP).

Figure 5. Masonry wall-to-wall connection enhancement (credits: STAP).

can strengthen the out-of-plane behaviour of masonry walls and overall structural stability.

2.4.1 *Wall-to-floor connections*

Low intrusive and removable devices are shown in Figure 4 (and 11). There are different types of connection devices, either to be installed in the longitudinal, or in the transversal direction of the beams. In the longitudinal direction, the L-shape plates are connected to the beams, usually with threaded bolts. In the transversal direction, the plates are bonded to the pavement. In order to assure the adhesion, a layer of glass fabric and epoxy resin composite may be put between the beam and the plate. The bars are anchored in the thickness of the walls (Cóias, 2007).

As the execution of aligned holes in the thickness of the walls is difficult to assure, the devices have a special semi-spherical hinge anchor that enables the fitting of important misaligning holes. In order to reduce the visual impact of the anchoring, the semi-spherical hinge and the nut are lodged in a semi-spherical cup with a lid. The set of pieces can be hidden in a small concavity on the wall, and then rendered (Cóias, 2007).

2.4.2 *Wall-to-wall connections*

Installation of horizontal steel ties corresponds to a usual technique to connect opposite masonry walls and enhance its lateral bracing. Currently, the ties are put at the floor levels, in two perpendicular directions, and are anchored in the masonry walls (Fig. 9 and Fig. 10).

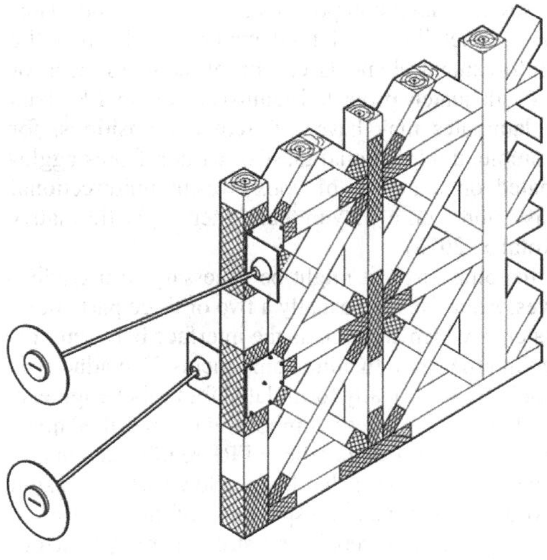

Figure 6. Wall-to-wall connecting devices (timber framed wall to masonry wall) (credits: STAP).

Another method that increases global stability and strengthening capacity is the improvement of the connection between intersecting walls. Figure 5 shows the stitching of intersecting orthogonal masonry walls through tie bars.

The internal timber framed walls of the original structure of the buildings in *Baixa Pombalina* represented a fundamental element in the bracing systems of the buildings. An important intervention in this kind of walls might be the reestablishment, and even the reinforcement, of the connection conditions to the main stone masonry façade walls. This can be achieved by means of the installation of a set of reversible connectors, as shown in Figure 6 (and 12). Similarly to the previously presented devices that improve the connections between the timber beams and the masonry, the reinforcing bars are inserted into the thickness of the main masonry walls, and anchored on the timber-framed wall by special devices (Cóias, 2007).

The L-shape plates are connected to timber with threaded bolts. Preferably, a GFRP fabric may be put between the timber and the plate. Ties bars are anchored in the masonry wall through semi-spherical hinge anchors like wall-to-floor connecting devices.

3 REHABILITATION OF A CASE STUDY: THE POMBALINO BUILDING

3.1 *The building*

The building is part of a *Pombalino* block, located in the downtown of Lisbon. *Pombalino* buildings correspond to the typology of construction that was used in the reconstruction of Lisbon after the devastating earthquake of 1755. One of the most remarkable features of the buildings is their structural concept, the distinctive *gaiola* system, which was designed to

Figure 7. Preservation of *frontal* walls. Replacement of decayed timber elements (credits: STAP).

provide the buildings with adequate seismic behaviour, enabling them to resist horizontal loads.

The *gaiola* system consists of an interior timber 3D grid, mainly formed by the timber framed walls (*frontal* walls) and the floors. These structural elements are connected to the principal stone masonry exterior walls and to the ashlars around the openings, by a set of timber members embedded along the inner face of the masonry walls. Further bracing to the two-directional vertical bracing system of the timber framed walls is provided by the timber floors roof timber trusses. The *gaiola* designation was coined because the building seemed like a big cage, with the carpentry work high up in the air.

3.2 *Conservation and strengthening*

The criteria for the rehabilitation were based on the assumption that the historic features and materials of the building were of primary importance. The works had to preserve the distinguishing quality of the structure and to respect, as far as possible, the integrity of the designer's original structural concept. The removal or alteration of any historic element should be avoided. The strategy defined for the intervention followed two basic lines: preservation and rehabilitation of all of the existing structural elements, and structural reinforcement, in view of increasing lateral stability.

First objective was achieved through maintaining the existing resistant masonry walls, as well as all the timber structural elements of the roof, floors and walls. For these elements, decayed and missing sections were restored to their original configurations. The figures 7 and 8 show the selective substitution of the deteriorated elements of the floors and of the *frontal* walls.

The reinforcement solutions comprised the installation of horizontal steel ties, to connect opposite masonry walls, and the strengthening of the joints of the timber beams over the intermediate support (*frontal* wall) – Figure 9.

The steel ties were put at the top level (ceiling) of each floor. Different exterior and interior anchor

Figure 8. Preservation of original timber structural elements of the floors. Renovation of deteriorated timber elements (credits: STAP).

Figure 9. Structural reinforcement with steel ties anchored in the masonry walls. Reinforcement of the timber beams joints (credits: STAP).

devices were used for the fixation of the horizontal ties to the façade and interior walls. In the interior, to avoid interfering with the adjacent building, specific anchor devices were applied. These devices were fixed to the wall throughout injected rods that were put inside the thickness of the masonry wall, thus not being noticed in the other face of the wall.

In the exterior, ductile anchors were used as presented in Figures 10 and 13. These anchors are less

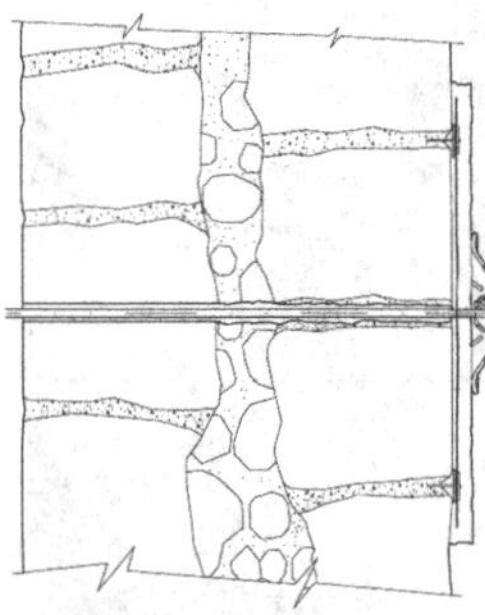

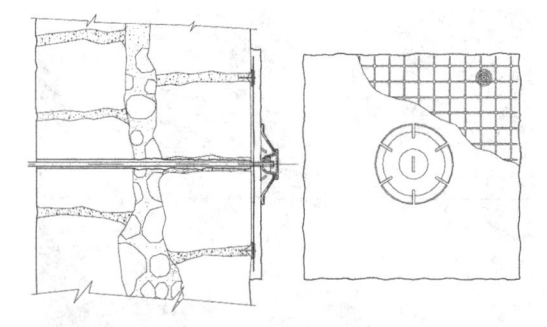

Figure 10. Exterior anchorage of the steel ties. Ductile anchors (credits: STAP).

Figure 11. Installation of wall-to-floor connection devices (credits: STAP).

Figure 12. Wall-to-wall connection devices (credits: STAP).

rigid that the traditional ones, thus improving the distribution of tensions to the walls, the deformation capacity and the ductility of the anchoring system.

At an experimental level, wall-to-floor and wall-to-wall connections' enhancement devices were also installed as shown in Figures 11 and 12.

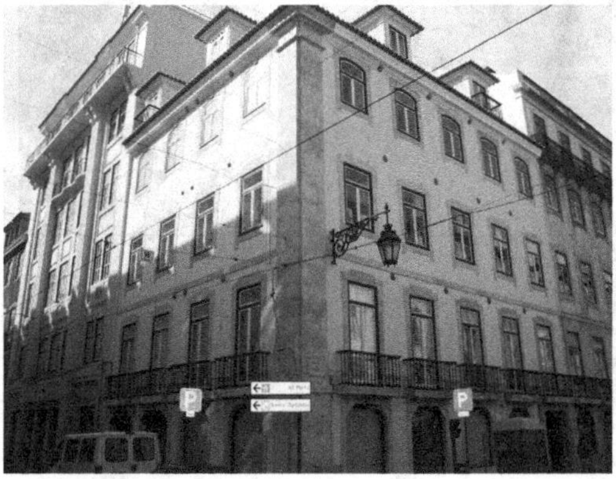

Figure 13. Final aspect of the building after intervention (credits: STAP).

4 CONCLUSIONS

In what concerns heritage buildings, minimum level interventions consistent with the need for safety and durability should be accomplished. Low intrusive and compatible solutions to improve seismic performance can be successfully applied, as briefly illustrated in this paper. In order to preserve and respect the original support, the products and systems that are used should be similar to, or compatible with the existing ones. The retrofit and restoration strategy should preserve, as much as possible, the existing structure by repairing and strengthen the elements rather than replacing them.

REFERENCES

Cóias, V. (2007). *Reabilitação Estrutural de Edifícios Antigos*, Lisbon: Argumentum.

Cruz, H. & Custódio J. (2004). Execução e Controlo de Qualidade da Reparação de Estruturas de Madeira com Colas Epoxídicas e FRPs, *CIMAD 2004*, pp.569–578.

Guerreiro, J. (2014). Out-of-plane flexural behavior of masonry walls reinforced with UHPPl, *9th International Masonry Conference, Proc. intern. conf., Guimarães, 7–9 July 2014*. Portugal.

ICOMOS (2003). Principles for the Analysis, Conservation and Structural Restoration of Architectural Heritage. Available at: http://www.icomos.org/en/charters-and-texts

Lopes, M. (2010). Construção Pombalina: Património Histórico e Estrutura Sismo-Resistente. At *8° Congresso Nacional de Sismologia e Engenharia – Sísmica 2010, Aveiro, 20–23 October 2010*.

Rotafix (2015). Available at: www.rotafix.co.uk.

Ross, P. (2002). Appraisal and repair of timber structures, London: Thomas Telford Ltd.

Paula, R., Cóias, V. & Cruz, H. (2006a). Sistema pouco intrusivo de reabilitação de madeira, Revista Pedra & Cal, Ano VII, N.° 29, Jan.–Mar. 2006.

Paula, R. & Cóias, V. (2006b, November). Rehabilitation of Lisbon's old 'seismic resistant' timber framed buildings using innovative techniques, *International Workshop on 'Earthquake Engineering on Timber Structures'*, Coimbra.

Seismic Retrofitting: Learning from Vernacular Architecture – Correia, Lourenço & Varum (Eds)
© 2015 Taylor & Francis Group, London, ISBN 978-1-138-02892-0

Criteria and methodology for intervention in vernacular architecture and earthen heritage

M.R. Correia
CI-ESG, Escola Superior Gallaecia, Vila Nova de Cerveira, Portugal

ABSTRACT: The assessment of criteria and methodology for intervention definition, its significance and the fact that there is a tendency for a non-rigorous application of the terms contributes for its inaccuracy in the field, having therefore a relevant negative impact on the decision-making process, regarding heritage intervention. This paper addresses the concepts associated with criteria and methodology for intervention in vernacular architecture and earthen heritage. A specific literature review is entailed for criteria and methodology for intervention; their frameworks are established; and key-criteria for intervention; and methodology key-components to consider for intervention are addressed. This assessment aims to contribute for a further well-defined and clear ground of work in conservation practice, leading to a more coherent and consistent intervention in heritage.

1 INTRODUCTION

When studying more rigorously the specific literature review, it was found that there was a serious lack of understanding amongst several experts, on the meaning or the need for a methodology of intervention in conservation; and the meaning or the need for devising criteria for conservation intervention.

To address the concept of criteria and methodology, a research was entailed based on a case study strategy, and using qualitative methods. Data was collected using documentary, questionnaire and interview methods from three sources: the case studies, the stakeholders and a selected group of international key-experts in built heritage conservation (Correia, 2010). The correlation of the results brought a clear contribution of the findings, concerning the identified research problem.

2 DEFINING TERMS

To contribute for a more accurate, consistent and clear approach some terms are to be defined. However, it is also important to clarify that the understanding of the meaning of the term, and its application to the structure or site can vary depending on the site context and the entailed assessment.

According to the New Shorter Oxford English Dictionary, the definition of 'criteria' entails 'a principle, standard, or test by which a thing is judged, assessed, or identified' (Brown, 1993, p. 551). This definition can help understand that criteria can be established through distinguished principles to facilitate and estimate impartial judgment.

On the other hand, according to the New Shorter Oxford English Dictionary, 'methodology' means 'the branch of knowledge that deals with method and its application in a particular field', but also as, 'the study of empirical research or the techniques employed in it' (Brown, 1993, p. 1759). Additionally, 'method' is considered 'a mode of procedure; a (defined or systematic) way of doing a thing' (ibid, p. 1759). Therefore, methodology of intervention in vernacular heritage conservation will be considered in this case, as the procedural process applied in order to prevent decay.

Finally, the definition of 'intervention', according to the New Shorter Oxford English Dictionary is 'the action or an act of coming between or interfering, esp. so as to modify or prevent a result' (Brown, 1993, p. 1401). In the present context, intervention is applied when trying to prevent decay of vernacular architecture and earthen structures and sites.

3 CRITERIA FOR INTERVENTION

Following the analysis of the research findings, it was observed that several international experts had a mixed interpretation regarding the meaning of criteria for intervention, and even mixed it with methodology for intervention. This illustrates that several of these concepts have not been clearly understood by the experts that work in a constant basis with conservation practice.

To have a clear notion on the use of criteria for intervention, it is important to develop a careful literature review on the concept; and then analyse and interpret the collected data. As a result, a criteria framework

Figure 1. The principles of authenticity and of integrity are key-criteria for intervention at World Heritage Sites In this case, they were certainly, also key-criteria for intervention at the San Estevan del Rey Mission Church, at Acoma city, NM, USA (credits: Mariana Correia, 2001).

is established, as well as the key-issues addressing criteria and key-criteria for intervention.

4 LITERATURE REVIEW REGARDING CRITERIA FOR INTERVENTION

When rereading the literature review concerning criteria, it was recognised different criteria for intervention, confusion regarding its meaning and a specific approach of the concept.

4.1 *Different criteria for intervention*

According to Correia & Walliman (2014), the decisive factors or reasons for intervention in earthen vernacular heritage are:

- *Bioclimatic criteria*, observed in Giardinelli and Conti (2000, p. 239);
- *Analytic methods criteria*, recognized in Shekede (2000, p. 170);
- *Design criteria*, identified in Guerrero Baca (2007, p. 198);
- *Conservation principles criteria*, detected in Morales Gamarra (2007, p. 262);
- *Criteria associated with the space use*, perceived in Guillaud *et al.* (2008, p. 63);
- *Sustainable principles criteria*, identified in Correia and Merten (2000, p. 229);
- *Criteria associated with values*, detected in Aguilar and Falck (1993, p. 250).

When addressing the body of literature, it was perceived that in several scientific papers, the term criteria was applied with different meanings. It was also often used to justify the intervention, not to contribute for a grounded proposal of intervening with consistency and coherency.

4.2 *Ambiguous meaning of intervention criteria*

In some of the revision of the literature, it was observed that there was confusion of concepts between:

- *Criteria & recommendations* detected in Calarco (2000, p. 22);
- *Criteria & intervention objectives*, recognised in Hoyle *et al.* (1993, p. 224);
- *Criteria & program to follow*, identified in Pujal (1993, p. 244);
- *Criteria & methodology*, observed in Orazi (2000, p. 89).

This confirms a recurrent use of the term 'criteria', not always acknowledging its accurate meaning, but even sometimes using it as a trend.

4.3 *An example of a specific criteria approach*

When addressing the identification of criteria, it was also recognised through the literature review, that Goldberg and Larson defined three types of criteria approach (1975, p. 145): *"Designative or Descriptive"; "Evaluative"*; and *"Prescriptive and Appraisive"*. In these terms, specifically in the field of earthen vernacular conservation, these approaches are also considered:

a) A *descriptive approach* occurs when it is tried to understand the site or structures by describing its different historical and technical parts, the different components and methodology applied, etc. This occurs in most of the papers presenting case study descriptions, as it is the case of Lassana Cisse (2000) and Correia and Merten (2000);
b) An *evaluative approach* takes place when it is tried to comprehend the reason for the site or the structures certain physical condition; or when it is tried to question the significance attributed to it, or to evaluate the best methodology to apply. This was recognised in Michon and Guillaud (1995);
c) A *prescriptive approach* happens when the evaluator points out immediately what to do, or how to intervene, without sometimes addressing: a comprehensive approach, an integrating documentation collection, recording, etc. This was the case of Mesbah *et al.* (2000) and Fahnert and Schroeder (2008).

Figure 2. Principle of authenticity and value of use were the key-criteria to rehabilitate the Kasbah of Ait Ben Moro, in Morocco (credits: Mariana Correia, 2006).

Figure 3. Physical condition protection was the key-criterion for intervention at Haft Tappeh, Iran. However, if the values inherent to the site (e.g. the sense of place, or the architectural value) are not recognized as criteria too, it is difficult for visitors to recognize the site's significance (credits: Mariana Correia, 2008).

The different approaches can coexist and could even interrelate on the same team work, as they complement each other. To recognise and integrate these approaches will also contribute for the interdisciplinary team to have a more proactive conservation intervention.

5 CRITERIA FRAMEWORK

The recognition of criteria is a relevant element contributing to decision-making. It can be based on indicators of quality and more than one set of guiding standards, such as **conservation principles** from conservation theory, **values** assigned by local communities, etc.

Throughout the analysis of the international expert questionnaires, 15% of the questioned experts separated conservation criteria into values based-decisions

Figure 4. Physical condition protection was the key-criterion, at Zchoga Zambil, Iran (credits: Mariana Correia, 2008).

or material based-decisions, clearly dividing into conservation through **value-assessment** of the site and preservation through **physical condition assessment** of the site. Another 15% of the questioned experts agreed that both approaches could interconnect (Correia & Walliman, 2014).

It is important to underline that criteria should be consistently maintained throughout the years, to avoid paradigmatic interventions, as it is the case of Chan Chan, where intervention criteria changed frequently and different conservation intervention trends can be recognised. A vernacular architecture site can have different criteria of intervention regarding different structures. This can result in distinct degrees of intervention concerning the different structures. However, it is fundamental that the principal of unity is present and balances the overall approach to the vernacular site.

6 KEY-CRITERIA FOR INTERVENTION

A third of the questioned international key-experts agreed that there are no universal criteria, as it will depend on the specifics of each structure or site. However, a framework of criteria was always considered important.

Following the analysis, two complementary notions of criteria were established. Only key-criteria for intervention are mentioned in the following list:

6.1 *Explicit criteria / tangible criteria*

Explicit criteria relate to guiding principles and more extrinsic characteristics of the built heritage, which means it reports to tangible issues (Correia & Walliman, 2014):

a) Conservation principles:
Authenticity; Compatibility; Uniqueness; Minimum intervention; Integrity; Reversibility of the

intervention; To consider balance between historical and aesthetical aspects.

b) Sustainable standards:
Economical sustainability; Availability of materials; Resources availability; Environmental sustainability.

c) Standards related to tangible requirements:
Minimum risk situation; To address pathologies; To try to preserve as much as possible; To consider life safety; To consider threats; To consider accessibility.

d) Preventive standards:
Continued maintenance; Continued use requirements; Maintenance capabilities; To consider functional use; Improve living conditions.

6.2 *Implicit criteria/intangible criteria*

Concerned the values that are inherent to the site (e.g. sense of place, spirituality, etc.) these are embedded in intrinsic characteristics of vernacular architecture (local building cultures, etc.). This means that it relates to intangible issues.

a) Values that define criteria:
Educational value; Historical value; Material document; Traditional value; Community value; Aesthetical value; Architectural value.

b) Intangible heritage significance criteria:
Cultural context; Sense of place; Sense of belonging; Knowledge inherent to the structure/site; History of the structure/site; Local building culture knowledge; and know-how.

There are other types of intervention criteria, which can be considered, as it is the case of design criteria, bioclimatic criteria, etc. The fundamental issue is that the use of criteria related to the intervention aims to the recognition of guiding-standards (as it is the case of principles, values, etc.), which contribute to a grounded an impartial judgment, when assessing actions required for conservation intervention.

7 METHODOLOGY FOR INTERVENTION

The consideration of methodology is important as it proposes different components to address before intervention. These components cannot only contribute to a more consistent and rigorous approach, but also to professional responsibility when carrying out conservation intervention (Correia & Walliman, 2014).

It becomes then essential, to first address a brief literature review regarding methodology of intervention, and then to address the perceptions that can be identified through methodology framework and methodology key-components.

8 BRIEF LITERATURE REVIEW CONCERNING METHODOLOGY FOR INTERVENTION

It can be observed through the assessment of the literature review that several of the different experts that intervene in heritage consider distinct phases.

On earthen heritage sites Matero (1995; 2000, 2003) addresses the process of intervention through different phases: documentation, stabilisation, interpretation and maintenance (1995, p. 7–8). In Casa Grande Earthen Ruins National Monument, Matero et al. defined four phases of intervention: documentation and condition survey, material characterisation, structural analysis, and treatment (2000, p. 54). In Mesa Verde, for instance, Matero intervene through the following components: survey, analysis, stabilisation and interpretation phases (2003, p. 39). All the case studies hereby mentioned presented a methodological approach based on the site physical condition assessment.

The Canadian Code of Ethics considers that a part of the conservation process consists on the following steps: inquiry (examination), documentation, preventive conservation, preservation, treatment, restoration and reconstruction (Earl, 2003, p. 191).

In the earthen site of Huaca de la Luna, Trujillo, Peru, Morales Gamarra proposed four different levels of approach for the conservation methodology procedure: i) preliminary recognition of the document and general characterisation; ii) preventive conservation; iii) integral conservation; iv) post conservation monitoring and systematic maintenance (2007, p. 264).

In the Casa dos Romeiros restoration intervention, a Portuguese vernacular architecture site, Correia and Merten addressed the following phases: documentation, rapid assessment, survey, intervention for the site safety, physical condition assessment, study drawing, plaster removal inspection, archaeological drawings of the façades, assessment of the social needs of the project targeting public, interpretation, project proposal, planning stage, structural consolidation, re-evaluation of the site needs, conservation and restoration intervention, reporting and maintenance, monitoring (2000, p. 227).

What clearly emerges from the revision of literature is that different sites, distinct programs of work and project aims define different methodological approaches to the site intervention.

9 METHODOLOGY FRAMEWORK

For several experts, the non-existence of methodology for conservation intervention is what the majority of projects have in common. In addition, following the literature review and the analysis of open interviews, site survey questionnaires and international expert questionnaires responses, it is generally acknowledged that there are few cases that have a comprehensive methodology of intervention. For instance, in Aït Ben Haddou, site in Morocco, there was no explicit methodology mentioned in the management plan, and there is still the need for a more formal approach towards methodology of intervention. It is important that the methodology components are followed as part of a process, and these components should have a stronger impact on the conservation intervention approach. On several sites this does not happen, as the different components of the methodology process do not cross or interrelate. For instance, recording is entailed as part of the documentation requirements, but it is not considered for the interpretation, or the decision-making. This was observed on the citadel of Bam, in Iran. In this case, components are seen as a procedure for a task to be completed and are not driven by the operated process. This means that the different components of the methodology process should be driven, looking for the achievement of a common desired outcome related with the intervention.

It is also interesting to notice that 15% of the international questioned experts stated that the same methodology approach should be followed in all types of structures and sites, as any conservation process shares the same underlying principles. Another 15% of the questioned international experts argued that a methodology should vary depending on the case, in order to be adequate. The fact is that both are accurate, as there are certain methodology components that are shared (e.g., documentation, recording), but the need for a particular study or not will depend on the specifics of the structure or site and the required priorities of intervention.

10 METHODOLOGY KEY-COMPONENTS

A comprehensive list of key-components for a qualitative methodological intervention was identified through the analysis of data (Correia, 2010). It is interesting to note that several of the components were generally agreed among questioned international experts, varying with an approval from 40% (to address the physical condition) to 5% (to address archaeological study). 10% of the questioned experts, in spite of not acknowledging the need for a methodological approach, understood the relevance of some of its components. Following, are the basic components identified for methodology of intervention (Correia, 2010). These key-components are not limited, as proper consideration of the contextual aspects has to be considered:

a) *Documentation and studies*
Collection of preliminary documentation for better understanding; Historical study of the building or site; Structural study of the building or site; Study of the construction techniques (the way they are used); Archaeological study; Stratification study of historical architectonic levels; Context study of the surroundings; Socio-cultural study; Functional (use) study; Collection of government policies for the site; Collection of all the information concerning availability of funding; Specific in-depth analysis for each study; Deep study and understanding of the historical technology; Other studies (e.g. local know-how).

b) *Recording and surveying*
Architectural survey, Detailed metric survey; Survey of the materials used; Survey of the materials added to the original fabric; Record of the historical technology; Registration (archaeological, etc.); Other records and surveys, when needed.

c) *Interpretation*
Continuous approach as it is addressed after the data collection and the documentation analyses but also following the phase of recording and surveying. This interpretation phase allows a grounded: Planning definition; Establishing programs; Addressing management; Addressing conservation implementation and practice; Addressing implementation, when needed.

d) *Assessment of significance*
Study on the structure and site significance. It entails the assessment of the community significance for the site, as well as its stakeholders.

e) *Assessment of physical condition:*
Identification of conditions (intrinsic and extrinsic) that affect the site; Analysis of the condition/deterioration: material pathologies; Analysis of the condition/deterioration: structural pathologies; Study and test of materials (including laboratory analysis); Technical diagnosis of the site/object (following the condition analysis); Other relevant components, when needed.

f) *Criteria for intervention*
It was addressed on chapters 5 and 6.

g) *Definition of an intervention proposal*
This phase entails the project proposal with its conceptual approach, and the definition of criteria, degrees, principles of intervention and assessed values.

h) *Intervention project*
This is the operative phase to address the intervention action on the heritage structure or site.

i) *Evaluation and monitoring*
Report assessment of the research data phase and of the intervention phase. It includes all the data collected throughout the intervention. Following the evaluation, the monitoring of the site should take place addressing: conditions of equilibrium; Routine Vigilance; Evaluation of conditions; Other components to be considered, when needed.

j) *Maintenance and follow-up*
After implementation of conservation approach, to include is also a maintenance plan; After maintenance plan implementation, there is the need for follow-up and monitoring.

11 CONCLUSIONS

In order to provide reliable and agreed criteria for intervention and for the methodology process in vernacular architecture and earthen heritage conservation, several relevant conclusions were established. Following are stated the main and overall contributions to the field, related to the research findings (Correia, 2010):

Figure 5–8. Methodology for intervention at Casa dos Romeiros, Alcácer do Sal, Portugal (credits: M. Correia, 1997–2001).

11.1 *Mix on the interpretation of Methodology for intervention and Criteria for intervention*

It is relevant to observe that 85% of the experts did not have a clear understanding, just a general idea, of the differences between methodology and criteria for intervention.

If there is mixture of interpretation and no full understanding of criteria for intervention, even among international experts that are experienced in conservation practice, then it is clear that in daily conservation actions, this mixture interpretation is even more likely to occur. This is an important conclusion and has a clear negative impact on the entailed conservation intervention process.

11.2 *Conservation intervention based on the expert's empirical experience*

Evidence from international key-experts questionnaires demonstrates that several of the experts carried out interventions based on their empirical experience. Nonetheless, a clear understanding of the different type of approaches is required, in order to promote a change towards built heritage conservation.

To avoid failure, conservation methodology and criteria for intervention have to be clearly discussed and further developed.

11.3 *Definition of criteria for intervention*

There are criteria for decision-making and clarification concerning procedures for intervention. Two complementary concepts of key-criteria were identified:

- ***Explicit criteria*** can be established through guiding principles, sustainable standards, tangible requirements and preventive standards.
- ***Implicit criteria*** can be defined by values and intangible heritage significance.

The combination of both with a social, physical and preventive approach can contribute to a more consistent and objective judgement for decision-making.

11.4 *Definition of indicators of quality and indicators of best practice*

The integration into the conservation process of indicators of quality and of best practice can provide clear evidence to accomplish high-standards in conservation.

It is important that these indicators are applied throughout conservation intervention, implementation of planning systems, and through the follow-up process of preventive conservation.

It is essential to acknowledge that the setting up of indicators of quality (relating issues such as conservation principles, values, interdisciplinary, community participation, etc.) is the main platform for the development of planning and the establishment of courses of action.

11.5 *Indicators for an Inclusive and Participative Conservation Process*

The inclusion of a combination of several indicators of best practice (such as balanced approach; capacity building; collaboration; commitment; communication; consistency; economic sustainability; expertise; holistic approach; long term approach; respectful practice; social aspects; and systematic approach) will promote an inclusive and more consistent conservation process. These indicators will also provide a reason to engage different actors, in order to achieve successful results. In fact, 60% of the international experts supported an increase in the involvement of the community and stakeholders, which demonstrates the will for more integrative and participative processes.

11.6 *Lack of agreement when identifying successful conservation intervention*

An important finding was the general lack of agreement among experts with reference to successful conservation intervention. 60% of the international key-experts mentioned isolated cases that others might not consider as positive examples.

This demonstrates disagreement about what is considered quality conservation interventions in earthen vernacular built heritage. Additionally, 60% of the experts referred to their own work, or their organisation's work, as exemplary conservation approaches. It is evident that there is lack of accuracy and impartial judgement in evaluating one's own work.

An agreement to nominate exceptional examples of quality in conservation practice could inspire higher standards in vernacular built heritage conservation.

11.7 *Final remarks*

When dealing specifically with intervention in vernacular architecture that recently suffered an earthquake impact, it is clear that the first criteria for intervention are the dwellings safety, to assure the survival of its inhabitants. This includes the standards related to tangible requirements (as mentioned in chapter 6.1c), which comprises life safety; to consider threats; to look for a minimum risk situation; to address the identified pathologies; to try to preserve, as much as possible, the authenticity and integrity of the structure; to consider accessibility.

Simultaneously, and whenever possible, intervention in vernacular structures should consider the use value, and the cultural and social value of the buildings where people live in. If not, the identity of the people and their culture is lost. Therefore, dwellings are safe, but inhabitants do not have a sense of place and of belonging to the house; even if the house recently went through a seismic retrofitted intervention.

ACKNOWLEDGMENTS

This paper is supported by FEDER Founding through the Operational Programme Competitivity Factors – COMPETE and by Nacional Funding through the FCT – Foundation for Science and Technology within the framework of the Research Project 'Seismic V – Vernacular Seismic Culture in Portugal' (PTDC/ATP-AQI/3934/2012).

REFERENCES

Aguilar, E. & Falck, N. (1993). San Antonio de Oriente conservation, restoration, and development. In Proceedings of the TERRA 93: 7th International Conference on the study and conservation of earthen architecture. Silves, Portugal, October 24–29, 1993. Lisbon, Portugal: DGEMN, 250–255.

Brown, L. (ed.) (1993). The New Shorter Oxford English Dictionary. Oxford: Clarendon Press.

Calarco, D. A. (2000). San Diego Royal Presidio. Conservation of an earthen architecture archaeological site [preprint]. In TERRA 2000: 8th International Conference on the Study and Conservation of Earthen Architecture. Torquay, UK, May 11–13, 2000. Torquay, UK: James & James, 20–25.

Correia, M.R.A.R. (2010). *Conservation Intervention in Earthen Heritage: Assessment and Significance of Failure, Criteria, Conservation Theory and Strategies*. PhD thesis. Oxford, UK: Oxford Brookes University.

Correia, M. & Merten, J. D. (2000). Restoration of the Casas dos Romeiros using traditional materials and methods—A case study in the southern Alentejo area of Portugal [preprint]. In TERRA 2000: 8th International Conference on the Study and Conservation of Earthen Architecture. Torquay, UK, May 11–13, 2000. Torquay, UK: James & James, 226–230.

Correia, M.R.A.R. & Walliman, N.S.R. (2014) Defining Criteria for Intervention in Earthen-Built Heritage Conservation, International Journal of Architectural Heritage: Conservation, Analysis, and Restoration, 8:4, 581-601, DOI: 10.1080/15583058.2012.704478

Fahnert, M. & Schroeder, H. (2008). The repair of traditional earthen architecture in Southern Morocco. In Lehm 2008, 5th International Conference on Building with Earth. Weimar, Germany: Dachverband Lehm e.V., 249–250.

Giardinelli, S. & Conti, G. (2000). The restoration of the D'Orazio house, Casalincontrada [preprint]. In TERRA 2000: 8th International Conference on the Study and Conservation of Earthen Architecture. Torquay, UK, May 11–13, 2000. Torquay, UK: James & James, 238–241.

Goldberg, A. & Larson, C. E. (1975). Group communication: Discussion processes & applications. Englewood Cliffs, NJ: Prentice-Hall Inc.

Guerrero Baca, L. F. (2007), July–December. Arquitectura en Tierra. Hacia la recuperación de una cultura constructiva. In Arquitectura en tierra. Revista Apuntes: Instituto Carlos Arbeláez Camacho para el Patrimonio arquitectónico y Urbano (ICAC)] vol. 20, number 2. Bogotá, Columbia: Pontificia Universidad Javeriana, 182–201.

Guillaud, H., Graz, C., Correia, M., Mecca, S., Mileto, C. & Vegas, F. (2008). Terra Incognita - Preserving European Earthen Architecture, No 2. Brussels, Belgium: Culture Lab Editions and Editora Argumentum.

Hoyle, A. M., Carcelén, J., & Saavedra, F. (1993). Conservation of the Tomaval Castle. Proceedings. In TERRA 93: 7th International Conference on the Study and Conservation of Earthen Architecture. Silves, Portugal, October 24–29, 1993. Lisbon, Portugal: DGEMN, 222–227.

Lassana Cisse, J. (2000). Architecture dogon. Problématique de la conservation des gin'na au regard des changementssociaux et religieux en milieu traditionnel-dogon [preprint]. In TERRA 2000: 8th International Conference on the Study and Conservation of Earthen Architecture. Torquay, UK, May 11–13, 2000. Torquay, UK: James & James, 218–225.

Matero, F. (1995). A programme for the conservation of architectural plasters in earthen ruins in the American Southwest. Fort Union National Monument, New Mexico, USA. Conservation and Management of Archaeological Sites. Volume 1, number 1. London: James & James, pp. 5–24.

Matero, F. (2003). Managing change: The role of documentation and condition survey at Mesa Verde National Park. Journal of the American Institute for Conservation, Vol. 42, No. 1, Architecture Issue. Washington DC: The American Institute for Conservation of Historic & Artistic Works, pp. 39-58.

Matero, F., del Bono, E., Fong, K., Johansen, R. & Barrow, J. (2000). Condition and Treatment history as prologue to site conservation at Casa Grande Ruins National Monument. Preprints. TERRA 2000: 8th International Conference on the study and conservation of earthen architecture. Torquay, England. 11-13 May 2000. Torquay, UK: James & James, pp. 52–64.

Mesbah, A., Morel, J. C., Gentilleau, J. M., & Olivier, M. (2000). Solutions techniques pour la restau- ration des ramparts de Taroudant (Maroc) [preprint]. In TERRA 2000: 8th International Conference on the Study and Conservation of Earthen Architecture. Torquay, UK, May 11–13, 2000. Torquay, UK: James & James, 266–271.

Michon, J. L. & Guillaud, H. (1995). Bahla Fort and Oasis Restoration and Rehabilitation Project. Follow-up report to the World Heritage Committee on a mission to the Sultanate of Oman. Paris, France: UNESCO–World Heritage Centre.

Morales Gamarra, R. (2007). Arquitectura Prehispánica de Tierra: Conservación y uso social en las hua- cas de Moche, Perú. In Arquitectura en tierra. Revista Apuntes: Instituto Carlos Arbeláez Camacho para el Patrimonioarquitectónico y Urbano (ICAC)]. vol. 20, number 2. Bogotá, Columbia: Pontificia Universidad Javeriana, 256–277.

Orazi, R. & Colosi, F. (2003). Integrated technologies for the study, documentation and exploitation of the archaeological area of Chan Chan, Peru [preprint]. In TERRA 2000: 8th International Conference on the Study and Conservation of Earthen Architecture. Yazd, Iran, November 19– December 3, 2003. Torquay, UK: James & James, 465–474.

Pujal, A. J. (1993). Restoration techniques for earthen buildings of historical value, actual cases. In Proceedings. In TERRA 93: 7th International Conference on the study and conservation of earthen architecture. Silves, Portugal, October 24–29, 1993. Lisbon, Portugal: DGEMN, 244–249.

Shekede, L. (2000). Wall paintings on earthen supports. Evaluating analytical methods for conserva- tion [preprints]. In TERRA 2000: 8th International Conference on the Study and Conservation of Earthen Architecture. Torquay, UK, May 11–13, 2000. Torquay, UK: James & James, 169–175.

Seismic Retrofitting: Learning from Vernacular Architecture – Correia, Lourenço & Varum (Eds)
© 2015 Taylor & Francis Group, London, ISBN 978-1-138-02892-0

Structural conservation and vernacular construction

P.B. Lourenço
ISISE, Faculty of Engineering, University of Minho, Guimarães, Portugal

H. Varum
CONSTRUCT-LESE, Faculty of Engineering, University of Porto, Porto, Portugal

G. Vasconcelos
ISISE, Faculty of Engineering, University of Minho, Guimarães, Portugal

H. Rodrigues
Polytechnic Institute of Leiria, Leiria, Portugal

ABSTRACT: Modern societies understand built cultural heritage, including vernacular construction, as a landmark of culture and diversity, which needs to be protected and brought to the next generations in suitable condition. Still, a large part of this heritage is affected by structural problems that menace the safety of buildings and people. In the case of vernacular construction, deterioration due to abandonment is often present, making the phenomena of urbanization one of the most important menaces to this built heritage. The developments in the areas of inspection, non-destructive testing, monitoring and structural analysis of historical constructions, together with recent guidelines for reuse and conservation, allow for safer, economical and adequate remedial measures, as discussed in this paper.

1 INTRODUCTION

1.1 *Cultural heritage and risk reduction*

Cultural heritage buildings are particularly vulnerable to disasters, because they are often deteriorated and damaged, or they were built with materials with low resistance, or even because they are heavy, and the connections between the various structural components are often insufficient. The main causes for damage are the lack of maintenance and water-induced deterioration (from rain or rising damp), soil settlements and extreme events such as earthquakes, see Figure 1. Extreme events often lead to disasters, in light of the high vulnerability, e.g. Neves et al. (2012) and Leite et at. (2013). Still, there are many other causes of damage, namely: high stresses due to gravity loading, alterations in layout or construction, cyclic environmental actions, climate change, physical attack from wind and water, chemical and biological attack, vegetation growth, fire, floods, vibration and micro-tremors, and anthropogenic actions. The built cultural heritage includes archaeological remains, monuments, dwellings and vernacular buildings, groups of buildings, ancient city centres, and historical urban texture, but also outstanding engineering works from antiquity to present, industrial heritage, 20th century heritage in steel or reinforced concrete, and even modern heritage. Despite the extension of cultural heritage legislation and protection to groups of buildings and urban spaces, and despite the listing (inventory) of complete town centres, the instruments and the application of monument protection is still fundamentally 'object' centred. The approach for risk reduction targeted to groups of buildings, urban spaces and isolated buildings is known, being necessary to: (i) characterise the existing built heritage; (ii) perform simplified analysis, at the territorial level, to estimate the vulnerability and risk of this heritage; (iii) in cases that are identified with higher risk in the previous step, perform detailed analyses to confirm the vulnerability and risk; (iv) define a plan with long-term intervention measures and their costs, taking into account the observed risk; (v) implement the plan, with periodic reviews of time and costs, considering the economic constraints, and the costs incurred in the interventions. Such a strategy requires political and societal commitment to become reality.

1.2 *Masonry and timber*

Most of the existing built heritage, particularly in the case of vernacular construction, is made with the so-called traditional materials (masonry, including earth, and timber). In many cases of vernacular construction, structural walls are made of masonry, while floors and roofs are made of timber. In some cases, structural walls are also made of half-timbered construction.

(a)

(b)

Figure 1. Collapse of vernacular construction: (a) Progressive damage due to water leakage from roof; (b) L'Aquila earthquake, 2009 (credits: ISISE).

The influence factors on construction practice were, mainly, the local culture and wealth, the knowledge of materials and tools, the availability of material, and aesthetical reasons. Ancient buildings are frequently characterised by their durability, which enabled them to remain in a good condition throughout relatively long time periods.

Innumerable variations of masonry materials, techniques and applications occurred during the course of time. The first masonry material to be used was probably stone. In addition to the use of stone, also earth brick was used as a masonry material, as it could be easily produced. Brick was lighter than stone, easy to mould, and formed a wall that was fire resistant and durable. The practice of burning brick probably started with the observation that the brick was stronger and more durable. Another component of masonry is the mortar, which, traditionally, was mostly clay or lime, mixed with sand and silty soil.

Wood is a largely available material in most regions of the world. Since ancient times, it has also been used by humans to build shelters. Even if it is not as durable as masonry, and it is combustible, it is possible to find several ancient buildings that use wood in their structures. Many of these structure, particularly those of hardwood, and protected from fire, exhibit remarkable longevity. Still, wooden construction often needs maintenance, and allows partial replacement of modules or damaged elements, without compromising the entire structure.

2 CONCEPTS

2.1 *Conservation, restoration and rehabilitation*

Conservation is defined in the Nara Charter (ICOMOS, 1994) as "all efforts designed to understand cultural heritage, know its history and meaning, ensure its material safeguard and, as required, its presentation, restoration and enhancement". A more technical oriented definition can be: all actions or processes that are aimed at safeguarding the character-defining elements of a cultural resource, so as to retain its heritage value and to extend its physical life.

A different concept is restoration, an action or process of accurately revealing, recovering, or representing the state of a cultural resource, or of an individual component, as it appeared at a particular period in its history, while protecting its heritage value. Restoration is a complex concept for the built heritage, as this heritage was hardly produced in any given period of time.

Rehabilitation is often defined as an action or process of making possible a continuing or compatible contemporary use, of a cultural resource or an individual component, through repair, alterations, and/or additions, while protecting its heritage value. The problem with this definition is that, making possible a modern use according to current standards and codes may be incompatible with sound protection of heritage value.

2.2 *Stabilization, repair and strengthening*

Other relevant technical concepts are stabilization, an action aimed at stopping a deteriorating process, involving structural damage or material decay (also applied to actions meant to prevent the partial, or total collapse of a deteriorated structure); repair, an action to recover the initial mechanical or strength properties of a material, structural component or structural system (also applied to cases where a structure has experienced a deterioration process, having produced a partial loss of its initial performance level); and strengthening, an action providing additional strength to the structure (needed to resist new loading conditions and uses, to comply with a more demanding level of structural safety, or to respond to increasing damage associated with continuous or long term processes).

In the context of conservation of historical structures, repair is not meant to correct any historical deterioration or transformation that only affects the appearance or formal integrity of the building and does not compromise its stability. Repair should be only used to improve structures having experienced severe damage, actually conveying a loss of structural performance, and thus causing a structural insufficiency with respect to either frequent or exceptional actions. Strict conservation will normally require stabilization or repair operations. Conversely, rehabilitation will frequently lead to strengthening operations.

3 THE ISCARSAH RECOMMENDATIONS

3.1 *Basis*

The first conservation attempts resulted often in significant negative experience accumulated, such as blind confidence in modern materials and technologies, mistrust towards traditional materials and original structural resources, devaluation of ancient structural features, and insufficient importance attributed to diagnostic studies before an intervention. On the contrary, modern conservation respects authenticity of the ancient materials and building structure, meaning that interventions must be based on the understanding of the nature of the structure, and the real causes of damage or alterations. Interventions are kept minimal, using an incremental approach, and much importance is attributed to diagnosis studies comprising historical, material and structural aspects.

ICOMOS, the International Council on Monuments and Sites, is a global non-governmental organisation, founded in 1965, dedicated to promoting the application of theory, methodology, and scientific techniques to the conservation of the architectural and archaeological heritage. ICOMOS shelters national committees in more than 100 countries, and more than 25 international scientific committees.

ISCARSAH is the International Scientific Committee on the Analysis and Restoration of Structures of Architectural Heritage. Founded by ICOMOS in 1996, it is a forum for engineers involved in the restoration and care of building heritage. These aspects were condensed in a document (ICOMOS, 2003), that recognises that conventional techniques and legal codes or standards, oriented to the design of new buildings, may be difficult to apply, or even inapplicable, to heritage buildings, stating the importance of a scientific and multidisciplinary approach involving historical research, inspection, monitoring and structural analysis.

3.2 *Principles and guidelines*

A multi-disciplinary approach is obviously required in any conservation or rehabilitation project; and the peculiarity of cultural heritage buildings, with their complex history, requires the organisation of studies and analysis in steps that are similar to those used in medicine. Anamnesis, diagnosis, therapy and controls, corresponding respectively to the condition survey, identification of the causes of damage and decay, choice of the remedial measures and control of the efficiency of the interventions.

The phases of the study involve:

- *Diagnosis:* Identification of the causes of damage and decay;
- *Safety evaluation:* Definition of the acceptability of safety levels by analysing the present condition of structure and materials;
- *Design of remedial measures:* Layout of repair or strengthening actions, to ascertain the required safety.

Diagnosis and safety evaluation of the structure are two consecutive and related stages, on the basis of which, the effective need for and the extent of treatment measures are determined. If these stages are performed incorrectly, the resulting decisions will be arbitrary: poor judgement may result in either conservative, and therefore heavy-handed conservation measures, or inadequate safety levels.

All phases should be based on both qualitative (such as historical research and inspection) and quantitative (such as monitoring and structural analysis) methods that take into account the effect of the phenomena on structural behaviour. It is stressed that the approach adopts a scientific method to reach conclusions on the condition of the building, and optimal interventions, resorting to sources such as historical information, inspection of current condition or monitoring. These provide empirical data, and structural modelling, which is based on a hypothetical representation of the reality. Certainly, those models are a very important contribution, even if they will not represent the full reality, and must be validated, while their possibilities are always limited to some extent. In a first step, the models are calibrated and validated against *in situ* testing or performance, while, in a second step, they are used for extrapolating the behaviour, and for defining the safety level.

Still, there are several difficulties, namely with respect to the limited applicability of available codes and subjectivity. Codes prepared for the design of modern structures are often inappropriately applied to historical structures. They are based in calculation approaches that may fail to recognise the real structural behaviour and safety condition of ancient constructions. The enforcement of seismic and geotechnical codes can lead to drastic and often unnecessary measures that fail to take into account the real structural behaviour. Nevertheless, recent standardisation advances have been made, e.g. in Italy (PCM, 2007) and USA (ASCE, 2013).

In addition, any assessment of safety is affected by two types of uncertainties: First, the uncertainty attached to data used (actions, geometry, deformations, material properties...); second, the difficulty of representing real phenomena in a precise way and with an adequate mathematical model. The subjective aspects involved in the study and evaluation of a historic building may lead to conclusions of uncertain reliability.

Modern legal codes and professional codes of practice adopt a conservative approach, involving the application of safety factors to take into account the various uncertainties. This is appropriate for new structures, where safety can be increased with a small increase in member size and cost. However, such an approach might not be appropriate in historical structures, where requirements to improve the capacity may lead to the loss of historical fabric, or to changes in the original conception of the structure. A more flexible and broader approach needs to be adopted for historical structures, to relate the remedial measures more clearly to the actual structural behaviour, and to retain the principle of minimum intervention, limiting, in any case, risk to an acceptable level.

4 METHODOLOGICAL ASPECTS

4.1 *Diagnosis*

Many developments have been recently made, namely on investigation procedures, for the diagnosis of historical fabric, e.g. Binda et al. (2000) and Kasal & Anthony (2004). Visual inspection is one of the most important tasks to be carried out for structural diagnosis, requiring adequate training and expertise. This often requires opening up the structure, if possible, and the use of additional equipment, such as a baroscopic camera (for internal vision), a laser scan, or a total station (for geometry and deformation definition), among others.

Several non-destructive tests (NDT) can be used for the experimental determination of the mechanical, physical or chemical properties of materials, or structural members. These tests do not cause any loss of, or damage to, the historical fabric and, therefore, sometimes the synonymous term non-invasive techniques is used, see Figure 2 for examples. These can be based on elastic waves (e.g. ultrasonic and sonic testing), in electromagnetic waves (e.g. ground probe radar) and in other concepts. Alternatively, minor destructive testing (MDT) causes minimal and easily reparable damage to the historical fabric. Among many examples, coring, flat-jack testing, or drilling resistance are popular techniques.

(a)

(b)

Figure 2. Non-destructive testing: (a) Ground Probe Radar testing at Monastery of Jerónimos, Lisbon, Portugal; (b) Dynamic identification at Famagusta, Cyprus (credits: ISISE).

4.2 *Safety evaluation*

Many methods and simulation tools are available for the assessment of the safety of historical masonry structures. The methods have different levels of complexity (from simple graphical methods and hand calculations, up to complex mathematical formulations and large systems of equations); different availability for the practitioner (from well disseminated structural analysis tools, accessible to any consulting engineer office, up to advanced structural analysis tools, only available in a few research oriented institutions and large consulting offices); different time requirements (from a few seconds of computer time, up to a number of days of processing) and, of course, different costs. Still, many structural analysis techniques can be adequate, possibly for different applications, if combined with proper engineering reasoning, see Lourenço (2002) for a review.

Seismic assessment of historical built heritage is rather complex, as the safety assessment techniques, used for modern buildings, usually fail to accurately replicate the true behaviour of such structures. Many advances have been made in the last decades, namely with respect to macro-block and macro-element analysis, see Lourenço et al. (2011) and Figure 3.

Masonry is a heterogeneous material that consists of units and joints. Usually, joints are weak planes and concentrate most damage in tension and shear. The

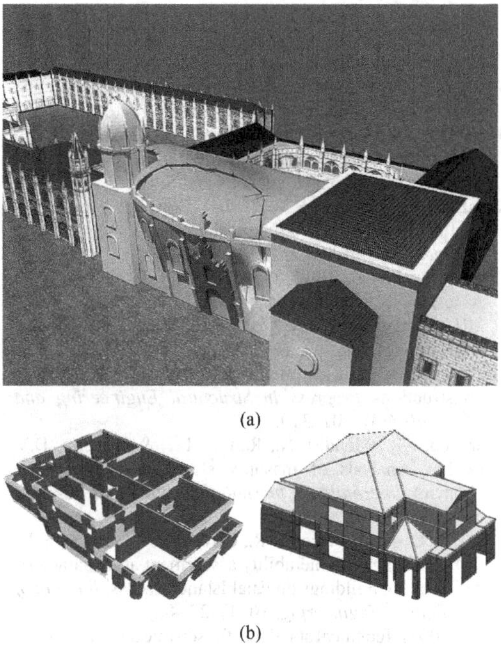

Figure 3. Advanced numerical simulations for seismic safety assessment, using: (a) Finite elements for Monastery of Jerónimos, Lisbon, Portugal; (b) Macro-elements for residential masonry structures (credits: ISISE).

Figure 4. Examples of application of remedial measures: (a) insertion of stainless steel bars; (b) injection of lime based grout; (c) textile reinforced mortars (credits: ISISE).

use of modern structural analysis techniques requires a thorough experimental description of the existing materials. This information is available in great extent, given the recent investment in studying the existing materials. In particular, the Italian normative provides a wealth of information regarding aspects, such as compressive strength, shear strength, Young's modulus, shear modulus and specific weight (PCM, 2003). In addition, several aspects can be taken into account, such as the quality of the mortar, the thickness of the joints, the presence of regular courses of masonry, the regular presence of through stones, the presence of an internal infill layer, the strengthening with grout injection, or the strengthening with reinforced plaster.

4.3 *Remedial measures*

Any remedial actions should respect existing materials and structure, and are expected to have minimal impact. The basis for design includes safety, compatibility, least invasion, durability, reversibility and controllability. Injection grouts, for example, are a much-used remedial technique, which can be durable and mechanically efficient, while also preserving historical values. Still, the selection of a grout for repair must be based on the physical and chemical properties of the existing materials. Parameters such as rheology, injectability, stability, and bond of the mix should be considered to ensure the effectiveness of grout injection. The insertion of bars (ideally stainless steel, or composite) within the masonry, using coring, has also been a popular technique to enhance structural capacity.

Increasing consideration has been given, in recent years, to the development of innovative technologies that apply externally bonded reinforcement systems using composite materials for strengthen, see Figure 4a,b. Applications of fibre reinforced polymers (FRP) to vaults, columns, and walls have demonstrated their effectiveness in increasing load-carrying capacity, and in upgrading seismic strength, even if concerns on durability exist. During the past decade, in an effort to alleviate some drawbacks associated with the use of polymer-based composites, inorganic matrix composites have been developed. This broad category includes steel reinforced grouts (SRG, unidirectional steel cords embedded in a cement or lime grout), fabric-reinforced cement-based matrix (FRCM) composites / textile reinforced mortars (TRM) (a sequence of one or more layers of cement-based matrix reinforced with dry fibres in the form of open single or multiple meshes). Currently, natural fibres are becoming increasing popular, as a green research field for crack control and strengthening, see Figure 4c.

5 CONCLUSIONS

Earthquakes are, and will remain, one of the most powerful sources of destruction for cultural heritage buildings. Many developments have recently been made, namely on methodological aspects, investigation procedures for the diagnosis of historical fabric, structural analysis techniques, or remedial measures. The application of these developments to vernacular construction is possible and needed, in order to retain its heritage value, and to reach cost-effective interventions.

Cracking occurs at early stages of loading, and adequate approaches for safety assessment are available, together with a wealth of information on mechanical characterisation. Recent developments in intervention techniques that better confine and tie together building parts, thereby reducing the possibility of separation of parts and disintegration of individual elements during a seismic event, are also significant.

ACKNOWLEDGEMENTS

The authors gratefully acknowledge the partial support by the research project 'SEISMIC-V – Vernacular Seismic Culture in Portugal' (PTDC/ATP-AQI/ 3934/2012), from the Portuguese Science and Technology Foundation (FCT).

REFERENCES

ASCE (2013). Seismic Evaluation and Retrofit of Existing Buildings. *ASCE Standard ASCE/SEI,* 41–13.

Binda, L., Saisi, A. & Tiraboschi, C. (2000). Investigation procedures for the diagnosis of historic masonries. *Construction and Building Materials,* 14(4), 199–233.

ICOMOS (1994). The Nara Document on Authenticity, International Council on Monuments and Sites.

ICOMOS (2003). Recommendations for the Analysis and Restoration of Historical Structures, ISCARSAH, International Council on Monuments and Sites.

Kasal, B. & Anthony, R.W. (2004). Advances in in situ evaluation of timber structures. *Progress in Structural Engineering and Materials,* 6(2), 94–103.

Leite, J., Lourenço, P.B. & Ingham, J.M. (2013). Statistical Assessment of Damage to Churches Affected by the 2010–2011 Canterbury (New Zealand) Earthquake Sequence. *Journal of Earthquake Engineering,* 17(1), 73–97.

Lourenço, P.B. (2002). Computations of historical masonry constructions *Progress in Structural Engineering and Materials* 4(3): 301–319.

Lourenço, P.B., Mendes, N., Ramos, L.F. & Oliveira, D.V. (2011). Analysis of masonry structures without box behavior, *International Journal of Architectural Heritage* 5(4–5): 369–382.

Neves, F., Costa, A., Vicente, R., Oliveira, C.S. & Varum, H. (2012). Seismic vulnerability assessment and characterisation of the buildings on Faial Island, Azores. *Bulletin of Earthquake Engineering,* 10 (1), 27–44.

PCM, (2003). Technical standards for seismic design of structures. Directive of the Prime Minister, 20/03/2003. G.U. n.252 of 29/10/2003. Modified by OPCM 3431, 3/5/2005.

PCM, (2007). *Guidelines for evaluation and mitigation of seismic risk to cultural heritage. Directive of the Prime Minister, 12/10/2007. G.U. n.24 of 29/01/2008.* Rome: Gangemi Editor.

Seismic Retrofitting: Learning from Vernacular Architecture – Correia, Lourenço & Varum (Eds)
© 2015 Taylor & Francis Group, London, ISBN 978-1-138-02892-0

Seismic retrofitting of historic earthen buildings

C. Cancino
The Getty Conservation Institute, Los Angeles, California, USA

D. Torrealva
Pontificia Universidad Católica del Perú, Lima, Peru

ABSTRACT: During the 1990s, the Getty Conservation Institute (GCI) carried out a major research and laboratory testing program – the Getty Seismic Adobe Project (GSAP) – to investigate the seismic performance and develop effective retrofit methods for historical adobe structures. In April 2006, the GCI's Earthen Architecture Initiative (EAI) hosted an international colloquium, in order to assess the impact and efficacy of the GSAP. The participants concluded that the GSAP methodology was reliable and effective, but its reliance on high-tech materials and professional expertise was a deterrent to its wider implementation. In response to these conclusions, the EAI initiated, in 2010, the Seismic Retrofitting Project (SRP) with the objective of adapting the GSAP guidelines, so that they better matched the equipment, materials, and technical skills available in many countries with earthen buildings. Peru was selected as the project's location, due to its current and historical knowledge and professional interest in the conservation of earthen sites.

1 BACKGROUND

For millennia, humans have constructed buildings of earth. In places, ranging from ancient archaeological sites, to living cities, from the vernacular to the monumental, earth is used, both as a structural, and as a decorative material. The remarkable diversity of earthen heritage presents equally complex conservation challenges.

For nearly two decades, the GCI has developed methodologies, and set standards for the conservation of earthen architectural heritage worldwide.

Earthen buildings, typically classified as unreinforced masonry structures, are extremely vulnerable to earthquakes and subject to sudden collapse during a seismic event – especially, if a building lacks proper and regular maintenance. Historical earthen sites located in seismic areas are at high risk of being heavily damaged and even destroyed.

1.1 *The GSAP*

During the 1990s, the GCI carried out a major research and laboratory testing program – the Getty Seismic Adobe Project (GSAP) –, which investigated the performance of historical adobe structures during earthquakes, and developed cost-effective retrofit methods that substantially preserve the authenticity of these buildings. Results of this research have been disseminated in a series of publications, both in English and Spanish (Tolles et al. 2002 & 2005).

Figure 1. GSAP Colloquium participants (credits: Getty Conservation Institute).

In 2006, the EAI convened two meetings: the Getty Seismic Adobe Project Colloquium, and New Concepts in Seismic Strengthening of Historic Adobe Structures (Fig. 1). Held at the Getty Center, the meetings focused on implementation of the GSAP. Papers presented at the colloquium, as well as the main conclusions of colloquium's round table discussions, were published as part of the colloquium proceedings (Hardy et al., 2009). The participants in the colloquium concluded that the GSAP methodology was excellent and effective. However, the methodology's reliance on high-tech materials and professional expertise was a deterrent to it being more widely implemented.

1.2 *The Pisco Earthquake Assessment*

The MW 7.9–8.0 magnitude inter-plate 2007 *Pisco* earthquake occurred off the coast of central Peru. It had a maximum local Modified Mercalli Intensity (MMI) of VII-VIII; its epicentre was located at 13.35S and 79.51W, at a depth of 39 km (USGS) and a total duration of approximately 300 seconds (Tavera et al., 2009). There have been numerous studies to define the geology of the affected area (Lermo et al., 2008, & CISMID et al., 2008, p. 88–91).

Post-earthquake assessments offer an opportunity to understand why buildings fail, and to provide information that can serve as the basis for the improvement of seismic performance. Lessons learned from earthquakes and other natural disasters are used to advance construction techniques. More recently, such lessons have fostered the development of the engineering and historic preservation disciplines, as well as the testing and review of current building codes and disaster management policies.

The *Pisco* earthquake tragic human losses resulted from the collapse of buildings in the states of *Ica, Lima, Huancavelica, Ayacucho* and *Junín*, among others (Johansson et al., 2007). The damages have been described by several national and international organizations that travelled to the affected region immediately after the earthquake.

From October 28 to November 2, 2007, the GCI, in collaboration with other Peruvian institutions, led a multidisciplinary team of national and international earthquake engineers, preservation architects and conservators, visiting a total of 14 buildings. The main objective of the GCI rapid assessment was to visually evaluate the damaged sites, while recording pre-existing conditions (abandonment, deterioration or structural interventions) that might have affected their seismic performance. Results of that assessment have been published, both in English and Spanish (Cancino, 2011, 2014).

2 THE SEISMIC RETROFITTING PROJECT

2.1 *The project and its location*

The GSAP colloquium conclusions, and the lessons learned from the *Pisco* earthquake assessment, prompt the GCI to create the Seismic Retrofitting Project (SRP). The main objective of the SRP is to adapt the GSAP guidelines in countries, where equipment, materials and technical skills are not readily available, by providing low-tech seismic retrofitting techniques and easy-to-implement maintenance programs for historical earthen buildings, in order to improve their seismic performance, while preserving their historical fabric.

Peru was selected as the project's location due to its current and historical knowledge and professional interest in the conservation of earthen sites, as well as its potential partners that could implement retrofitting techniques through model conservation projects.

2.2 *The SRP objectives*

From its conception the SRP's several activities have tried to achieve the following objectives:

- Design seismic retrofitting techniques using locally available materials and expertise for selected Peruvian building prototypes, which have potential for application in other Latin American countries;
- Validate the techniques using numerical modelling and testing;
- Obtain the recognition, approval and promotion of the techniques by local authorities;
- Develop guidelines for those responsible for the techniques implementation; and,
- Develop a model conservation project to demonstrate the execution of the techniques.

2.3 *The SRP methodology and partners*

In 2009, the GCI developed the SRP methodology, which involved a number of phases: 1) identifying prototype buildings that represent key-earthen historical buildings found in South America; 2) undertaking structural and material assessments of each prototype, followed by laboratory testing of key-building elements, and developing numerical models for each of the prototypes; 3) designing, testing, and modelling of potential retrofitting strategies for each prototype building; 4) implementation of the retrofit strategies on selected prototypes; and 5) dissemination of the results and methods.

In order to carry the phases on, the GCI joined forces with the *Escuela de Ciencias e Ingeniería* of the *Pontificia Universidad Católica del Perú* (*PUCP*), the *Ministerio de Cultura del Perú* and the School of Architecture and Civil Engineering of the University of Bath, in 2010. In 2013 and 2014, the Civil, Environmental and Geomatic Engineering School at University College London (UCL) continued the work started by the University of Bath; and in 2015, the University of Minho is further advancing the modelling of the prototypes and potential retrofitting strategies.

3 THE SELECTION OF BUILDING PROTOTYPES

3.1 *The process*

As part of the first phase, up to four building typologies were identified for study. Each selected typology was to represent buildings that are priorities for seismic retrofitting, based upon their level of historical and architectural significance; the current lack of - and thus greater need for - retrofitting solutions; and their demonstration of typical modes of failure, so that their reinforcing techniques would be able to be more widely applied to other earthen sites.

At the beginning of 2010, the GCI in collaboration with the *Ministerio de Cultura del Perú* and *PUCP*,

Table 1. Evaluation of matching buildings with the defined building typologies (credits: authors).

Sites & Criteria	Original structure	Historical value	Significance	Availability (Owner)	Security (safe for project staff)	Representative pathologies	Accessibility (location)	Historical/architectural data	Visibility	TOTAL
Typology 1: Coastal houses										
Hospicio Ruiz Dávila	6	7	8	1	4	6	6	4	4	46
Quinta Heren	8	8	8	8	4	7	5	5	5	58
Hac. San Juan Grande	5	7	6	2	2	6	6	4	5	43
Hotel Comercio	6	8	8	9	8	8	8	9	6	70
Typology 2: Coastal churches										
Cathedral of Ica, Ica	9	7	7	6	9	9	8	6	8	69
San Antonio, Mala	5	5	8	9	7	9	4	4	3	54
San Juan, Ica	8	5	8	9	7	9	4	4	3	57
San Luis, Cañete	4	5	8	9	7	9	4	4	3	53
Santuario de Yauca, Ica	7	5	8	8	7	9	4	4	3	55
Typology 3: Andean churches										
Andahuaylillas	7	9	8	7	7	3	5	7	9	62
Canincunca	6	9	8	7	7	3	4	6	8	58
Marcapata	9	8	8	8	6	8	2	6	8	63
Kuchi Wasi	9	8	7	8	6	5	2	6	7	58
Kuño Tambo	9	8	8	9	8	9	3	6	8	68
Rondocan	7	8	8	9	8	6	2	6	5	59
Typology 4: Andean houses										
Garrido Mendivil	6	8	6	5	8	5	9	7	6	60
Arones	8	8	9	9	8	9	9	8	6	74
Alonso del Toro	6	8	6	4	4	5	8	7	6	54
Serapio Calderón	7	8	6	4	3	6	8	7	6	55
Concha	5	9	7	8	8	4	8	8	8	65

identified four building typologies that met these criteria. A total of 20 sites were pre-selected by Peruvian architects, architectural historians and engineers. The evaluation of the sites was made by all project partners followed the criteria listed in Table 1.

3.2 *The prototype buildings*

The buildings receiving the highest scores, and thus selected for further study, were:

- The *Hotel Comercio* (Typology 1, Fig. 2), a XIX century residential building in the historical centre of Lima, constructed with adobe walls in the first floor, flat roofs, and quincha panels in the second and third floors;
- The Cathedral of *Ica* (Typology 2, Fig. 3), a XVIII ecclesiastical building constructed with thick adobe walls, quincha pillars, vaults and domes, and damaged during the 2007's earthquake;
- The Church of *Kuño Tambo* (Typology 3, Fig. 4), a XVII century building constructed with thick adobe walls, decorated with mural paintings and truss roof in *Acomayo, Cusco*; and,
- *Casa Arones* (Typology 4, Fig. 5), a XVII century residential building in the historical centre of Cusco, constructed with two-storey adobe walls and a truss roof.

Figures 2-5. The selected prototypes (from top to bottom): *Hotel Comercio*, Cathedral of *Ica*, Church of *Kuño Tambo* and *Casa Arones* (credits: Getty Conservation Institute).

4 THE CONSTRUCTION ASSESSMENT

As part of the architectural and structural research and studies phase, the selected prototype buildings were documented and assessed in two field survey campaigns in 2010. For those, the following documentation techniques were developed and tried on site.

4.1 *Conditions recording and survey forms*

As the four prototype sites have different structural configurations, the project team decided to carry out a general survey, consisting of recording forms to gather information of the structure as a whole, and detail assessments to obtain critical information, regarding construction techniques, as well as the severity of structural conditions of the different structural elements. The survey forms were accompanied by detailed drawings, so as to graphically indicate the location of construction materials, and the conditions described in the forms.

4.2 *Non-destructive research - thermal imaging*

The structural assessment of an existing building requires an understanding of the materials and connections that are usually not readily apparent through visual analysis. Even when some areas of the building are exposed, it cannot always be assumed that these sections of the building were constructed in the same manner, as those that are intact, necessitating of further research.

Non-destructive research techniques were adapted from other fields for use on this project. Two non-destructive assessment methods, infrared photography, and thermal imaging, were evaluated and considered for use; and of these, thermal imaging was identified for on-site trials. Thermal imaging is a way of visually capturing information on surface temperature. Since the materials within a wall can affect surface temperature, thermal imaging can reveal where different materials have been placed within a building. Particularly useful from a structural standpoint, was the ability to locate the quincha posts in the third floor of *Hotel Comercio*, since one of the questions raised during the survey was whether the posts are aligned with the floor joists. Also useful were the scanning of the wooden structural elements at the vaults and domes of *Ica* Cathedral.

4.3 *Prospections*

To complement the detailed structural assessment, annotated detail drawings, illustrating structural elements, systems, and connections were prepared for the four prototype buildings. These drawings were prepared by opening up select areas of the building foundations, wall, and roof structure, for further research – a process referred to as "prospection."

The areas to be opened up were selected by the SRP partners, and the work was carried out by GCI consultant Mirna Soto, a Peruvian architect and architectural conservator. The number and location of prospections did not require the permanent removal of original building materials. Each opened prospection area was recorded through field sketches, photographs, and narrated videos (Fig. 6). After that, any necessary repair work was carried out to return the prospection areas to their original configuration and appearance. All findings of the construction report have been published, both in English and Spanish (Cancino et al., 2012 & 2013).

The data collected for this assessment report was used for the construction of partial and global numerical models and seismic analyses; as well as for the experimental testing of materials, connections and components for the four historical earthen prototype buildings.

5 THE TESTING AND MODELLING PROGRAM

The testing and modelling phase intends to provide quantitative information for seismic behaviour of the prototypes, through a comprehensive testing program and numerical analysis of the four buildings.

The models of the four prototypes are now in progress. *In situ* analysis in 2015, consisting of extensive sonic testing and dynamic identification, has allowed University of Minho to obtain indirect measurements of the elastic mechanical properties of the masonry-like materials, to validate the numerical models of the four prototypes. Results of previous partial modelling analysis developed by UCL and the University of Bath, have been published in several international conferences (Ferreira et al., 2013), (Ferreira & D'Ayala, 2014), (Fonseca & D'Ayala, 2012a, b), (Quinn et al., 2012), (Quinn & D'Ayala, 2013 & 2014).

In 2011, a testing program to obtain material and mechanical properties of each site was designed, evaluated by all SRP partners and external peer reviewers, to be later developed at *PUCP*. From 2012 to 2014, *PUCP* conducted over 300 material, mechanical and static testing for all four building prototypes. Some of these tests were performed for the first time on earthen materials and/or structural components, providing valuable information to the field.

The performed tests included:

- Characterisation of historical soil material;
- Characterisation of historical wood material;
- Mechanical properties of historical masonry;
- Structural components: In plane cyclic shear tests on new and historical *Quincha* panels, and cyclic tests of connection timber assemblies;
- Traditional retrofitting techniques: Pull out test of wooden tie beams, and corner keys embedded on adobe walls.

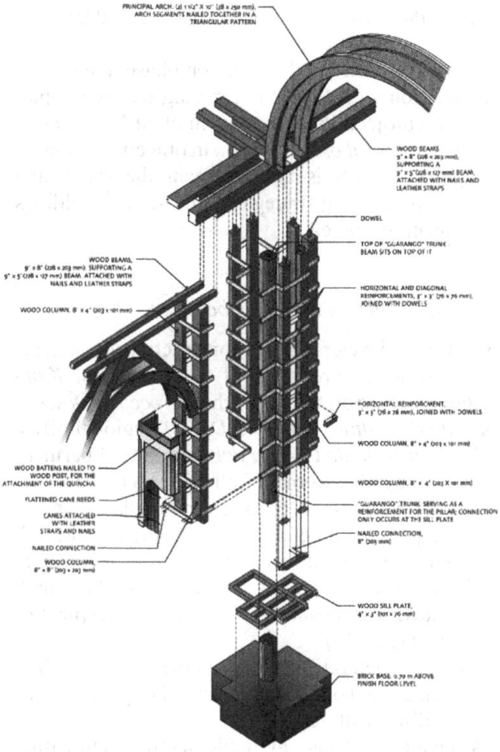

Figure 6. Prospection drawings showing an isometric detail of wooden structure of one of the pillars at Cathedral of *Ica* (credits: Getty Conservation Institute).

The tests also provided valuable information to the partial and global models of each of the prototypes. Partial Results of the testing program have been published in several international conferences (Torrealva & Vicente, 2012 & 2014), and will be published as part of the SRP publication series at the end of 2015.

5.1 *Findings of the testing program*

The main findings of the extensive experimental program can be summarised as follow:

Wood identification and mechanical properties of the timber: With the participation of the *Facultad de Ciencias Forestales* of the *Universidad Nacional Agraria La Molina*, the wooden species of the two prototype buildings located in the coast, *Ica* Cathedral and *Hotel Comercio*, were identified and studied. Five foreign and one native species were identified in the buildings located in the coast of Peru. The Peruvian coast, as an arid environment, didn't have available construction wood, and, in the time of the colony, most of the wood for construction was brought in in ships from North and Central America. The mechanical properties of these species were determined by testing samples in good condition, obtained from the original elements.

Mechanical properties of original masonry: Original material - adobe bricks, burned clay bricks and

Figure 7. Wattle and daub ("*Quincha*") walls being tested to analyze their seismic behavior at *PUCP* (credits: Getty Conservation Institute, 2011).

lime and mud mortar - was obtained from partial collapsed parts of *Ica* Cathedral and *Hotel Comercio*. Small masonry specimens were then built and tested in the *PUCP* laboratory, in order to determine the mechanical properties of the adobe and brick masonry of both buildings. The soil material composition in both buildings was also identified.

Cyclic shear behaviour in *Quincha* panels: It is generally accepted by the engineering and professional community that *Quincha* buildings perform well during earthquakes. Twelve full scale new *Quincha* panels were built reproducing the same construction details, and vertical loads found in the second and third stories of *Hotel Comercio,* and then subjected to horizontal cyclic force to simulate the lateral displacement produced by earthquakes (Fig. 7). The results showed that the panels can undergo big lateral displacements, without losing its vertical load carrying capacity. Furthermore, one original and historical panel was dismantled and extracted from an area already collapsed of the *Hotel Comercio* and then re-assembled at the *PUCP* laboratory. This historical panel was then subjected to the same cyclic test as the new ones. The results of this test showed that the original panel performed almost the same than the new ones, with great resilience to lateral displacements, in spite of the deterioration of its elements.

Cyclic tests on timber connections: Three types of connections were reproduced in full scale, using similar wood species, mortise and tenon connections, as well as nailed connection at the *PUCP* laboratory.

The specimens were then subjected to cyclic force, in order to determine their rotational stiffness. The transversal pinned connection was tested to obtain its maximum shear capacity. The cyclic test showed an almost null rotational stiffness after the first cycle, and the transversal pinned connection test showed a high shear capacity on the transversal pinned connection.

Experimental testing on traditional techniques: Wooden tie beams and corner keys are two common techniques for improving the lateral stabilization of adobe walls. Tie beams connect two parallel walls,

Figure 8. Pull out test of wooden corner key embeded in adobe Wall (credits: *PUCP*, 2012).

Figure 9. View of Kuño Tambo wall paintings at *Sotocoro* (credits: Getty Conservation Institute).

so each one helps the other, to avoid lateral overturning. Corner keys are timber assemblies, inserted in the corner of two perpendicular adobe walls, in order to maintain the connection, when the walls are subjected to lateral forces.

Full scale assemblies of both systems were reproduced at the *PUCP* laboratory, and subjected to pull out tests, in order to determine their failure mode and force capacity. The results showed that the maximum pull out force depends on the shear strength of the masonry wall (Fig. 8).

6 THE IMPLEMENTATION MODEL PROJECTS

As part of the SRP implementation phase, the GCI in collaboration with *PUCP* is carrying out two implementation projects in Peru: the Church of *Kuño Tambo* and the *Cathedral of Ica*. The main objective of developing the projects is to implement the tested and designed retrofitting strategies in those two buildings as model case studies.

6.1 *The Church of Kuño Tambo*

The SRP is developing the construction documents for the seismic retrofitting of the Church of *Kuño Tambo*, in collaboration with the *Dirección Desconcentrada de Cultura-Cusco (DDC-C),* regional office of the *Ministerio de Cultura del Perú*, a GCI partner.

Owned by the Roman Catholic Archdiocese of Cusco, the church has been in continuous use as a place of worship, since its construction in the seventeenth century and, remarkably, retains much of its original configuration and materials. The interior of the church is decorated with beautiful wall paintings, typical of that period (Fig. 10).

Further structural studies, diagnosis and modelling have been carried out, to understand the specific areas of need of retrofitting.

A major challenge in implementing retrofitting techniques in historical buildings with decorated surfaces is avoiding removal of the wall paintings - removal having been fairly common in Latin America, when the walls behind the paintings were repaired. To circumvent this practice at *Kuño Tambo*, the GCI has developed and carried out a series of interventions to consolidate, and then protect, the wall paintings during the construction work. The interventions were designed to be compatible with the characterisation of the original wall paintings, and adobe materials – analyses performed by personnel of the *DDC-C*.

Based on the results of this analysis, and a detailed wall paintings condition assessment performed by GCI staff and consultants, using rectified photography developed by the Carleton Immersive Media Studio of Carleton University, Ottawa, the project designed, tested, and conducted *in situ* interventions to reattach and consolidate the wall paintings, and to protect them during the construction.

Two campaigns in February and May 2015 were carried out to consolidate the wall paintings prior to retrofitting. An earthen-based grout was used to reattach all wall painting interfaces, in conjunction with the application of facing material to consolidate them. Once the paintings were stabilised, their physical protection included a lightweight and resistant mesh fastened to the top of the wall, and temporarily fixed to the church floor. The mesh will have an incline of 45° to prevent any contact between the paintings and material that might fall from the walls or ceiling during construction.

This strategy for protecting the wall paintings during construction can serve as an alternative model to the common practice of removing the wall paintings during retrofitting, and can help preserving the paintings' historic, aesthetic, and material values.

6.2 *Ica Cathedral*

The Cathedral of *Ica* was originally built in 1759 by the Society of Jesus. Presently owned by the Roman Catholic Diocese of *Ica*, the cathedral was used as a place of worship, until it was damaged in the 2007 *Pisco* earthquake.

The thick lateral walls are constructed with mud brick masonry, over a fired brick base course and stone foundations. The side aisles are separated from the central nave by a series of hollow *Quincha* pillars and arches, covered with painted earth mortar and gypsum plaster. The barrel vault and domes are also constructed with wood arches or ribs and *Quincha*.

The SRP is currently developing, in collaboration with local consultants, the construction documents for the seismic retrofitting of the site, as part of a comprehensive conservation project.

The preliminary results of the structural diagnosis of the Cathedral have shown that the extensive damage on the wooden elements was the principal cause for the partial collapse of the roofing structure after the 2007 earthquake. Detail modelling of structural components has proven that connections between the structural wooden frame and the adobe walls minimised displacement in the transept. On the contrary, the inelastic behaviour of the large adobe lateral walls didn't reduced displacement, stressing the pillars in the central nave.

6.3 *Retrofitting strategies*

The retrofitting strategies for both sites are now being refined and evaluated, trying to preserve the elements with highest significance and developing detail designs of the interventions, to minimise loss of the historical fabric.

Following the definitions of the Structural Engineers Association of California (Poland et al. 1995), the main objective of the structural interventions is to improve the seismic performance of both buildings, from "Close to Collapse", after an occasional earthquake, to "Operational", after a frequent earthquake, and avoiding the collapse in rare earthquakes.

In *Kuño Tambo*, the preliminary results of the church structural diagnosis has shown that minimal interventions repairing and improving existing reinforcement techniques, such as proper anchoring of the tie beams, reconstruction of damaged and/or collapse buttresses, grouting and stitching of structural cracks and proper connection between the truss roof and the adobe walls, would seismically retrofit the building. In the case of *Ica* Cathedral, the repairing of the severely damaged wooden structural elements is being further studied, to secure the authenticity of the architectural configuration and structural system, which provided value to the site. Again, a value-based analysis of the site, and all its elements, is being developed to help the team taking decision about conservation interventions.

Figure 10. Personnel of the *DDC-C* injecting an eaerthen grouts for consolidation of wall paintings (credits: Getty Conservation Institute).

7 NEXT STEPS

The SRP methodology designed in 2009 included the analysis of the retrofitting strategies for all for buildings using finite modelling analysis and testing. The project is now in the process of developing those strategies, and results of their performance will be published soon.

In unison with the implementation of retrofitting techniques at *Ica* Cathedral and *Kuño Tambo*, the SRP intends to organise a peer review meeting, to discuss the final conclusions of the testing and modelling phase. Following that meeting, the SRP will start working on guidelines and manuals for those responsible for the techniques implementation (architects, engineers, and conservators).

REFERENCES

Cancino, C. (2011). *Damage Assessment of Historic Earthen Buildings after the August 15, 2007 Pisco, Peru Earthquake.* Los Angeles, CA: Getty Conservation Institute.

Cancino, C., Lardinois, S., D'Ayala, D., Fonseca Ferreira, C., Torrealva Dávila, D., Vicente Meléndez, E., & Villacorta Santamato L. (2012). *Seismic Retrofitting Project: Assessment of Building Prototypes.* Los Angeles, CA: The Getty Conservation Institute

Cancino, C., Lardinois, S., D'Ayala, D., Fonseca Ferreira, C., Quinn, Q., Torrealva Dávila, D. &Vicente Meléndez, E. (2013, December) "Seismic Retrofitting of Historic Earthen Buildings. A Project of the Earthen Architecture Initiative, The Getty Conservation Institute". *Earthen Architecture in Today's World.* UNESCO International Colloquium on the Conservation of World Heritage Earthen Architecture. Paris, France.

Cancino, C., Lardinois, S., D'Ayala, D., Fonseca Ferreira, C., Torrealva Dávila, D., Vicente Meléndez, E., & Villacorta Santamato, L. (2013). *Proyecto de Estabilización Sismorresistente: Estudio de edificaciones tipológicas.* Los Angeles, CA: Getty Conservation Institute.

Cancino, C. (2014). *Estudio de daños a edificaciones históricas de tierra después del terremoto del 15 de agosto del 2007 en Pisco, Perú*. Los Angeles, CA: Getty Conservation Institute.

Centro Peruano Japonés de Investigaciones Sísmicas y Mitigación de Desastres (CISMID), Facultad de Ingeniería (FCI) & Universidad Nacional de Ingeniería (UNI) (2008) "Informe Final del Estudio de Microzonificación Sísmica y Zonificación de Peligro de Tsunami en las Ciudades de Chincha Baja y Tambo de Mora.". Available at: http://www.cismid-uni.org/component/k2/item/69-articulos-de-interes

Ferreira, C. F., D'Ayala, D., Fernandez Cabo, J.L., & Díez, R. (2013) Numerical Modelling of Historic Vaulted Timber Structures. Advanced Materials Research, 778, 517–525.

Ferreira, C. F. & D'Ayala D. (2014). "Structural analysis of timber vaulted structures with masonry walls". In F. Peñan & M. Chavez (Eds.), *Proceedings of the SAHC2014-9th International Conference on Structural Analysis of Historical Constructions*. Wroclaw, Poland.

Fonseca, C. & D'Ayala, D. (2012). Seismic Assessment and Retrofitting of Peruvian Earthen Churches by Means of Numerical Modelling. *Proceedings of The Fifthteenth World Conference on Earthquake Engineering; Lisbon, Portugal.*

Fonseca, C. & D'Ayala, D. (2012). "Numerical modelling and structural analysis of historical ecclesiastical buildings in Peru for seismic retrofitting ", *8th International Conference on Structural Analysis of Historical Constructions, Wroclaw, SAHC: Wroclaw, Poland.*

Hardy, M., Cancino C. & Ostergren, G. (2009). *Proceedings of the Getty Seismic Adobe Project 2006 Colloquium*. Los Angeles: The Getty Conservation Institute

Johansson, J., Mayorca, P., Torres, T. & León, E. (2007) "A Reconnaissance Report on the Pisco, Peru Earthquake of August 15, 2007." p. 1. Available at: http://www.jsce.or.jp/report/45/content.pdf

Lermo, J., Limaymanta, M., Antayhua, Y. & Lázares, F. (2008). "El Terremoto del 15 de Agosto de 2007 (MW+7.9), Pisco, Perú. Mapas de Clasificación de Terrenos con fines de diseño sísmico para las ciudades de Pisco, Ica y Lima-Callao" In El Terremoto de Pisco (Perú) del 15 de Agosto de 2007 (7.9 M_W), edited by Hernando Tavera. Lima: IGP, pp. 233–268.

Poland, C. D., Hill, J., Sharpe, R. L. & Soulages, J. R. (1995). Performance based seismic engineering of buildings Sacramento, California. In: Structural Engineers Association of California, 1995-04-03. Available at: http://nisee.berkeley.edu/documents/elib/www/documents/200806/vision-2000.pdf

Quinn, N., D'Ayala, D. & Moore, D. (2012). Numerical Analysis and Experimental Testing of Quincha under Lateral Loading. Proceedings of the 8th International Conference on Structural Analysis of Historical Constructions, 15-17 October 2012, Wroclaw, SAHC: Wroclaw, Poland,

Quinn, N. & D'Ayala D. (2013) "Assessment of the Realistic Stiffness and Capacity of the Connections in Quincha Frames to Develop Numerical Models." Advanced Materials Research 778 (2013): 526–533.

Quinn, N. & D'Ayala D. (2014) "In-plane expertimental testing on historic quincha walls" In F. Peñan & M. Chavez (Eds.), *Proceedings of the SAHC2014-9th International Conference on Structural Analysis of Historical Constructions*, Wroclaw, Poland.

Tavera, H., Bernal, I., Strasser, F.O., Arango-Gaviria, M.C., El Alarcón, J. & Bommer, J.J. (2009) "Ground motions observed during the 15 August 2007 Pisco, Peru, earthquake". *Bulletin of Earthquake Engineering* 7 (1), pp. 71–111.

Tolles, E.L., Kimbro, E.E. & Ginell W.S. (2002). *Planning and Engineering Guidelines for the Seismic Retrofitting of Historic Adobe Structures.* Los Angeles: The Getty Conservation Institute.

Tolles, E.L., Kimbro, E.E. & Ginell W.S. (2005) *Guías de planeamiento e ingeniería para la estabilización sismorresistente de estructuras históricas de adobe.* GCI Scientific Program Reports. Los Angeles, CA: Getty Conservation Institute.

Torrealva, D. & Vicente, E. (2012). "Numerical modelling and structural analysis of historical ecclesiastical buildings in Peru for seismic retrofitting". *8th International Conference on Structural Analysis of Historical Constructions, Wroclaw, Poland.*

Torrealva, D. & Vicente E. (2014) "Experimental behavior of traditional seismic retrofitting techniques in earthen buildings in Peru". In F. Peñan & M. Chavez (Eds.), *Proceedings of the SAHC2014-9th International Conference on Structural Analysis of Historical Constructions*, Wroclaw, Poland.

United States Geological Survey (USGS). *Magnitude 8.0 – Near the coast of central Peru*. Earthquake Summary.

Seismic Retrofitting: Learning from Vernacular Architecture – Correia, Lourenço & Varum (Eds)
© 2015 Taylor & Francis Group, London, ISBN 978-1-138-02892-0

Local building cultures valued to better contribute to housing reconstruction programs

T. Joffroy & P. Garnier
CRAterre – ENSAG, AE&CC research unit, France

ABSTRACT: This article presents the summarised results of a study conducted by Haitian and international organizations that joined their expertise and efforts to respond effectively and sustainably to issues relative to housing reconstruction, following the January 2010 earthquake in Haiti.

1 A CASE STUDY IN HAITI

1.1 *Introduction*

This paper intends to focus on the potential of the presented approach, derived from field observations and from exchanges that took place at several international meetings, having already been introduced in a manifesto entitled *«Promoting local building cultures to improve the efficiency of housing programmes»* (http://craterre.org/diffusion) co-authored by Misereor, Caritas France and Caritas Bangladesh, IFRC, CRAterre and Fondation Abbé Pierre.

As stated in this manifesto, an introduction to the building cultures approach, «societies all over the world have developed specific local building cultures, resulting in the establishment of recognizable "situated" architectures and building systems respectful of their local environment», which is actually what the concept of «sustainable development» leads us to seek today.

In order for this principle to be applied in reconstruction programs, it was suggested to adopt participatory approaches, and to value holders of local knowledge and know-how. It was also suggested to identify local types of expertise and organisation patterns in tune, with resilience strategies; and the protection of built structures, and to integrate these elements when defining and designing (re)construction programs to be implemented, so as to allow affected populations to return to dignity, while strengthening social ties.

Rather accustomed to derogatory speeches of such cultures, a number of Haitian partners were initially surprised by these proposals, while others were against them. This situation evolved rather quickly, as early interventions were completed and were deemed satisfactory by local populations. Later, other partners from the same platform, but also from other NGOs involved in reconstruction related activities, decided to adopt this approach in their projects, aware of the efficiency of such solutions. Indeed, beyond meeting immediate basic needs, these solutions are reproducible, as they are adapted to the technical and financial means of local populations and artisans. This allows for housing recipients to extend basic housing structures, and for non-recipients to employ trained artisans to meet their construction needs. Moreover, these solutions contribute to efforts in the fight against poverty, due to the good return on investment in the community, as most expenses are directly injected into local economies.

Thanks to all stakeholders in this project, which has been conducted in Haiti over the last 5 years, these recommendations have been tested, corrected and applied on a large scale. This has allowed to achieve clear and tangible results, in the field, for local populations, but also for the organizations and institutions involved in reconstruction processes, through now better structured elements which, undoubtedly, will allow to better prepare for the future and, together, better cope with similar situations.

2 CONTEXT AFTER THE EARTHQUAKE OF JANUARY 12, 2010

The earthquake of January 12, 2010 heavily hit Haiti, and particularly its capital, Port-au-Prince, and the towns of Leogane, Jacmel, Petit Goave, as well as their adjoined peri-urban and rural areas, causing more than 300 000 casualties, and injuring an equal number. In total, almost 1.5 million people were affected, finding themselves homeless or displaced.

The terrible damage caused resulted from both, the extremely violent natural hazard, and the degree of vulnerability in the area, linked to the high human density, fragile buildings, non-compliance to construction rules, poverty and a «deconstructed» society, etc. In addition, every year nature takes a heavy toll on the country during the hurricane season, with significant flooding, which delays all «development» processes, or even annihilates any efforts that had been made over several years.

Haiti, ranked among the poorest countries in the world before this earthquake, was already in need of substantial support for rebuilding. The government of Haiti stated that "rebuilding Haiti does not mean returning to the situation that existed on January 11, on the eve of the earthquake, but to address all factors of vulnerability, so that natural disasters may never inflict such suffering or cause such loss."

It is in this perspective that CRAterre came to work with several organizations, national and international, and to develop research activities, both basic and applied. The first encouraging results obtained in the field, especially in rural areas, in the framework of projects supported by Misereor and Caritas and PADED – PAPDA platforms, enabled the gradual establishment of various partnerships.

In addition, the support of the ANR (French National Research Agency) in the context of the ReparH research project facilitated some of the aspects of these operations, and especially allowed to draw a number of useful lessons linked to the pursuit of reconstruction efforts in Haiti, as well as in the framework of similar projects, which may be useful in the future, in areas at risk.

3 ANALYSIS OF LOCAL CONSTRUCTION PRACTICES

The development of a qualitative method for analysing the building practices in a specific region aims at addressing the existing weaknesses, both in terms of field practice and of technical knowledge (norms and standards), and taught knowledge (technical training).

The content of this analysis is presented around four main themes.

The natural environment, that includes the physical, climatic, geological and morphological analysis of the territory and, if applicable, the alterations, which it has undergone.

Buildings and their architectural and structural characteristics, designed as a response to the needs, aspirations and activities of their users, but also as the result of a set of specific technologies and materials. Knowledge and know-how as experience and information, developed by a given population in the development of housing and in response to specific challenges. Tangible and intangible available resources, including materials, means and capacities for the implementation and development of built environments; as well as for the management and preparation of crisis situations.

In 2010, a first analysis of local construction practices was conducted in collaboration with Haitian associations. This approach was continued later on, at the start of each new project, taking into account observations made in the field that provided additional information, deserving to be integrated to better adapt projects to local conditions, including observations regarding the implementation of projects.

In analysed rural areas, the wood-frame structure housing is the most common, but more importantly, such constructions are the ones that most successfully withstood the seism. They are composed of a regular grid of wooden poles constituting the frame of the building, featuring fillings of different techniques, depending on the materials available locally. *Clissage* is the oldest of such techniques. It consists of the horizontal braiding of palm slats forming panels, inserted between the vertical elements of the wooden frame. Sometimes left exposed, these panels are usually covered on both sides with a mortar made of earth, or a mixture of earth and lime. *Tiwoch* is another technique that characterizes most rural dwellings on analysed areas, and it is very popular for the construction of new buildings. It involves filling a main frame made of wooden poles with stone masonry, and a mortar made of earth, earth and lime or cement.

Planks nailed to the outside poles set apart a third technique, known as *palmiste*, formerly widespread but having recently become economically inaccessible, due to the scarcity and high cost of wood. Aside from an ease of implementation and expansion of the structure, in this technique the planks provide bracing, thus improving the uniformity and rigidity of the general structure.

Most of the assessments made allowed to see that although localised damage was done, these traditional buildings caused few casualties. Damages were mostly limited to secondary structures, preserving the main supporting structure, thus safeguarding lives, and securing the most expensive parts of buildings. Moreover, eventual reparations on such buildings proved more accessible, both technically and economically.

It is interesting to note that several of the measures or rules that appear to have been followed in the design of these housing units also improve their behaviour facing a more recurrent phenomenon: cyclones. In vernacular dwellings, low roof heights limit wind resistance and, by pushing the centre of gravity of the building down, also promote good resistance to earthquakes. As for wood framing, horizontal bracing in the four corners of the wall plate reinforce the beam structure (Fig. 1a), while at the level of the roof, the diagonal bracing between the ridge beam and the king post provide stability for the framework (Fig. 1b). Mortise and tenon joints are used to allow the building to withstand slight deformation, while maintaining structural integrity.

In addition to the specific characteristics of structural partitioning techniques, the gradual reduction in the thickness of the walls and the use of lighter materials on the upper portions help protect people inside the buildings in the event of a partial collapse of the walls (Fig. 1c).

To increase the resistance to the impact of strong winds, several devices feature the integration of hurricane-engineering principles currently advocated. A slope of about 30° for the roof, as well as reduced overhangs, helps reducing the capture of wind energy. The separation between the cover of the main portion

a) b) c)

Figure 1. Grande Rivière: (a) bracing of the framework, (b) and framing of the roof, (c) gradual reduction of wall thickness (credits: CRAterre).

of the building and that of the gallery follows the principle of structural disconnection between primary and secondary structures; the gallery roof becomes a detachable element, and even if wrenching occurs, this will not compromise the structure of the main roof (Fig. 2a).

Furthermore, some architectural elements will not only meet functional and aesthetic requirements, but also play a structural role: in addition to serving as a protected area to stock goods, closed granaries provide greater resistance to roofs, especially gable roofs. Similarly, the fencing of the gallery achieved with planks positioned crosswise secures the structure that, even in case of damage of the base of the masonry wall, will maintain some of its consistency and prevent a collapse (Fig. 2b). The decorative friezes adorning the gables allow breaking the wind flow and thus minimising the depressions that may blow the cover (Barré et al., 2011).

The application of vulnerability reduction strategies incorporated into the management of the environment (Fig. 2c), the architecture, the practices and the knowledge within the community, emphasises the awareness of risks and their taking into account, as part of vernacular solutions.

From a perspective of bringing support to programs for the reparation and reconstruction of housing, the investigative work carried out in Haiti has allowed to collect detailed data on vernacular buildings and know-how that can be used to contribute to improve technical and methodological solutions in ongoing

a) b) c)

Figure 2. Grande Rivière: (a) disconnection of the gallery structure, (b) bracing of the gallery and decorative frieze (c) Cap Rouge: reduced height and plant barrier to protect against strong winds (credits: ??).

programs, and to train local stakeholders to perform this type of analysis.

4 FROM IDENTIFICATION TO CHARACTERISATION

As part of the ReparH project supported by the ANR, collaboration was established between complementary disciplines: social sciences, architecture and engineering. Through the association of researchers in Haiti and their CRAterre-ENSAG counterparts, who were able to witness these local solutions, and those at 3SR-UJF laboratory, who could meet characterisation needs to cater to the demands of structural engineering and control departments, always necessary when projecting to implement programs at a large scale, but also to connect fieldwork and laboratory research, moving back and forth through the different dimensions of mutual enrichment.

The idea is not only to reuse constructive intelligences identified through research work, but also, if necessary, to bring improvements to existing methods, for instance, to overcome the main weaknesses of local construction systems by modifying ground anchoring and superstructure bracing systems, so as to improve sustainability.

Thus, the possibility for the replacement of a construction feature was considered: the post system

Figure 3. (a) Timber frame prototype under construction (credits: CRAterre).

Figure 4. Prototype ready to be tested on FCBA vibrating table at FCBA in Bordeaux (credits: CRAterre).

directly anchored in the ground, which eventually and rather quickly ends up rotting, to be replaced with an anchoring system of the posts over a foundation, with the addition of bracing through the implementation of St Andrew's crosses.

The results of the study described above were used to develop several types of models for the reconstruction of houses, offering a good compromise in terms of costs, the respect of local cultures and para-seismic performances, mostly based on the use of wooden structures braced with St. Andrew's crosses, and filled with stones or earth where possible.

In order to understand and model the behaviour of such structures, the study was based on a multi-scale approach for analysing the global and local behaviours of housing structures subject to an earthquake (assembly, individual unit, wall, full structure). The study of behaviours at each scale provided the necessary information to predict behaviours at the following scale. All of these studies could be verified through testing on a vibrating table, implemented by the FCBA in Bordeaux in April, 2013 (Fig 3–4).

As a result, a construction system has been certified by the Ministry of Public Works, Transport and Communication (MTPTC), paving the way for multiple achievements, with the adoption by more and more organizations of the proposed process/method, accompanied by further efforts dealing with training programs, including the training of trainers.

Figure 5. EdM housing unit in Port-au-Prince (credits: CRAterre).

Figure 6. EPPMPH housing unit in Rivière Froide (credits: CRAterre).

5 CONCLUSIONS AFTER 5 YEARS

The work undertaken in Haiti in partnership with many local and international stakeholders morphed into a variety of complementary projects, and generated important synergies. Nearly five years after the disaster, results are tangible and measurable, both quantitative and qualitative, in the field, as well as in terms of a reflection on methods and practices.

The various projects, in which the team has been involved to date, have resulted in the construction or rehabilitation of about 3 000 dwellings (Fig. 5–6–7).

However, the adoption of the «building cultures» approach by other organisations has also brought about indirect effects, with a wider circulation of models (technical and methodological), and a level of achievement that is already estimated at over 1000 additional basic housing models. This is at once small and great: small compared to the needs of the country, but great

Figure 7. New timber frame Housing unit in Ducrabon. (credits: CRAterre).

as regards to the number of permanent residents that could be found by their families. Indeed, this production should soon surpass the 5 000 unit mark, representing a significant portion of the sustainable buildings that could be constructed as part of the «national reconstruction» process.

After the necessary step of running diagnostics and making preliminary studies to understand the context and dynamics, it soon became possible to launch the first experimental constructions that have allowed to demonstrate and convince, but also to identify the needs for adjustments, to stakeholders. It was also necessary to prepare, in collaboration with local partners, the implementation of projects, from a technical point of view (architecture and construction requirements, training, etc.) and from an organisational point of view (sensitisation, logistics, coordination, administration, financial management).

In this process, the certification of the building system by the MTPTC, made possible through a partnership with the organisation entrepreneurs du Monde was a decisive step, allowing to validate scientific hypotheses and to reassure a number of local partners and actors on the capacity of the construction systems developed to withstand hazards. Later, the excellent results from the vibrating table tests have helped erasing any remaining doubts.

With a growing recognition, gradually, other organisations have expressed their interest in this «building cultures» approach, PADED member organizations at first, and other organizations later on, with training and sensitisation efforts (Fig. 8–9), and an early structuring of the sector led by Entrepreneurs du Monde.

However, through this process, it is by implementing a training program with UN-Habitat and the Atelier-école de Jacmel that this outreach effort could specifically be developed, through the establishment of a pedagogical engineering aimed at 7 organisations. These organisations are now able to carry out similar actions, and to train the professionals and trainers required at the national level to develop sustainable practices.

Figure 8. Training and sensitization workshop with communities and Kombit (credits: CRAterre).

Figure 9. National Seminar in Kenscoff, with all organisations involved in the reconstruction project (credits: CRAterre).

Another important result deals with unit costs, allowing to take into account more beneficiaries, but also to establish such building principles and processes on the longer term. The construction costs for basic models is very reasonable (around 150 US $\$/m^2$ for a new building – between 40 and 60 US $\$/m^2$ for rehabilitation). Another economic aspect has to do with local benefits. Indeed, the projects mainly involve local populations and local professionals, enabling them to start and benefit from income generating activities. These models are very accessible and allow some of the families to expand their homes by implementing the improved traditional techniques proposed. Kombits that have been sensitised get organised to pursue reconstruction efforts following the same principles. Such results are extremely important, given the fact that, currently, international assistance during major disasters hardly covers 20% of the real needs, and Haiti is unfortunately no exception to this «rule». This culture of solidarity is remarkable in hard to reach areas, where «construction» kombits pursue their efforts within their communities to maintain communication paths, or invest in replanting and reforestation projects to secure wood for building.

Moreover, an important and complementary set of publications has been produced: technical manuals and forms, competency frameworks and assessment tools, pedagogic materials, plans, etc., as well as articles that describe and popularise the approach implemented.

The project, or rather the projects that have been carried out in synergy, allowed making further significant progress. With the support of the ANR, it also allowed to question the social utility of science and its added value, and to articulate a dialogue between local knowledge and scientific knowledge, humanities and social sciences, in this very specific context of reconstruction. In this framework, in addition to two doctoral theses (one in architecture and the other in engineering), other research results have or will be achieved, and new research pathways have been opened and have already been presented and discussed at various international meetings.

Some significant data on PADED reconstruction programs, supported by Misereor (October 2014):

- Construction of 782 houses, 22 m^2 (basic module) and 85 reparations (35 m^2) performed over 3 years.
- Local capacity-building: 7 engineers, 15 construction foremen and 273 masons and carpenters, trained on construction sites; training validated following an evaluation protocol.
- Yield for the construction of a new building: 1 foreman and 6 artisans can build, together with the kombits, 4 small houses per month. 1 mason and 1 carpenter manage 1 site over 3 weeks, accompanied by 2 apprentices evaluated after 6 construction projects.
- Cost of 22 m^2 (new construction), excluding external technical assistance expenses: US $ 3,000 or US $ 135/m^2, of which US $ 1,700 is spent on materials, US $ 200 on transportation, $ 50 US on tools, US $ 450 on local skilled labour and 600 US $ on monitoring and coordination by a local NGO.
- On average, repairs on 35 m^2 units cost US $ 900 for labour, and the cost of materials and transport is reduced to US $ 1000. Valorisation of local expertise, diversified solutions.

This set of elements has undoubtedly contributed to achieve better and faster short-term goals in Haiti, but also to take into consideration alternative approaches to reconstruction and development, as well as different scales, other than that of the prototype, and, at last, to aim at sustaining an approach that has proven efficient.

6 OUTLOOK IN HAITI AND ELSEWHERE…

The reconstruction of Haiti is unfortunately far from being complete. While much has already been done, the needs are still immense, especially in terms of finding solutions for populations still living in camps or under unacceptable conditions. It is therefore necessary to encourage and facilitate the construction of suitable housing in urban and peri-urban areas (of various densities). Infrastructure needs are also important, especially with regard to the education sector.

Therefore, the government and a number of organisations try to remain involved with local communities still living in poverty, or who have no access to basic services. Beyond meeting these basic needs, it remains imperative that efforts can still be made, so as to sustainably change and replace harming practices, and in particular to transfer the knowledge generated during the reconstruction phase to professional, academic and research groups.

These are the major challenges that CRAterre and its partners, among others, are now invited to address. From this perspective, it is clear that new methods and additional strategies must be developed. This new phase will be useful to further examine the situation and produce new knowledge and skills, which will be useful to achieve further work in Haiti, including upstream intervention, prevention, also in other countries facing the consequences of natural hazards.

In conclusion, it is clear that the implementation of appropriate strategies requires the wide dissemination of useful information before disasters strike. It is indeed important that policy makers and particularly local stakeholders, who always play a critical role in reconstruction programs, may be sensitised on the importance of local cultures, so that the right decisions can be made early, as reconstruction efforts start to be developed. It is therefore necessary to consider the implementation of prevention and harm reduction programs integrating building cultures at all levels of decision-making, articulating training activities, fundamental and applied research, and the dissemination of supports for the enhancement of «good practices».

REFERENCES

Belinga Nko'o, C. (2014). Projet ReparH, essais sismiques sur table vibrante: rapport. Grenoble: AE&CC-ENSAG.

Caimi, A., Guillaud, H., Garnier, P. (2014). Cultures constructives vernaculaires et résilience. Entre savoir, pratique et technique: appréhender le vernaculaire en tant que génie du lieu et génie parasinistre [en ligne]. thèse de doctorat. Grenoble : Université de Grenoble. 539 p. Disponible sur : <https://tel.archives-ouvertes.fr/tel-01148207> (consulté le 18 juin 2015).

Correia, M., Dipasquale, L., Mecca, S. (2014). Versus: heritage for tomorrow: vernacular knowledge for sustainable architecture. Florence: Firenze University Press.

Dejeant, F., 2012. Construction en ossature bois et remplissage en maçonnerie: Bâtiments parasismiques et paracycloniques à 1 ou 2 Niveaux. [s.l.] : CRAterre-ENSAG & Entrepreneurs du monde.

Frey, P., Bouchain, P. (2014). Learning from vernacular: towards a new vernacular architecture. Arles : Actes Sud.

Garcia, C., Trabaud, V., (2015). La reconstruction d'habitats en Haïti: enjeux techniques, habitabilité et patrimoine. Available at http://urd.org/IMG/pdf/La_reconstruction_d_habitats_en_Haiti_Final_compresse.pdf

Garnier P., Moles, O., Caimi, A., Gandreau, D., Hofman, M. (2013). Natural Hazards, Disasters and Local Development. Integrated strategies for risk management through the strengthening of local dynamics: from reconstruction towards prevention. Grenoble: CRAterre ENSAG

Guillaud, H., Moriset, S., Sanchez, N., Sevillano, E. (2014). Versus: lessons from vernacular heritage to sustainable architecture. Villefontaine : CRAterre

Hofmann, M. (2015). Le facteur séisme dans l'architecture vernaculaire: un décryptage entre éléments déterminants culturels, typologies structurelles et ressources cognitives parasismiques (PhD Thesis, Ecole polytechnique fédérale de Lausanne). Available at http://infoscience.epfl.ch/record/207731/files/EPFL_TH6578.pdf

JIGYASU, Rohit, 2002. Reducing disaster vulnerability through local knowledge and capacity: the case of earthquake prone rural communities in India and Nepal. (PhD Thesis, Norwegian University of Science and Technology, Trondheim). Available at www.diva-portal.org/smash/get/diva2:123824/FULLTEXT01.

Joffroy, T., Garnier, P., Douline, A., Moles, O. (2014). Reconstruire Haïti après le séisme de janvier 2010: réduction des risques, cultures constructives et développement local. Villefontaine : CRAterre.

The Sphere Project. (2011). The sphere project: humanitarian charter and minimum standards in humanitarian response. Available at http://www.ifrc.org/PageFiles/95530/The-Sphere-Project-Handbook-20111.pdf.

Vieux-Champagne, F., Caimi, A., Garnier, P., Guillaud, H., Moles, O., Sieffert, Y., Grange, S., Daudeville, L. (2014). " Savoirs traditionnels et connaissances scientifiques pour une réduction de la vulnérabilité de l'habitat rural face aux aléas naturels en Haïti ". M. Oriol (Ed.), Innovations locales et développement durable en Haïti. Haïti : Editions de l'université d'Etat d'Haïti.

Vieux-Champagne, F., Daudeville, L. (2013). Analyse de la vulnérabilité sismique des structures à ossature en bois avec remplissage. (PhD Thesis). Grenoble : Université de Grenoble.

Part 2: Local seismic culture around the world

Seismic Retrofitting: Learning from Vernacular Architecture – Correia, Lourenço & Varum (Eds)
© 2015 Taylor & Francis Group, London, ISBN 978-1-138-02892-0

Local seismic culture in Latin America

L.F. Guerrero Baca
Universidad Autónoma Metropolitana, Ciudad de Mexico, Mexico

J. Vargas Neumann
Pontificia Universidad Católica del Perú, Lima, Peru

ABSTRACT: This article analyses some of the main Latin American vernacular architecture strategies in order to address the effects of earthquakes, which are widespread among the region. Most of this constructive knowledge results from the legacy of pre-Columbian indigenous cultures generated from ancestral processes of trial and error, which was combined with Western building work, brought by European conquerors from the sixteenth century. Among the preventive measures in vernacular construction, one important aspect is finding ways to face earthquakes from the joints and flexibility of structures, instead of establishing a rigid opposition against telluric forces. Building components of plant origin such as wood, canes and ropes are used to reinforce masonry and therefore, control movements between structural parts.

1 INTRODUCTION

Despite the cultural and natural diversity that characterizes Latin America, many of the common features of vernacular architecture in this area have derived from the seismic condition of this huge geographic region.

The continent lies on three plates of the Earth's crust (North American, Caribbean, and South American) that according to the Theory of Continental Drift move and generate in their contacts with great Oceanic and Antarctic plates, a permanent tectonic activity that accumulates energy, generating cyclical earthquakes.

There are other minor plates (Cocos, Caribbean and Nazca) interacting between the large plates previously mentioned. Finally, in different parts of the continent, there is a presence of active volcanic chains that release energy, producing tremors of diverse intensities.

These factors forced ancient cultures that lived in these territories to develop constructive strategies, which allowed their buildings to resist movements of the Earth's crust, through the sustainable use of the most abundant and accessible natural materials available to them, including earth, stone, wood, canes, and various plant fibers.

Upon arrival of the European conquerors from the sixteenth century, local building cultures experimented different changes resulting from the incorporation of a building tradition of Gothic and Renaissance heritage, based mainly on the static work forces of gravity within the structures.

It took European builders several decades to understand the motivation of constructive concepts of native civilizations, since the seismic experience of the Old Continent was much less intense, and thus allowed the development of structural systems based on the use of relatively slim stone pieces, with building components that worked in compression, such as the masonry arches, vaults and domes, which did not exist in American cultures.

Attempts to replicate in America, such stone structures and heavy roofs produced diverse failures, disasters and even deaths. To mention an example, the old city of Santiago de los Caballeros of Guatemala, former capital, collapsed many times due to cyclic earthquakes.

It is also worth mentioning the heavy adobe or stone constructions of Lima, capital of the Viceroyalty of Peru, which were catastrophically destroyed by the great earthquakes of 1687 and 1746. These disasters led to a Spanish Royal Ordinance that prohibited earth constructions more than one story high.

Vernacular architecture that emerged from both constructive visions led, in some regions, to a setback of the ancestrally proven knowledge, when materials such as pebbles, earth mortars, adobes and rammed earth were used in different dimensions and proportions than those that characterized pre-Columbian structures.

Changes in use, shape and size of livable spaces, along with the implementation of building components intended to be used massively, as it happened with the pre-Colombian pyramids and platforms, generated architectural typologies that, while possessing high cultural significance, presented a certain vulnerability to seismic events.

Various conservation and restoration problems that traditional architecture suffers, and even some monumental buildings of the continent, are linked to the

Figure 1. Wattle and daub or *quincha* wall, made around 3000 BC. Caral, Peru (credits: Julio Vargas, 2010).

Figure 2. Archaeological remains of stonewall bagged with plant fibers ropes, made around 3000 BC in Caral, Peru (credits: Julio Vargas, 2010).

collision between opposite building logics: one static and other dynamic.

Pre-Columbian public buildings were not roofed with heavy materials due to seismic activity. Only small areas roofed with wood, cane and vegetables were used. Figure 1 shows a space with vestiges of wattle and daub walls, wooden ceiling and central wooden columns at the top of the pyramid "La Galería" of Caral, Peru, which was built by stable platforms made of stones wrapped with vegetal fiber ropes (Vargas et al, 2011).

Instead, most of the vernacular constructions associated with communities where European influence had a minor impact, still maintain current seismic design strategies that have been passed from generation to generation.

2 PRE-COLUMBIAN BACKGROUND

Vernacular architecture that remains in various regions of Latin America surprises for its resemblance with discovered archaeological evidence, as well as graphic representations found in ceramics, sculptures and pre-Columbian codex.

The civilizations that inhabited the continent realized, from atavistic trial and error, that it was impossible to make rigid and heavy structures to resist earthquakes; thus they developed different resources to seek that living places had flexible and light involutes, also able to move in harmony with the ground.

In regions with forest resources or bamboo species, their components were used integrally or by sections, since wood, canes and ropes possess remarkable flexibility and deformation tolerance in different directions.

If there were not enough wood resources, another option was to combine elastic materials, such as sticks, branches or canes, that were tied with plant fibers, and combined with earth mortars or stone structures, which also ensured the possibility of building walls and roofs, both resistant and flexible. There are remarkable archaeological remains of wattle and daub building technique, locally known as *quincha*, *bahareque*, *bajareque*, *fajina*, *embarrado* or *tequezal*, in different regions of the continent.

However, in different seismic regions in Latin America both types of vegetal sources were scarce or nonexistent, like in desert areas of Northern Mexico, Peru, Chile and Argentina. These conditions led to the development of sophisticated building systems made by ancient cultures, based on the combination of earth and stone in the constructive logic of masonry (Guerrero et al., 2014).

In order to generate stairs, ramps, platforms, stepped pyramids and walls that confined ceremonial, governmental or open spaces, inventive earthquake-resistant solutions were created.

It is worth mentioning in more detail the example of the stable nuclei of the stepped pyramids of Caral previously cited, made from rope bags filled with stones, technique that dissipated seismic energy by impact and friction (Fig. 2).

Figure 3. Discontinuous arrangement of the adobe courses to form layers and blocks that function as articulated elements, in Huacas de Moche, Peru (credits: Luis Guerrero, 2007).

Figure 4. Pre-Columbian vestiges of a foundation platform, in Joya de Cerén, El Salvador (credits: Luis Guerrero 2014).

Other important example is the adobe masonry volumes of different shapes and dimensions of Huaca del Sol in Trujillo, Peru, which were set fulfilling two conditions: to be placed by overlapping layers in both vertical and horizontal planes, and to be built in large unit blocks with "seismic joints" between them.

Therefore, instead of producing massive units evenly built, articulated 'packages' were generated and, although they had the necessary strength to form platforms and support structures for lightweight roofs, they also had certain deformability qualities so that seismic forces would not make them collapse. The separate joints of the building components allow seismic energy to dissipate through cracking, impact and friction between materials, which causes a much less destructive effect (Fig. 3).

This combination of requirements is highly complex, since the elements that form the structures must be strong enough, in order to not suffer crushing or breaking effects. They must also be arranged alternatively to form superimposed layers, in order to transfer both gravitational and lateral thrusts derived from diagonal loads or horizontal forces that characterize earthquakes.

Many of these buildings also had one or more overlays that were made for ritual purposes, set as loose layers on the pre-existing ones, in order to help deviate vertical shearing stresses. Sometimes, even within the same building phase, builders intentionally overlapped materials by layers, which reacted like plates with horizontal movements that were relatively free, stabilizing the system.

3 BUILDING RESOURCES

3.1 *Foundation platforms*

Among the most frequently used components in vernacular architecture built in adobe or wattle and daub, foundation platforms were very important because they not only prevented rising damp from ground water below, as they also stabilized the built spaces they supported.

Furthermore, these platforms commonly had small elevations to form continuous baseboards that supported walls, so that the axial loads were evenly transmitted.

But in the event of an earthquake, these elevations of the foundation interrupt the continuity of the potential failures coming from the ground, giving the structure a more efficient response (Fig. 4).

These platforms were usually built by superimposed layers, which were compacted as they were erected. The compacting of the surfaces was a part of the overall strategy of the system.

Hypothesis have been made around the idea that these platforms known as 'stone embankment' (pedraplén), which were already present in pre-Columbian houses of primitive periods, gradually evolved over time until becoming the pyramidal, platforms over which palaces and temples of the great American civilizations were built.

Even though most of these great platforms associated with pre-Columbian urban settlements stopped being built after the Spanish conquest. Many of them continue fulfilling their original functions, specially their defense purpose against earthquake effects.

The geometry, as well as the building procedure of those platforms work as a cushion element towards seismic forces, because they are made of several pieces that, as already mentioned, discontinue and divert possible subsoil failures.

3.2 *Wooden walls*

In regions with forest resources, complex architectural systems with outstanding responses to earthquakes were developed.

Wooden vernacular architecture has been generally built with joinery system of planks and beams, in order

Figure 5. Wooden vernacular house in Michoacán, Mexico (credits: Luis Guerrero, 2006).

Figure 6. Traditional wattle and daub or *taquezal* wall, in Granada, Nicaragua (credits: Luis Guerrero, 2006).

to allow free movement that this material develops as a result of daily and seasonal changes in humidity and temperature. In addition to the flexibility that wood possesses, this construction criteria results in structures that are fully articulated with an excellent response to seismic shocks (Fig. 5).

In some areas, vernacular architecture is built entirely of wood, so that the stresses and seismic movements are distributed on floors, posts, walls and ceilings without damaging the structures.

3.3 *Meshed structures*

One of the main building types in vernacular architecture of Latin America is meshed earthen walls, or wattle and daub with natural fibers.

This mixed technique (earth, fibers and branches or canes), which has remained largely unchanged for millennia, has the quality of presenting, a very appropriate response to seismic events.

These mixed structures, in which the flexibility of the vegetal material constitutes the core of walls, floors and roofs, functions as reinforcement mesh of the system, providing an outstanding response towards multidirectional stresses (Fig. 6).

The earth that covers and confines the vegetal material prevents excessive displacement, while the structure moves, but without making it too rigid. This gradually dissipates energy, and the buildings remain standing. This energy dissipation in walls causes cracking, which can be easily repaired.

Wattle and daub vernacular architecture is usually built one story high, so the tallness also provides stability to the property.

However, there are several buildings made in two or more levels, or only one but with greater heights, such as churches, monasteries or public buildings, which remain with a favorable response towards earthquakes.

3.4 *Small windows*

Due to the fact that an important part of earth, stone and wooden vernacular houses are located in rural areas, many communities that inhabit them carry out most of their daily activities outdoors. Thus, the indoor spaces, for resting or cooking, usually do not have windows, or those that do, have small dimensions, since there is no greater need to look outside.

In climates with very high or low temperatures, the wide earthen walls with small windows are very convenient for their thermal mass, and their respective cooling or heating qualities.

This condition is also highly suitable as a seismic solution, since the walls transmit stresses continuously, without concentrating forces, and thus developing more stable responses (Fig. 7).

Figure 7. Variant of a rustic *quincha* house made of wood, canes and earth. Even though, it has larger windows than usual, it resisted the 2005 northern earthquake of Lamas, Peru (credits: Teresa Montoya, 2015).

Furthermore, the absence of dissimilar materials, such as those used for lintels and window frames, prevents competition between components of different mechanical strength and flexibility that may weaken the structure.

Often lintels have perforating effects that crack walls with wide windows during earthquakes.

3.5 *Architectural layouts with regular geometry*

Rural households tend to occupy the properties in a dispersed manner through functional units of compact volume. These often have architectural layouts with a regular shape, tend to be symmetrical, and have small differences between their relative wall lengths.

Thus, when earthquakes occur, each unit resists due to the balance of the thrusts, its wide and low walls, minimum number of wall openings, small and centered windows, almost squared layouts for rooms, the existence of buttresses and finally, the control of movements that are achieved by timber or canes and ropes (Vargas, 2013).

As a result of using building systems, in which the support for rooftops is made with straight and long sections of wood or bamboo, it is very common for squared or slightly rectangular floor plans to be developed. When the original forms are modified by growths derived from functional needs, with imbalances between the longitudinal and transversal walls, differences in mass and thrusts can lead to their collapse.

Chronicles of the Spaniards after the conquest describe that in Andean regions they found circular plan houses made of stone and earth, almost without windows with very low height and straw roofs.

The circular shape prevents the stress concentration in the joint of perpendicular walls that tend to turn in different directions.

Figure 8. Adobe circular plan house, in Chacarilla, Bolivia (credits: L. Guerrero 2011).

Nowadays, circular households of shepherds are still located in the highlands of Peru, Bolivia, northern Chile and Argentina (Fig. 8).

3.6 *Weight loss to greater heights*

A design factor that characterizes vernacular building lies in the care of the hierarchy and mass distribution of the building components. Low building areas are made with higher density and weight materials, while the upper areas have light and flexible ones.

In this way, it is possible to move the building's center of gravity to the bottom of the structure, thus decreasing possible affectation of seismic movements that usually have greater intensity in horizontal directions (Dipasquale et al., 2014).

3.7 *Lightweight roofs*

Another key feature of traditional houses is the predominant use of roofs made with lightweight materials, always supported by bamboo or wooden beams. Logically, the ceiling slope and the number of slants vary, depending on the rainfall conditions in different regions.

In arid areas, such as those prevailing in northern areas of Peru, Mexico, Argentina and Chile, roofs have been noticeably flat and with a single slope. In rainy regions, higher ceilings are dominant, with inclinations above 45°, with two or four slopes.

The cover surfaces of the oldest roofs were made with straw materials, such as grasses or palm leaves. These components were intertwined and tied to the wooden structure of the roof, and it also contributed to the proper distribution of the forces that lateral thrust exerts in buildings, and that become critical in the upper components of the structures.

The lightweight and flexible closures tend to move with greater impulse than the lower parts of the structure, that are more solid and rigid, which dissipates the energy of the building, as a whole (Fig. 9).

Figure 9. Thick *quincha* walls with lightweight roofs resisted the 2005 Lamas earthquake, in Peru (credits: T. Montoya 2015).

4 CONCLUSIONS

Vernacular architecture is the materialization of endless knowledge generated in the past, and whose efficiency has been proven after millennia of trial and error experimenting.

Therefore, it is an unlimited source of knowledge that can be retaken for repair, or design, both contemporary and future spaces that help improve society's quality of life and, above all, provide security for its inhabitants.

There is a lesson from this accumulated knowledge that is worth mentioning: the characteristics of buildings emerged from a local and vernacular seismic culture, which have generally been developed with design criteria based on resistance and stability, are valuable, but are not enough to face large earthquakes in the future.

They must be enhanced with design criteria based on seismic behavior, considering reinforcements to prevent loss of lives, and help reduce property damage (Fig. 10).

Wall reinforcements are essential, as well as the connection between walls and ceilings, in order to control

Figure 10. Houses that have been affected the most by the 2010 earthquake, in Maule, Chile. There was a lack in conditions of seismic forecast derived from the vernacular building cultures and require constructive cultures (credits: Julio Vargas 2010).

the cracking and displacement among pieces separated by fissures.

Modern technology, seismic testing and earthen simulation models, in addition to local seismic experience gathered since ancient times, help the reinforcement in advancing common vernacular heritage, before a future earthquake occurs, rather than after, in order to avoid deaths and further damage (Vargas, 2014).

REFERENCES

Dipasquale, L., Omar, D. & Mecca, S. (2014). Earthquake resistant systems. In M. Correia, L. Dipasquale & S. Mecca, (eds), *VERSUS. Heritage for Tomorrow*: 233–239. Firenze: Firenze University Press.

Guerrero, L., Meraz, L. & Soria, F. J. (2014). Cualidades sismorresistentes de las viviendas de adobe en las faldas del volcán Popocatépetl. In L. Guerrero (Coord.), *Reutilización del patrimonio edificado en adobe*: 194–215. México: U.A.M.

Vargas, J., Iwaki C. & Rubiños A. (2011). Evaluación Estructural del Edificio Piramidal La Galería. Proyecto Especial Arquelógico Caral-Supe. Fondo del Embajador EEUU.

Vargas, J. (2013). Consideraciones para incluir la técnica del tapial en la normativa de tierra Peruana. In 13° Seminario Iberoamericano de Arquitectura y Construcción con Tierra. 13° SIACOT. Valparaíso, Chile: Duoc UC.

Vargas, J. (2014). El patrimonio cultural en tierra del Perú y la influencia de los desastres en su historia. Una propuesta de conservación. In L. Guerrero (Coord.), Reutilización del patrimonio edificado en adobe: 268–303. México: U.A.M.

Seismic Retrofitting: Learning from Vernacular Architecture – Correia, Lourenço & Varum (Eds)
© 2015 Taylor & Francis Group, London, ISBN 978-1-138-02892-0

Local seismic culture in the Mediterranean region

L. Dipasquale & S. Mecca
DIDA, Department of Architecture, University of Florence, Italy

ABSTRACT: In contexts of high seismic activity, such as the Mediterranean area, many local communities have developed strategies for managing such a risk, adapting all available resources for creating earthquake-resistant rules, shaping not just a particular building culture, but a complex local seismic culture. Over the centuries, the Mediterranean area has known an unequalled variety of building experiences thanks to the continuous exchanges and the dissemination of innovative solutions. The paper investigates and analyses the contribution of Mediterranean local building culture in the strategies of defence against earthquakes, through their conditions, logic and specific devices. This text presents those technical building devices, which are strictly connected to the local seismic culture, describing the techniques used to reinforce and to absorb horizontally loads in earthen, stone, and brick masonries.

1 REGIONAL SEISMICITY IN THE MEDITERRANEAN CONTEXT

The Mediterranean Basin is a complex environment, where over the centuries widely diverse cultures and communities came into contact, through trade and the exchange of goods and ideas. Indeed the circulation and the mingling of cultures and beliefs have shaped a composite civilization, consisting of tangible as well as immaterial factors, which is deeply rooted in our social and cultural imagination as Mediterranean culture.

Despite of local bioclimatic variations – with outstanding contrasts between the geographical extremities of the Mediterranean area – the territories surrounding the Mediterranean sea share a mild climate, with hot and dry summers, soft and wet winters, and rainy springs and autumns. Two thirds of the lands of the Mediterranean area are constituted of limestone, which characterise the natural landscape, and provides the most widely used building material throughout the Mediterranean area (AA.VV, 2002). The alluvial surfaces of the Mediterranean banks produce instead soil that is rich in clay, which has played many roles and functions in the vernacular architecture: it has been used for the body of the wall (brick or compacted earth), the mortar, and the protection rendering.

The environmental context, the climatic condition and the available resources and materials have influenced the development of building typologies and techniques, which present in all the Mediterranean basin common elements and attributes (i.e. heavy-weight fabric, with an high thermal inertia, building types with a simple and compact geometric shape, compact urban fabric, etc.), and an amazing amount of variations, depending on the local environmental cultural and socio-economic characters.

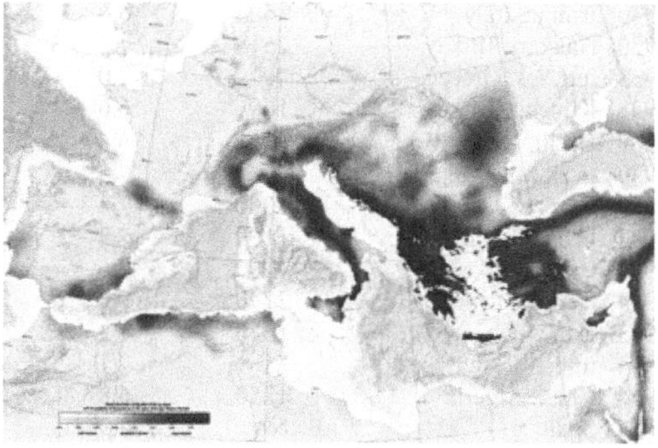

Figure 1. European – Mediterranean seismic Hazard map. Available at: http://preventionweb.net/go/10049.

A relevant factor that has influenced over the centuries the evolution of building typologies, techniques and specific devices, is the environmental risks, such as earthquakes.

North Africa and Southern Europe represent a region, around the Mediterranean, that is often prone to earthquakes. Seismicity in this area is especially due to the interaction along the boundary between the African and Eurasian plates.

Seismic events are not uniformly distributed along these boundaries. The oceanic crust of the Mediterranean basin is divided into two parts by the Italian peninsula, with the western part belonging to the Eurasian plate, and the eastern to the African plate. Typically, earthquakes with moderate magnitudes are widespread in all the area, while large earthquakes take place mostly along the Hellenides, and around the Aegean. Other segments of the boundaries are seismically less active, such as the Tunisian part of

Table 1. List of the larger earthquakes (magnitude >6.5) of the last 300 years (Levent Erel & Adatepe, 2007; Marturano, 2002; Utsu, 2002; USGS, 2012; USGS, 2014).

Year	Location	Casualities	Magnitude
1802	Vrancea region, Moldavia	3 in Bucharest	7,9
1810	Crete, Greece	2.000	7,5
1837	Galilee, Palestine	6.000–7.000	>7
1857	Basilicata, Italy	11.000	6,9
1881	Chios, Turkey	7.886	7.3
1894	Gulf of Izmit, Turkey	1300	7,9
1903	Malazgirt, Muş, Turkey	600	6,7
1903	Southern Greece	NA	8,4
1905	Calabria, Italy	557–5.000	6,7–7,9
1908	Messina, Italy	75.000–200.000	/,1
1909	Provence, France	6000	7,3
1912	Mürefte, Turkey	2800	7,3
1912	Mürefte Tekirdağ, Turkey	216	7,3
1914	Burdur, Turkey	4000	7,0
1915	L'Aquila, Italy	30000+	7,5
1920	Garfagnana, Lunigiana, Italy	171	6.5
1924	Horasan, Erzurum, Turkey	60	6,8
1928	İzmir, Turkey	50	6,5
1930	Irpinia, Italy	1404	6,6
1930	Hakkari, Turkey	2514	7,2–7,5
1932	Ierissos, Greece	491	7,0
1938	Kırşehir, Turkey	160	6,6
1939	Dikili, İzmir, Turkey	60	6,6
1939	Erzincan, Turkey	32.700	7,8
1940	Vrancea, Romania	1000	7,3
1942	Erbaa, Tokat, Turkey	1100	7,3
1943	Ladik, Turkey	4000	7,6
1944	Gerede, Turkey	2790	7,4
1953	Yenice-Gonen, Turkey	1070	7,3
1954	Ionian Islands	31	7,1
1954	Chlef, Algeria	1250	6,8
1954	Spain	NA	7,9
1956	Amorgos Island, Greece	53	7.8
1960	Agadir, Morocco	15000	5,7
1962	Bu'in Zahra, Qazvin, Iran	12225	7,1
1963	Skopje, Macedonia	1100	6,0
1964	Western Turkey	7,0	36
1966	Varto, Turkey	2529	6,8
1967	Mudurnu Valley, Turkey	173	7,3
1968	Western Sicily	231	6,5
1969	Portugal-Morocco area	13	7,8
1970	Gediz, Turkey	1086	6,9
1971	Bingöl, Turkey	1000+	6,9
1975	Eastern Turkey	2300	6,7
1976	Northeastern Italy	1000	6,5
1976	Turkey-Iran border region	5000	7,3
1978	Greece	50	6,6
1980	El Asnam, Algeria	5000	7,7
1980	Southern Italy	2735	6,5
1981	Greece	16	6,8
1983	Erzurum and Kars, Turkey	1342	6,9
1992	Erzican, Turkey	498	6,8
1999	Izmit, Turkey,	17.000+	7.6
1999	Düzce, Turkey	894	7,2
2003	Northern Algeria	2266	6,8
2006	Southern Greece	3	6,7
2011	Van Province, Turkey	604	7,2

Northern Africa, or the boundary between the Mid Atlantic Ridge and the Strait of Gibraltar (Udías, 1985, Vannucci et al, 2004; Marturano, 2002) (Fig. 1). In such regions, affected by frequent and medium to high intensity earthquakes, local communities have developed vernacular strategies to protect themselves from the risk, such as building systems or specific devices that are the result of a natural selection, dictated by the climatic context and locally available materials (Dipasquale et al., 2014).

Seismic vernacular reinforcements in the Mediterranean area are numerous, and often depend on available materials, local building cultures and the skills of the builders. The local seismic cultures include the earthquake-resistant regulations, which have not been formally laid out in written code, but which are still visible in the building characteristics, in the choice of the site, and in the general layout of the territory (Ferrigni et al., 2005).

Table 1 includes the larger earthquakes (magnitude >6.5) of the last 300 years. However, to understand the reason for the development of a seismic culture, seismic episodes of lower intensity should be also considered. In fact, the development and improvement of seismic retrofitting systems in a building culture is not only linked to the intensity of the earthquakes, but also to their recurrence, which is the factor that changes communities' practices and behaviours. Indeed, the origins and persistence of a local seismic culture can be determined, both by the scale of intensity and the frequency with which the earthquakes occur, and the economic and social conditions, including resource availability and the cultural traditions (Ferrigni et al., 2005). In Italy, for instance, violent catastrophic earthquakes have led most of the time to spend all the human and material resources for quick reconstructions (Marturano, 2002). After the disaster, often new seismic code have been improved, but only in few areas – e.g. in Lunigiana and Garfagnana, where earthquakes are endemic – a real seismic culture has been developed. On the other side, in the Balkans or in Turkey, where small and large seismic episodes occur frequently, local communities have developed behaviours and buildings with a good seismic resilience.

2 HISTORICAL SEISMIC RETROFITTING BACKGROUND IN THE MEDITERRANEAN AREA

In the Mediterranean area, amongst the ancient cultures, the Cretan (2000–1200 BC) and Mycenaean (in the 14th century BC) had developed a great sensitivity towards earthquakes. Archaeological excavations have revealed the systematic use of timber elements to reinforce bearing structures of palaces and villas in the Minoan Crete of the late Bronze Age (Tsakanika, 2006). Even if wooden elements have not survived to the present day, their location and dimensions are evidenced by vertical, horizontal or oblique holes in the masonry (Vintzileou, 2011). Another seismic retrofitting device found in the Minoan palace in Crete – which can be considered as an ancient system of seismic isolation – is the predisposition of alternate layers of sand and gravel under the foundation plan, useful to dampen the vibrations transmitted from the ground (Rovero & Tonietti, 2011). The use of timber reinforcement has been revealed by archaeologists as well in Akrotiri (Santorini, Greece): inside the large stone blocks, they found the housings of large pins crossing the rocks to accommodate wooden connecting elements, with the purpose of keeping the various blocks connected, and to give a strong plasticity to the whole structure (Touliatos, 2005).

In Anatolia the technique of timber frame structures with adobe blocks infill is employed from the Late Chalcolithic Age onwards. In general, that use of timber frames and adobe has a long documented tradition in nearly all Anatolia, in Northern Syria and in Egypt; all of them regions with a constant seismic activity (Didem Aktaş, 2011; Lloyd and Müller, 1998).

Wooden frame systems were documented also in many buildings of the ancient Roman Provinces. One of the most ancient examples in Italy of timber-frame reinforced buildings techniques is the *opus craticium* by Vitruvius; today visible in some of the surviving houses of the archaeological sites of Herculaneum and Pompeii (Fig. 2a). The structural grid of the *opus craticium* consists of squared uprights (*arrectarii*, of 8–12 cm per side) and horizontal elements (*trasversarii*, of 6–8 cm in width), arranged so as to form frames of 60 × 70 cm. The connections are made through joints and rivets (Langenbach, 2007).

Furthermore, in ancient Roman building traditions, rows of bricks were set down horizontally, through the conglomerate wall section, functioning, not only to connect and reinforce, but at the same time, serving to interrupt the possible spreading of cracks. This technique is still visible in many stone walls in Italian historic cities (Fig. 2b), and its wide use by the Romans is confirmed by the presence of continuous belts of red bricks in the old stone walls of Istanbul (Langenbach, 2007, Omar Sidik, 2013).

Figure 2. Roman Seismic retrofitting in ancient Roman building tradition. a) Opus craticium in Pompei, Regio I Insula XII. b) Random rubble masonry with horizontal layers of bricks, Lamezia (credits: L. Dipasquale).

3 SEISMIC-RETROFITTING TECHNIQUES IN MEDITERRANEAN VERNACULAR ARCHITECTURE

3.1 *Masonry reinforcement measures*

In the Mediterranean region the traditional construction technology is commonly based on load-bearing masonry walls.

Masonry buildings are vulnerable to seismic events, since they are constituted by the assembly of heterogeneous elements and materials (such as stones, or earth), whose characteristic is the not tensile strength. The seismic forces are transmitted through the soil at the base of the building, as horizontal actions. These forces give rise to a rotation out of lane, at the base of the wall perpendicular to the direction of the earthquake, which tends to an overturning and a consequent partial or general collapse of the masonry building. If the bonds between the orthogonal walls are effective, the building shows a box-like behaviour: the overturn is avoided, and the horizontal actions are transferred onto the walls in the direction of the earthquake, as shear actions that produce diagonal cracks, primarily distributed along the mortar joints.

Under seismic action, in order to avoid the first kind of damage, the structure has to guarantee a box-like behaviour, which can be achieved through the structural and dressing quality of the wall, such as good connections in the corners between perpendicular walls, and effective horizontal tying elements. Indeed certain parts, such as corners, lintels, jambs of openings and base plates, are more solicited than others, and they must better perform. For this reason ashlar blocks or bricks are often used to strengthen the structure in these more solicited parts.

In seismic areas where stone, earth or bricks masonry is the prevalent building technique, the most frequent prevention and/or reinforcement measures consist of adopting the mechanism of mutual contrast between parts of the buildings, to counteract horizontal forces.

The most common traditional devices used with the purpose of contrast and seismic reinforcement in Mediterranean vernacular masonry buildings are described below.

- *Buttresses*, or *counterforts*, are elements used both in vernacular and in monumental architecture in almost all the Mediterranean area. They are made of strong materials, such as brick or stone, and present a rectangular or trapezoidal cross-section. These elements are placed against and embedded to the wall in more stressed areas, to resist the side thrust created by the load on an arch, or a roof. Buttresses can be added to existing walls, or they can be built at the same time as the building, with the purpose of reinforcing corners or walls (Fig. 3). The traditional *loggia,* used mostly in Italy (interesting examples

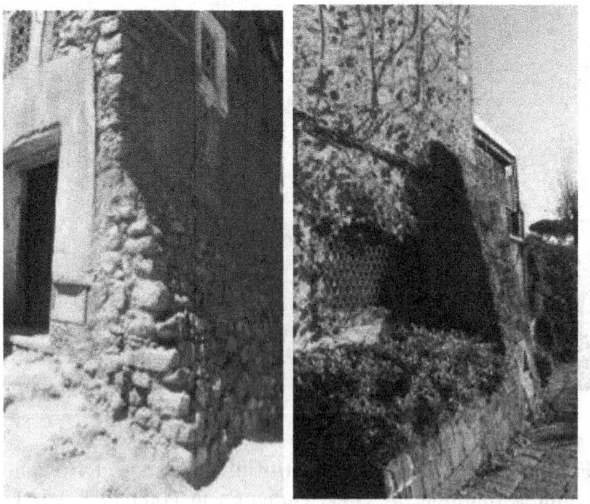

Figure 3. Buttress of random rubble masonry in Asni, Morocco, and buttress made of ashlar masonry in Naples, Italy (credits: L. Dipasquale).

Figure 4. Loggias in Ischia, Italy (photo: A. Picone).

can be seen in Sicily, the Amalfi coast, Ischia and Procida), in Croatia, and in southern France –, can be seen as an evolution of the buttress system (Ferrigni et al., 2005). It is used as reinforcement at the base of the building, and at the same time it provides a shading space at the entrance of the house

- *Anchor or patress plates* are metal plates connected to a tie rod or bolt, the purpose of which is to assemble the braces of the masonry wall against lateral bowing, holding the exterior wall from bowing out. Tie rods are stretched across the building from wall to wall, creating a horizontal clamping between the outer walls of the building. Anchor plates are made of cast iron, and sometimes of wrought iron or steel, and can be also fixed to the ends of the wooden floor beams (Pierotti & Ulivieri, 2001). The plates are mostly placed at the height of the floors, and are used in brick, stone, rammed earth, adobe or other masonry-based buildings (Fig. 5). There are many different designs of end-plate: square, circle, cross, double C, S or I shaped, etc. The shape is often an element characterising the local building culture. The dimension of the plate is bigger when the masonry is made of less resistant materials, such as earth or of smaller pieces as bricks. Following the 1909's earthquake in Provence, the use of wall tie plates to reinforce building walls on each floor was officially recommended (Ferrigni et al., 2005).

Figure 5. Anchor plates connecting two adjacent buildings in Florence, Italy (credits: L. Dipasquale).

Figure 6. Contrast arch in Marrakech, Morocco (credits: L. Dipasquale).

Figure 7. Vaulted street in Lunigiana, Italy (credits: L. Dipasquale).

- *Reinforcement, contrast or discharging arches* are stone or bricks masonry arches, set between two opposed buildings separated by a small street or a narrow passage. They allow the transmission of horizontal constraints to the opposite building at the level of the floor. In this way, buildings behave as an ensemble of dynamic blocks, and not as isolated elements. Reinforcements arches are visible in many historical towns situated in seismic areas, where stone and bricks are the prevalent building material (Pierotti & Ulivieri, 2001; Giuliani, 2011) (Fig. 6).

Figure 8. External staircase in Sicily and Apulia, Italy (credits: L. Dipasquale).

- *Lowering the centre of gravity.* Several techniques were used to increase the stability of buildings, by concentrating their mass closer to the ground. The most common solution is the use of increasingly lighter materials. The ground floor walls are often made by strong and compact stone (able to resist water pounding the base of the building), heavier and deeper, while the upper floors' walls are thinner, and often made of a combination of lighter elements and materials (in the seismic areas of the western Mediterranean they are generally made of timber, filled with small stones, bricks, or earth blocks). A frequent strategy to lower the centre of gravity of the building is the use of vaulted spaces at the ground floor. An example can be seen in some villages in the Lunigiana region (Italy), where many streets are vaulted, forming a single block with the adjacent buildings (Pierotti & Uliveri, 2001) (Fig. 7). Another example can be found in South-East Sicily, where after the earthquake of 1793, almost all the ground levels of reconstructed buildings have been covered by a masonry vaulted structure, while the intermediate floor structures are made of wood.

Also buttresses and staircases at the base of walls contribute to lower the centre of gravity of the building (Fig. 8).

3.2 *Seismic-retrofitting systems using wooden elements*

The difference between the solution described before and the following, is that the first group of devices aims to counteract horizontal forces; while the systems using wooden reinforcement are developed to metabolise, rather than counteract, these shearing motions, by favouring the deformation of single parts and junctions (Ferrigni et al., 2005). In areas where earthquakes are endemic, the use of wooden elements as seismic-retrofitting is a recurring strategy, and sometimes it defines an architectural typology, which characterises the local building culture.

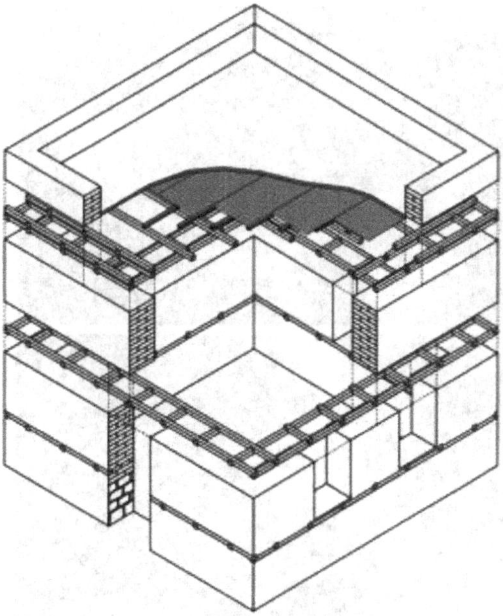

Figure 9. Building structure of hooping wooden system as seismic-retrofitting (credits: D. Omar Sidik).

The great elastic properties of wood, its characteristics of flexibility, lightness and deformability without reaching the breaking point, offers good resistance capacity against horizontal loads, and enables the dissipation of substantial amounts of energy. Moreover, timber elements divide the structure into sections, which prevent the spread of cracks occurring in portions of the masonry. By creating horizontal and vertical connections, wooden devices applied to structures with good compression behaviour (such as stone, adobe or brick masonry), can improve the resistance to shearing, bending and torsion forces. Therefore, in the case of an earthquake, whereas the rubble-stone walls may collapse, the wood is ductile enough to ensure that the building does not, thereby, saving lives (Dipasquale et al., 2015).

There can be various uses of wood as earthquake-proof reinforcement material: timber may be embedded in stone masonry walls to tie the stone units together, reinforce corners or, if braced, offer lateral resistance. Two main categories of systems that use timber structures have been identified: the *hooping* and the *frame systems*.

The first – *hooping*– provides the arrangement of circular or square section wooden beams, horizontally disposed, within the load-bearing masonry during the construction phase (Fig. 9). In many cases two beams are used, one on the inner side of the wall, and the other on the outer; connected by transverse wooden pieces. The empty spaces between the beams are filled with fragments of brick or stone. Interlocking systems of nailing are used for the connections between perpendicular elements. The ring beams can be inserted at the height of the floors, in correspondence to the openings

Figure 10. Two storeys adobe building with timber ties in Kastaneri, Greece (credits: S. Mecca).

and lintels or regularly distributed along the height of the construction, and they could prevent the propagation of a diagonal cracks in the wall during seismic actions.

This system can be found elsewhere in seismic regions in the Mediterranean: from the Balkans to Turkey, the Maghreb and Greece (it was systematically used in houses in Akrotiri on the island of Santorini). In Algeria wooden chaining is also used in the portico of the courtyards, as it can be seen, for example, in the Bey Palace in Algiers, where three beds of acacia logs have been inserted between the capitals and the beginning of the arch structure (Rovero & Tonietti, 2011).

In eastern Anatolia (in the area near Erzurum and Arapgir) runner beams (*hatil*) are aligned with the edges of the walls, at intervals of 50–100 cm, and they are crossed and overlapped at the corner with a scarf joint or half lap joint (Inan, 2014, Langenbach, 2007). In the town of Mut (Southern Turkey) wooden horizontal beams (*köstek*) – are placed every 50–150 cm inside and outside of the 50–75 thick masonry walls. These two beams are bounded with thin smaller wooden elements (Çelebioğlu & Yergün, 2014).

In northern Greece horizontal timber ties are almost always used to reinforce adobe bearing masonry: the ties are spaced at 70–100 cm intervals, passing at the level of the floor, or at the openings (Bei, 2011) (Fig. 10).

This building system has been documented also in Albania: the historical towns of Gjirokastra and Elbasan (fig. 11), for instance, conserve traditional

Figure 11. Stone masonry reinforced by horizontal wooden timber ties (credits: L. Rovero).

dwellings with limestone bearing masonry, reinforced by horizontal wooden timber ties, located on both sides of the wall, and ensured with a diagonal tie element at the corner (Pompeiano & Merxhani, 2015; Rovero & Tonietti, 2011).

In Algeria, wood logs of *tuja*, about 2 m in length, are embedded in the masonry thickness at intervals of 80/120 cm. Only in some cases logs are overlapped at the corners, more often they are used to connect adjacent buildings (Omar Sidik, 2013).

The second category of seismic-retrofitting systems using wooden elements includes *wooden frame systems*, which are articulated in round or square section beams and pillars, and frequently, diagonal bracing elements. The constructive system is always based on a grid of wooden poles making the main structure, while infill techniques vary depending on the locally available materials (stones, bricks, adobe, cob, daub or mixed materials).

If the beams are not as long as the entire wall, timbers are connected together through elaborate interlocking systems. In some cases, the longitudinal beams are held together in the thickness of the wall by transverse elements that are wedged or nailed, and the corners present additional reinforcement.

Wooden frame construction systems are widespread, both in rural houses, and townhouses in all the countries that have been subjected to the influence of the Ottoman Empire. The design principles of this building tradition, known worldwide as the Ottoman house, has been well-established as a building tradition since the 16th century in Turkey: from the Western Aegean Region it was successfully applied to a vast area, from Southern Middle Anatolia to the Black Sea Coasts of Romania and Bulgaria, as well as in Macedonia, Bosnia Herzegovina, Croatia and the north of Hungary (Didem Aktas, 2011; Langenbach, 2007).

The Ottoman house is based on the use of masonry laced bearing wall constructions on the ground floor level, and lighter infill-frames for the upper stories (Fig. 12). The intervals between the vertical, horizontal and diagonal timber elements of the upper floors

Figure 12. Ottoman house building system (credits: D. Omar Sidik).

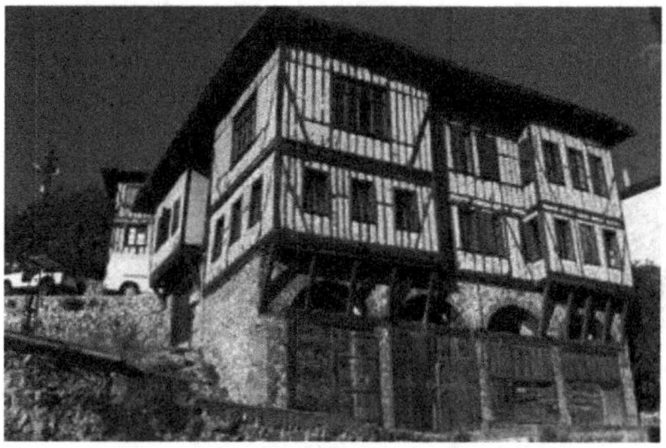

Figure 13. *Hımış* building Safranbolu (credits: Uğur Başak).

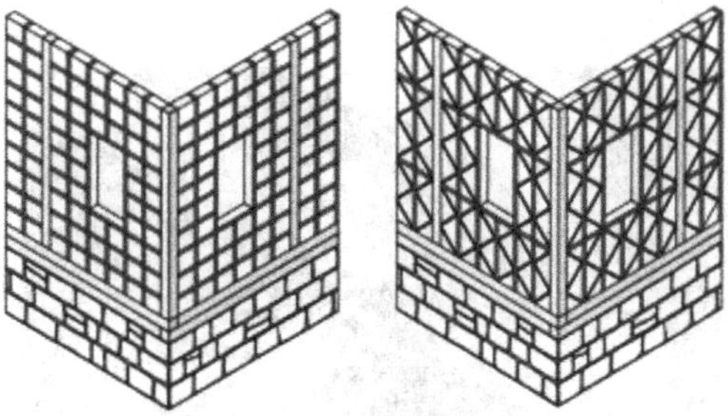

Figure 14. Building structure of *goz dolmas* and *muskali dolma* (credits: D. Omar Sidik).

are filled with bricks, adobe or stone – depending on the material availability of the region–, which are generally set with an earthen mortar. The ground floor masonry walls are often laced with horizontal timbers; these elements can be thin timber boards laid into the wall or squared wooden beams.

There are numerous reports demonstrating the validity of the Ottoman house as a seismic-proof building system. Many reports are based on observation, made after historical or contemporary earthquakes, and show that such buildings have been more resistant than other construction types, such as reinforced concrete and/or masonry structures (Gulkan, & Langenbach, 2004; Langenbach, 2007, Homan, 2001). Scientific researches have validated the empirical assessment, showing the different performance of structures with different infill materials, and also demonstrating that the location of failure nearly often coincides with the connections. (Aktas et al., 2010; Dikmen & Er Akan, 2005; Didem Aktas, 2011).

Turkish buildings called *hımış*, *bağdadi*, *muskali dolma* and *goz dolmasi*, can be considered as variations of the Ottoman house. All of them have a regular plan, two or three floors; the timber frame is often left in sight; and the upper plans projecting over the ground floor. The substantial differences between these techniques reside in the warping of the frame, as well as in the filling system opted.

The *hımış* timber skeleton, composed of horizontal, vertical elements and diagonal braces, is fastened to the ground floor masonry walls via the wooden beams embedded within the upper part of masonry (Gulkan & Langenbach, 2004) (Fig. 13). The cross-sections of the timber elements are approximately from 9 × 9 to 15 × 15 cm for the main elements, and 5 × 10 cm for the secondary ones. The interval between the studs varies between 50 cm (in Ankara) and 150 m (in North-western Anatolia) (Didem Aktaş, 2011; Omar Sidik, 2013). Diagonal braces were used in a variety of configurations and inclination angles (in North-western Anatolia, the angle is 30 to 45°). A single frame, forming the façade of a room, is the smallest module forming the timber frame section (Didem Aktas, 2011). The walls of the ground floor are generally 50–70 cm thick, and they can be made of rubble stone, cut stone or alternating courses of stone and brick or adobe, with timber lintels (hat*ı*ls) at regular intervals (Gulkan & Langenbach, 2004; Tanac Zeren & Karaman, 2015).

Typological variation of the *hımış* house are *muskali dolma* and *goz dolmasi* buildings (Fig. 14), where the secondary elements of the timber structure generate a very dense frame, made of triangular or square hole, of 15–30 cm width, which are filled with little stone or earthen mortar (Fig. 15). These techniques require a large use of wood, which is available in great quantities in the region. The joints between elements are accurately made, without the use of metal connectors (Omar Sidik, 2001).

The *bağdadi* typology consists of light exterior wooden laths, which are nailed onto the timber frame, filled with a mortar of straw and earth, sometimes gravel, and then covered with an external plaster (Dikmen & Er Akan, 2005; Şahin, 1995). This type of construction is typical in northern Greece and in Turkey

Figures 15 and 16. Building structure of *bağdadi* and *bondruk* (credits: D. Omar Sidik).

Figure 17. *Tsatma* structure in Antartiko. (credits: Luca Lupi).

(mostly in Ankara) (Tampone et al. 2011) (Fig. 15). In Lesbos (Greece), and other areas in Greece, it is used for the upper stories of townhouses, which project out over the streets, creating shaded areas (Jerome, 2014).

In Macedonia, timber frame buildings are called *bondruk*, and their structural configuration is comparable to the *himis* house (Namicev & Namiceva, 2014) (Fig. 16). Half-timbered architecture, mostly combined with earth, is also present in the south of Hungary, in the North-West and Central regions of Bulgaria, and in Macedonia, with high-quality architectural form and construction, revealing the influence of the Ottoman building culture (Vegas et al., 2011). In Bulgaria the main timbers consist of horizontal beams (*tabani*), vertical posts (*diretsi* or *mertetsi*) – placed every 60–70 cm, and diagonal braces (*payanti*). The houses are usually rendered and white-washed, so it is difficult to distinguish, at first sight, the material used for the filling (brick, adobe, wattle and clay daub or, rarely, stone).

In northern Greece (regions of west Macedonia), in mid-mountain areas, the second floor, upon the stone bearing walls, is traditionally made with timber framed walls, infilled with adobe, known as *tsatma* (fig. 17).

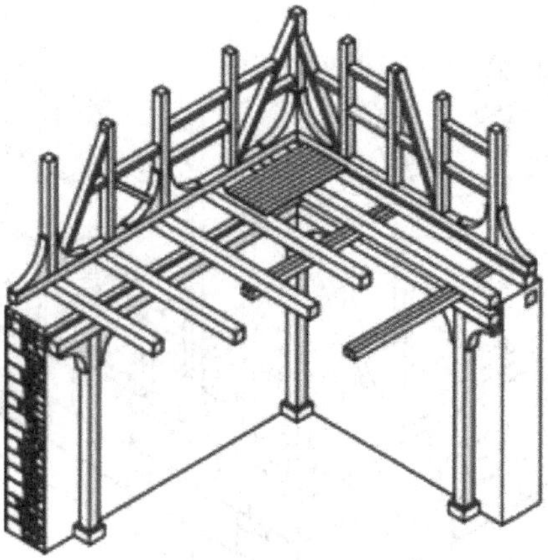

Figure 18. Building structure of Lefjkada traditional house. (credits: D. Omar Sidik).

The walls had a top layer of earthen plaster, which covers all empty voids (Bei, 2001).

In some Greek islands timber frame was firstly used as reinforcement to repair buildings affected by earthquakes, and only from the 19th century on, has it formed part of the new building structure (Ephessiou et al., 2005).

In the island of Lefkada stone and wood are combined in a unique indigenous dual system, where the timber frame is independent from the masonry, (Karababa & Guthrie, 2007). Buildings have masonry walls on the ground floor, while the upper floors are infilled timber frames; the walls of the upper floor are supported by both the ground wall masonry, and by a secondary structural system (offset at 5–10 cm from the masonry walls), consisting of timber columns, with a typical cross-section between 15 and 20 cm. This secondary system of support – locally called *pontelarisma* – makes the upper floors statically independent from the stone masonry, and it is able to safely sustain vertical loads, in case the masonry is severely damaged owing to a seismic event, giving the time that is necessary for repair or reconstruction of the damaged masonry system (Fig. 18) (Vintzileou, 2011; Ferrigni et al., 2005).

In Italy, the inheritance of the Roman system *Opus Craticium* has been almost lost over the centuries, although there are testimonies of vernacular buildings, widespread in the regions of Basilicata, Campania, Calabria and Sicily until the early 18th century, called *baracca*. These present a wooden frame structure, hidden by the exterior wall covering: this system was the model for designing seismic proof buildings in the phase of reconstruction after the disaster earthquake of 1783 in Calabria. The new building system called *casa baraccata*, designed by Giovanni Vivenzio (Fig. 19), presents a more rigorous architectural scheme, where specific devices act to create solid connections and to develop a good box-like action between all the elements of the building. The system consists of timber

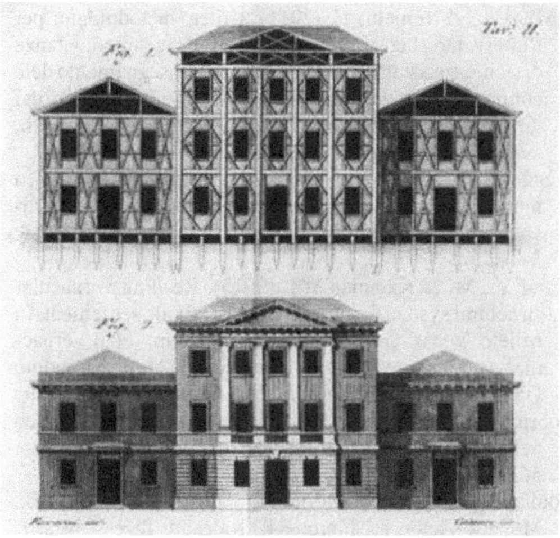

Figure 19. Casa *baraccata* in drawings (credits: Giovanni Vivenzio).

frame structures with stone or adobe infill. The external walled structure is made of straight vertical and horizontal pieces, with a square section of 10–12 cm. The internal load bearing walls include sloping timbers as braces, giving extra support between horizontal or vertical members of the timber frame. The connections between the wooden beams and pillars should take the form of snaps and rivets. The frame elements are covered externally with mortar, thus protected from the deterioration caused by atmospheric agents and by insects (Ruggeri, 1998; Tobriner, 1997; Dipasquale et al., 2015)

The good anti-seismic performance of this system was tested during the earthquakes that struck Calabria in 1905 and 1908: the buildings suffered few significant damages, and limited portions of masonry have collapsed. In the following decades the *baraccata* system has not been implemented with the original rigor, and it often presents insecure timber connections. In the last decades it was definitely abandoned. In 2013, a research conducted by the Italian National Research Council (CNR-Ivalsa) and the University of Calabria, scientifically demonstrated the validity of this building system as an effective seismic resistant solution.

4 CONCLUSION

The built environment of the Mediterranean, characterised by a balance between established tradition and continuous transformation, in an area with a high seismic activity, is a resource of knowledge, from which the scientific community could glean to better assess the seismic vulnerability of the existing buildings, as well as to identify strategies to improve the seismic safety of our building heritage.

Observing the traditional earthquake resistant structures, are understood some common rules that can improve the seismic inertia of the buildings, such as: a good execution of the work, a good connection between the elements of the buildings (walls, floors, roof, façade elements, etc.), a progressive reduction of the weight of materials, from the bottom to the top of the building, devices capable of counteracting horizontal forces, and systems able to increase the ductility of the buildings.

The awareness of the extraordinary quality of many traditional solutions, and the interest in the preservation of this heritage and the building culture, represents essential achievements, through which models for appropriate effective rehabilitations, future sustainable architectures and settlements can be composed.

REFERENCES

AA.VV. (2002). Architecture traditionelle méditerranéenne. Barcelona: CORPUS

Aktas, E., Didem, Y. A., Turer, B., Erdil, U., Akyüz, G. N., Sahin (2010). Testing and Seismic Capacity Evaluation of a Typical Traditional Ottoman Timber Frame. Advanced Materials Research, 133–134, 629–634.

Bei, G. (2011). Earthen architecture in Greece. In Correia, M., Dipasquale, L. Mecca, S. Terra Europae Earthen Architecture in the European Union. Pisa, Italy: ETS, pp. 124–127.

Çelebioğlu B. & Yergün U. (2014). An example of vernacular architecture in Central Anatolia: The mut houses. In Correia, Carlos & Rocha (eds) Vernacular Heritage and Earthen Architecture: Contributions for Sustainable Development. London: Taylor & Francis Group, pp.65–70.

Didem Aktaş, Y. (2011). Evaluation of seismic resistance of traditional ottoman timber frame houses. PhD Thesis in Restoration. Graduate School of Natural and Applied Sciences of Middle East Technical University.

Dikmen, N & Er Akan, A. (2005). Structural behavior of traditional timber buildings against natural disasters from different regions of Turkey. In Acts of 5th International Postgraduate Research Conference in the built and human environment, pp. 297–238.

Dipasquale, L., Omar Sidik, D. & Mecca, S. (2014). Earthquake resistant structures. In Correia, Dipasquale & Mecca (eds). VERSUS: Heritage for Tomorrow. Vernacular Knowledge for Sustainable Architecture. Firenze, Italy: University Press.

Dipasquale, L., Omar Sidik, D. & Mecca, S. (2015) Local seismic cultural and traditional earthquake-resistant devices: The case study of "Casa Baraccata". In Mileto, Vegas. García Soriano, &. Cristini (eds) Vernacular Architecture: Towards a Sustainable Future. London, UK: CRC Press, Taylor & Francis Group, pp. 255–260.

Ephessiou, I. Gante, D.J., Mitropolous, M. (2005). Levkàs . In Ferrigni, F., Helly, B., Mendes Victor, L., Pierotti, P., Rideaud, Teves Costa, P. Ancient Buildings and Earthquakes: the Local Seismic Culture Approach: Principles, Methods, Potentialities. Bari, Italy: Edipuglia, pp. 144–150.

Ferrigni, F., Helly, B., Mendes Victor, L., Pierotti, P., Rideaud, Teves Costa, P. (2005). Ancient Buildings and Earthquakes: the Local Seismic Culture Approach: Principles, Methods, Potentialities. Bari, Italy: Edipuglia.

Georgieva, D. & Velkov, M. (2011). Earthen architecture in Bulgaria. In Correia, M., Dipasquale, L. Mecca, S. Terra

Europae Earthen Architecture in the European Union. Pisa, Italy: ETS. 124–127.
Giuliani C. F. (2011) Provvedimenti antisismici nell'antichità. Rilievo Archeologico. JAT XXI, 25–52.
Gulkan, P. & Langenbach, R. (2004). The earthquake resistance of traditional timber and masonry dwellings in Turkey. In Acts of 13th World Conference on Earthquake Engineering Vancouver, B.C., Canada. Paper n. 2297.
Homan, J. & Warren J. E. (2001). The 17 August 1999 Kocaeli (Izmit) Earthquake:Historical Records and Seismic Culture. Earthquake Engineering Research Institute (EERI), Earthquake Spectra, Vol. 17, 4, 617–634.
Inan, Z. (2014). Runner beams as a building element of masonry walls in Eastern Anatolia, Turkey. In Correia, Carlos & Rocha (eds) Vernacular Heritage and Earthen Architecture: Contributions for Sustainable Development. London: Taylor & Francis Group, pp. 721–726.
Jerome, P. (2014). Vernacular houses of Lesvos, Greece: Typology and construction technology. In Correia, Carlos & Rocha (eds) Vernacular Heritage and Earthen Architecture: Contributions for Sustainable Development. London: Taylor & Francis Group.
Karababa, F. S. & Guthrie P. M. (2007). Vulnerability reduction through local seismic culture. Ieee Technology and Society Magazine, 26(3), 30–41 doi: 10.1109/mts.2007.906674.
Langenbach, R. (2007). From opus craticium to the Chicago frame. Earthquake resistant traditional construction. International Journal of Architectural Heritage. Conservation, Analysis, and Restoration, 1:1, 29–59.
Levent Erel, T. & Adatepe, F. (2007). Traces of Historical earthquakes in the ancient city life at the Mediterranean region. J. Black Sea/Mediterranean Environment Vol. 13, 241–252.
Lloyd, S. & Müller, H.W. (1998). Architettura: origini. Milano: Electa.
Marturano, A. (ed) (2002). Contributi per la storia dei terremoti nel bacino del Mediterraneo (secc. V–XVIII). Salerno: Laveglia editore.
Namicev P. & Namiceva E. (2014). Traditional city house in Northeastern Macedonia. Skopje. Macedonia: 2ри Август, Штип.
Omar Sidik, D. (2013). Presidi antisismici nelle culture costruttive tradizionali. Prime validazioni sperimentali relative all'impiego del legno negli edifici in terra. Unpublished PhD thesis. Università degli Studi di Firenze.
Pierotti, P. & Uliveri, D. (2001). Culture sismiche locali, Pisa: Edizioni Plus-Università di Pisa.
Pompeiano, F. & Merxhani, K. (2015) Preliminary studies on traditional timber roof structures in Gjirokastra, Albania. In Mileto, Vegas. García Soriano, &. Cristini (eds) Vernacular Architecture: Towards a Sustainable Future. London, UK: CRC Press, Taylor & Francis Group, pp. 631–636.
Rovero, L. & Tonietti, U. (2011). Criteri metodologici per l'intervento sl costruito storico a rischio sismico: istanze di sicurezza, istanze di salvaguardia e l'insegnamento dele culture costruttive locali. In Nudo, F. (eds) Lezioni dai terremoti: fonti di vulnerabilità, nuove stretgie progettuali, sviluppi normativi. Firenze: University Press. 289–301.
Ruggeri, N. (1988), Il sistema antisismico borbonico muratura con intelaiatura lignea genesi e sviluppo in Calabria alla fine del '700. Bollettino Ingegneri, 2012, 10, 3–14.
Tanac Z., M. & Karaman Y. I. (2015). Reading vernacular structural system features of Soma-Darkale settlement. In Mileto, Vegas. García Soriano, &. Cristini (eds) Vernacular Architecture: Towards a Sustainable Future. London, UK: CRC Press, Taylor & Francis Group 691–694.
Tobriner, S. (1997), La casa baraccata: un sistema antisismico nella Calabria del XVIII secolo. Costruire in laterizio, no. 56., 110–115.
Touliatos, P. (2005) Santorini. In Ferrigni, F., Helly, B., Mendes Victor, L., Pierotti, P., Rideaud, Teves Costa, P. Ancient Buildings and Earthquakes: the Local Seismic Culture Approach: Principles, Methods, Potentialities. Bari, Italy: Edipuglia. 159–162.
Udías, A. (1985) Seismicity of the Mediterranean Basin, In Stanley D.J, Wezel, F (eds) Geological Evolution of the Mediterranean Basin. New York, USA: Springer. 55–63.
Utsu, T.R. (2002). A List of Deadly Earthquakes in the World: 1500–2000. In International Handbook of Earthquake & Engineering Seismology, Part A, Volume 81A (First ed.), Massachusetts, USA: Academic Press.
USGS (2012). Historic World Earthquakes, At: *http://earthquake.usgs.gov/earthquakes/world/historical.php*.
USGS (2014). Earthquakes with 1,000 or More Deaths 1900–2014. At *http://earthquake.usgs.gov/earthquakes/world/world_deaths.php*.
Vannucci, G., Pondrelli, S., Argnani, A., Morelli, A., Gasperini, P.& Boschi, E. (2004). An atlas of Mediterranean seismicity. Annals of Geophysics, Supplement to vol. 47, n. 1, 2004, 247–306.
Vegas, F., Mileto, C. & Cristini, V. (2011). Earthen architecture in East Central Europe: Czech Republic, Slovakia, Austria, Slovenia, Hungary and Romania. In Correia, M., Dipasquale, L. Mecca, S. Terra Europae Earthen Architecture in the European Union. Pisa, Italy: ETS. 124–127.
Vintzileou, E. (2011) Timber-reinforced structures in Greece: 2500 BC–1900 AD. Structures and Buildings 164 June 2011 Issue SB3, 167–180 doi: 10.1680/stbu. 9.00085.

Seismic Retrofitting: Learning from Vernacular Architecture – Correia, Lourenço & Varum (Eds)
 ISBN 978-1-138-02892-0

The central and eastern Asian local seismic culture: Three approaches

F. Ferrigni
European University Centre for Cultural Heritage, Ravello, Italy

ABSTRACT: The central and eastern Asia are regions with a very high seismicity, and very rich in many old monuments too. In all monumental buildings, and in vernacular architecture too, it is possible to recognise seismic-proof techniques. These techniques are various: very large / very high buildings, massive/light structures, stone/wood materials. By analysing some monuments of the Asian region, this paper aims to show that there are two different approaches to resolve the same problem: how to realise structures able to "metabolise" the energy transmitted by the earthquake. A third approach, mainly adopted in vernacular architecture, is to accept the collapse of the building, on the condition that it can be easily rebuilt.

1 INTRODUCTION

Comparing the images of two famous Asian monuments the Potala Palace in Lhasa, and the Pagoda of Six Harmonies, in Hangzhou (Fig. 1) – a totally different look emerges: massive and larger than high, the first one; light and higher than large, the second one.

The same difference can be observed in vernacular architecture, comparing the mass of Tibetan houses with the lightness of the Japanese one (Fig. 2).

All these buildings are located in a very seismic region, all have survived to hundreds of earthquakes, and all present seismic-proof techniques, although different. We can easily consider them, as the result of the Local Seismic Culture (LSC).

Anyway, a question arises: why this deep difference?

Figure 1. The Potala Palace and the Pagoda of Six Harmonies': same seismicity, very different architectural language: massive masonry in the first one, very light wood structure in the second one (credits: "布达拉宫" by *Coolmanjackey*. Licensed under CC BY-SA 3.0 via Wikimedia Commons – https://commons.wikimedia.org; "Liuhe Pagoda" by *PericlesofAthens* at en.wikipedia – Transferred from en.wikipedia. Licensed under Attribution via Wikimedia Commons – https://commons.wikimedia.org).

It is banal to observe that in earthquake regions the temples and other public buildings were constructed with special care: monuments are intended to last in time. In the light of this, it would appear useful to analyse some of the technological approaches devised by the ancient communities, in order to reduce the impact of earthquakes on both listed buildings and vernacular architecture. First and foremost, this analysis may help understand how a local seismic culture (LSC) came about and took root, and how it might be recovered

Figure 2. The difference between the massive Tibetan houses and the light Japanese ones is impressive, although they stay in regions with similar very high seismicity (credits: S. Rijnhart, Creative Commons; Hamano House by *663highland*. Licensed under CC BY 2.5 via Wikimedia Commons – https://commons.wikimedia.org)

today, in order to reduce the vulnerability of the ancient architectural heritage.

In general terms, it is right to say that two methods have been developed, in order to make a building significantly more resistant to shear and torsional stress. One is to counteract horizontal forces (shear stress), by using materials with a stronger inherent frictional potential; and to counteract the effects of torsion forces, by strengthening the corners of buildings. The other is to absorb, rather than counteract, these shear or torsion motions, by favouring the (controlled) deformation of single parts and junctions.

In both these methods the overall shape of a building (a symmetrical plan, and masses decreasing from bottom to top) becomes an effective means of protection, since it reduces the impact of such horizontal forces. In simple terms, it is possible to say that two different seismic cultural approaches emerge in the Asian ancient architectural heritage: the 'rigidity' approach, i.e. a building of more rigid structures, capable of withstanding horizontal forces; and the "deformability" approach, i.e. protecting a building against the effects of earthquakes, by allowing its structures to deform.

In terms of seismic engineering, in the rigidity approach the energy generated by the earthquake is 'metabolised' by temporarily increasing the tensions inherent in single structural elements, while in the deformability approach it is "metabolised", thanks to displacements of the structure, to the friction between the components of the junctions, and to their elastic deformation.

In addition to these two cultural approaches there is also a third one, which is referred to as the 'passive' approach, i.e. the acceptance that the earthquake causes heavy damages to the buildings (mainly, houses), on condition that the structures are enough light to safe human lives when collapse; and that the building can be easily rebuilt.

A concise review of the traditional seismic-proof techniques, recognised in the Asian monuments and vernacular architecture, proves a well-rooted LSC, with some similarities and many differences.

Figure 3. The Taktsang Monastery, in Bhutan, and the Chinese Hanging Monastery appear very different, except the location, the safest one, despite the appearance (credits: "Taktshang" by D. McLaughlin – Licensed under CC BY 2.5 via Wikimedia Commons; "Xuankongsi" by E. Brunner – Licensed under Public Domain via Wikimedia Commons).

Figure 4. The Thikse Monastery, in Ladakh, and the Chinese Miao houses share the same "two side ground connection", very effective against EQs-(credits: Thikse Monastery by Sameer karania, licenced under CC, via Wikimedia Commons; Miao houses, autor Wu Liguan).

2 SIMILARITIES AND DIFFERENCES

Wherever they are located, whatever the used materials are, Asian temples and pagodas have a symmetrical, or nearly symmetrical, plan. This is a well-known seismic-proof feature, because it responds in the same way to the seismic motion, whatever its direction is. It also minimises the moment of the structure. Another general feature of Asian monuments is the regular decreasing of dimensions, from the bottom to the top. This is a seismic-proof feature too, because the larger base increases the strength to the horizontal components of the seismic shock.

The Taktsang Monastery, in Bhutan, and Chinese Hanging Monastery (Fig. 3) apparently show no similarities: heavy masonry in the first one, light wood in the second one; narrow windows in Bhutan, large glass walls in China. But they share a very effective seismic-proof similarity: the location. Appearances aside, the sites are the safest ones against earthquakes. In fact, both are founded on rocks; both have two sides connected to the mountain. These two features are very seismic-proof. First of all, rock is the safest soil in case of earthquake, because it minimises the site effect. Secondly, a building connected to the soil on two sides is totally protected against the shear forces, because foundations and floors move together.

It can be noted the same differences/ similarities above all relatively to the 'two sides soil contact' feature comparing the Thikse Monastery, in Ladakh, with Miao wood buildings, in China, Longsheng County (Fig. 4).

On the contrary the differences are numerous and eye-catching. The Mongolian Gandan-Monastery look

Figure 5. Massive masonry in Mongolian Gandan Monastery, light wood structure in Chinese Yingxian Wood Pagoda, but the same seismic-proof features: symmetrical plan and dimensions decreasing from bottom to top (credits "Gandantegchinlen Khiid Monastery" by Angelo Juan Ramos. Licensed under CC BY 2.0 via Wikimedia Commons; Yingxian Wood Pagoda, by Zhangzhugang, Licensed under GNU Licence, via Wikimedia Commons).

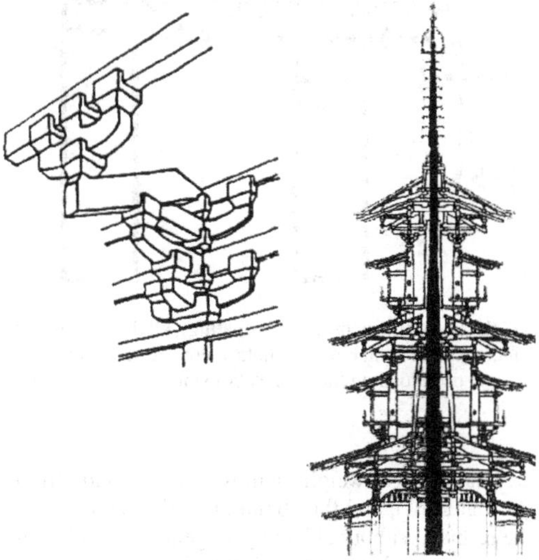

Figure 6. The structure of the Japanese XIV AD pagoda is isostatic: just a pole thrust into the ground. The roofs are hanged to the pole by a very jointed system of wood "bracket arms" (credits: modified from R. Tanabashi – Ferrigni et al, 2005).

is very unlike from the Chinese Yingxian Pagoda one (Fig. 5).

Of course main cultural approaches recurring in different parts of the world, depend on the locally available materials (the wood is scarce in Mongolia, abundant in China), as well as on the intended use of the artefacts themselves (the Mongolian building is allocated to monks' home, the Chinese pagoda to Buddha's statue and relics). These are the main factors determining the global aspect of the two buildings, but the construction techniques are totally different: massive masonry in Mongolia, and deformable wood in China.

Many and deep differences characterise the "deformability approach" of Japanese and Chinese Pagodas too. The structural scheme of the 14th century Japanese pagoda (Fig. 6) is very plain: a long pole thrust into the ground, and the various levels of the roof are hanged to the pole, by a very jointed sys- traditional "bracket arm", a very effective energy dissipator.

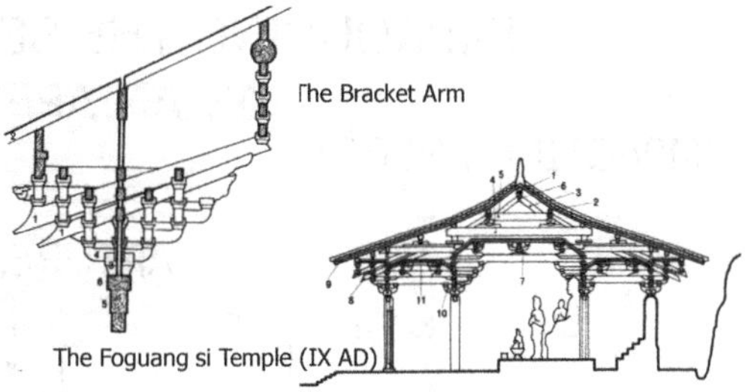

Figure 7. The same "bracket arm" in Chinese Foguang si Temple. The only difference with the Japanese one (see Fig. 6) is .that the roof is leaned on columns (credits: modified from N. Bouvier & D. Blum – Ferrigni et al, 2005).

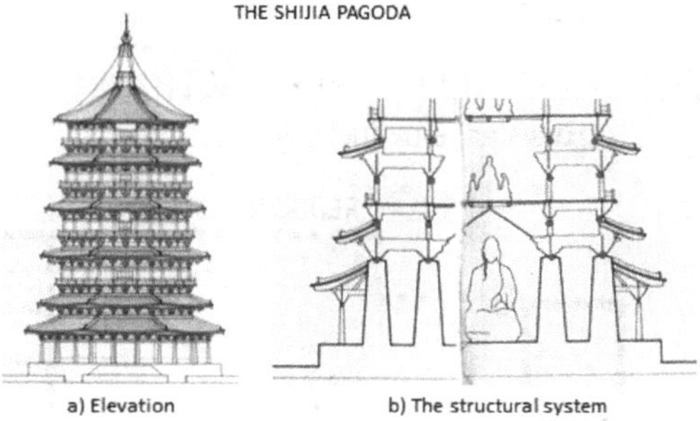

Figure 8. The very sophisticated system of "rolling floors" in the Chinese Shijia Pagoda (credits: modified from N. Bouvier & D. Blum – Ferrigni et al, 2005).

A more sophisticated "bracket arms" system is also present in Foguang si Temple on Mount Wutai (Shanxi, China) (Fig. 7). It is important to note that the "bracket arms" appear first in 3rd century BC, and have been used for all the following centuries.

The structural scheme of the 14th century Chinese Shijia Pagoda (Fig. 8) is totally different: floors are connected each other, by two roller systems.

The structural scheme of the 14th century Chinese Shijia Pagoda (Fig. 8) is totally different: floors are connected each other, by two roller systems.

3 THREE APPROACHES

The above short review of some examples of LSC in Central and Eastern Asia shows that even if seismic-proof techniques may be various and different, their aim is just one: to "metabolise" the energy transmitted by earthquakes.

The graphic in Fig. 9 schematises the problem of the metabolising seismic energy, and the consequent origin of the two different approaches. Under seismic forces, of the global energy transmitted by the ground

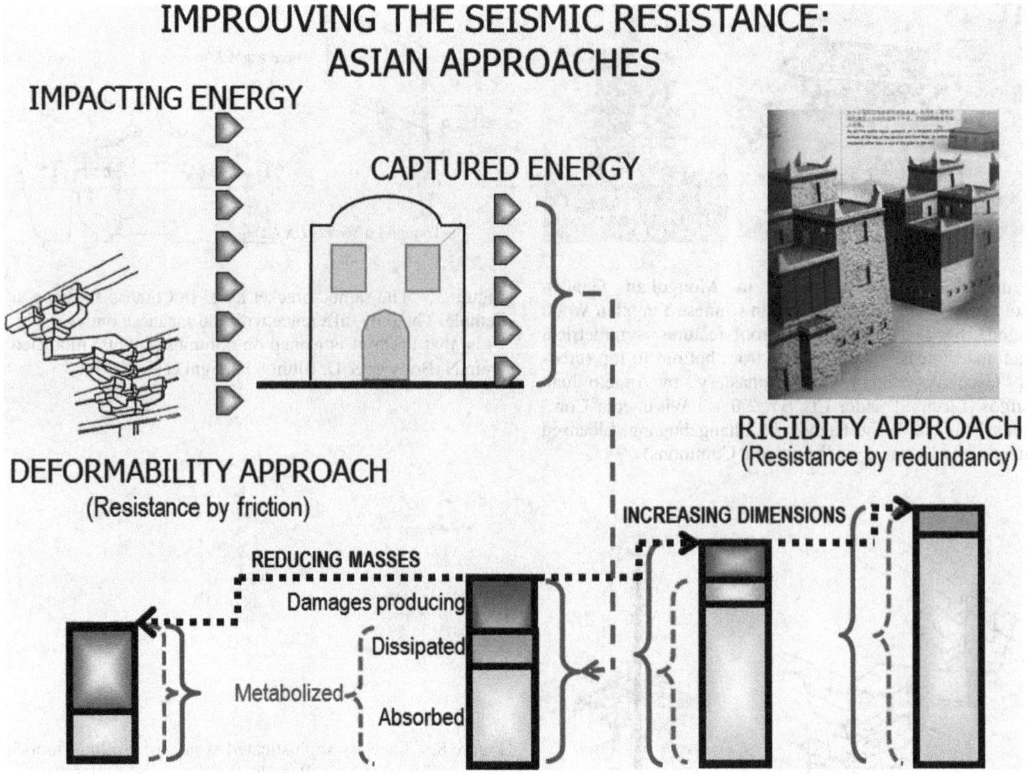

Figure 9. To resist to the seismic stress it is necessary to "metabolise" (by absorbing and/or by dissipating), all the "captured" quota of impacting energy transmitted by the ground motion. To increase the metabolised quota there are just two way: increasing the dimensions of the structures (metabolising by redundancy) or unloading them (metabolising by displacement, deformation and friction) (credits: F. Ferrigni).

motion, the building "captures" a quota, proportional to the mass of the artefact. This energy produces both, reversible (elastic) and irreversible (plastic) deformations. In seismic engineering the two quotas are defined as absorbed and dissipated energy. As these two quotas produce deformations but do not collapse, it is possible to define the addition of absorbed and dissipated quotas, as the "metabolised energy". The energy surplus produces damages.

To eliminate the gap between "captured" and "metabolised" energy there are only two ways: increasing the metabolising capability, by increasing the dimensions of structures, or reducing the captured energy, by lightening the structural elements. The first way produces an increasing of captured energy, so the dimensions of the structures have to be further increased, with the result of the redundancy of the structures. On the contrary, the second one involves an increase of deformations.

The two approaches co-exist in Asian LSC. But there is a third approach too. Alongside the development of flexible approaches to monumental building, some regions also witnessed the emergence of a LSC, which present no earthquake-resistant buildings, on the assumption that earthquakes were inevitable and frequent. In these systems, non-monumental buildings were mainly light-weight (almost entirely made from light materials), and thus bound to collapse, but able to be re-built without difficulty. In general this kind of seismic culture is to be found in regions exposed, not only to seismic risk, but also to hurricanes, as the Asian earthquake regions in the Pacific area. For example, traditional Japanese houses are one-storey structures made of wood (see Fig. 2). Thin wood and paper panels delimit the rooms. Should the structure collapse, their inhabitants are seldom at risk; and the constructions themselves can be re-built (or replaced) very rapidly, at a comparatively low cost.

We can define the houses built according to techniques, which offer no resistance to seismic shock as the result of a "passive LSC".

4 CONCLUSIONS

Cultures based on the flexibility approach are particularly fragile, in the sense that they are exposed, not so much to the impact of earthquakes, but to the contamination from different seismic cultures. In fact, the effectiveness of flexible techniques depends almost exclusively on the rupture threshold that is chosen. Builders, who adopt flexible approaches mainly use natural materials (wood, bamboo), and have recourse to traditional craftsmanship methods, handed down to them by word of mouth. Now that the empirical knowledge concerning such methods is forgotten, or has been contaminated, such thresholds are difficult to define because uneven, irregular-shaped materials are not easily represented in structural models, on which the required calculations can be based.

Thus, as no reliable parameters are available to predict the response of traditional flexible structures, in terms of their earthquake-resistance, and as, conversely, steady advancements in technology lead to the use of ever more rigid structural elements, local flexibility cultures are doomed to oblivion, and will inevitably take a back seat with respect to imported technologies and know-how. Therefore, very often, buildings with a light structure are 'reinforced' through the addition of rigid elements, which determine diametrically effects, opposed to those desired. Combinations of these two seismic approaches – the flexibility approach for monuments, and the passive approach for ordinary constructions – are typical of Oriental cultures. This combination is clearly the result of social calculations (monuments are intended to last, while houses can be re-built), but it also reveals a different relationship with nature, from that typical of Western civilisation. Here, nature is either feared or dominated. There, the local people attempt to draw maximum benefit from it.

REFERENCES

Ferrigni, F. (1990). À la recherché des anomalies qui protégent. Actes des Ateliers Européens de Ravello, 19–27 Novembre 1987. Ravello: PACT Volcanologie et Archéologie & Conseil de L'Europe.

Ferrigni, F., Helly, B., Mauro, A., Mendes Victor, L., Pierotti, P., Rideaud, A. & Teves Costa, P. (2005). Ancient Buildings and Earthquakes. The Local Seismic Culture approach: principles, methods, potentialities. Ravello: Centro Universitario Europeo per i Beni Culturali, Edipuglia srl.

Seismic Retrofitting: Learning from Vernacular Architecture – Correia, Lourenço & Varum (Eds)
© 2015 Taylor & Francis Group, London, ISBN 978-1-138-02892-0

The earthquake resistant vernacular architecture in the Himalayas

Randolph Langenbach
Conservation Consulting, Oakland, California, USA

ABSTRACT: This paper examines the traditional construction found in the Himalayan region in Indian and Pakistan Kashmir in comparison with Nepal, which has just at the time of this writing been subjected to the devastating Gorkha earthquake on April 25, 2015. The chapter describes the widespread tradition of the use of timber reinforcement of masonry construction in Kashmir in the context of the less common use of such features in Nepal, as shown by the widespread damage and destruction of traditional masonry buildings in Kathmandu. However, some of the heritage structures in Nepal do possess earthquake resistant features – most importantly timber bands – and there is now evidence many of those buildings have survived the earthquake without collapse.

1 INTRODUCTION

Earthquake! When this chapter was written, the dust had not yet completely settled from the April 25, 2015 earthquake in Nepal (Fig. 1). The body count continued to increase with each day and the rescue efforts to find and free people entrapped beneath the ruins continued with more distress on the part of the survivors, as hope for those lost from sight continued to fade until rescue efforts were terminated a little over two weeks after the earthquake.

'Seismic Culture'? From the evidence seen in the media images over the first weeks following the earthquake, "seismic culture" does not appear to have existed in Nepal. Considering that this has long been known as a very seismically active part of the globe, the pertinent, or perhaps impertinent, question is "*why not*?".

The Himalayan chain was created by the collision of continental plates, creating the highest mountains in the world, along with one of the world's most active earthquake hazard areas (Fig. 2). If any region would seem to have a reason for the emergence of a "seismic culture," one would think that Nepal would be close to the top of the list, along with neighboring Bhutan, Tibet, Indian and Pakistani Kashmir, and Afghanistan.

Historical records indicate that there was an earthquake in 1255 AD that killed a quarter to a third of the population of Kathmandu Valley (NSC, 2015). By comparison, the death toll of the 2015 earthquake is a little over 1,100 in Kathmandu city, a week and a half after the earthquake. This is but a small fraction of the city's population of 2.5 million (New York Times, 2015).

Even in the more heavily destroyed well populated rural area to the north of Kathmandu known as the Sindhupalchok District, which suffered more than double the fatalities in Kathmandu, this death toll represents less than 1% of the population of the district.

Figure 1. View of Bhaktapur, Kathmandu area, Nepal after the Gorkha Earthquake (Credits: Xavier Romão & Esmeralda Paupério).

While this may seem like evidence that substantially greater earthquake resistance has been achieved, one still can see that the destruction in some of the mountain villages near to the epicenter has been almost total and there is little visual evidence of pre-modern earthquake resistant features in the ruins (Fig. 3). If any such features did exist in the collapsed houses, they have proven to be ineffective. However, if indeed a third of

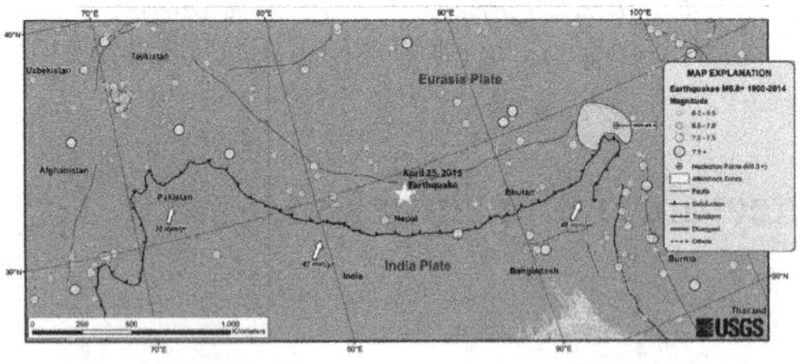

Figure 2. Himalayan and Eurasian Plate collision boundary with M6+ earthquakes since 1900 (Credits: USGS).

Figure 3. Typical view of ruins of destroyed stone masonry rural home, in Sangachowk Village, Sindhupalchowk District, Nepal (Credits: UNICEF/Chandra Shekhar Karki).

the population was killed in 1255, it is hard to argue that building safety has not somehow improved, however, at the same time, earthquakes from across almost a millennium of time are extremely hard to compare.

A similar dialectic exists in Italy, which like Nepal has frequently been subjected to damaging earthquakes throughout its multi-millennia of recorded history. Yet, certain features of traditional construction remain common in the country. These include rubble cores in the masonry walls that have long been known to make buildings vulnerable to collapse in earthquakes. However, there are many other features which have been identified by scholars as indicative of a pre-industrial era seismic culture, such as buttresses against masonry walls and corners, and arches between buildings; as well as iron ties connecting floor diaphragms and walls, and box-like building configurations.

The more important question is "What constitutes a seismic culture?" Is it simply a rise in construction quality and technological sophistication, or does it feature certain specific details the purpose of which can best be ascribed to resistance against earthquake forces? Or is the only proof of a seismic culture to be found in documents or in generations of knowledge and folklore of a known need for certain earthquake resistant details, such as was done so deliberately after the Great 1755 Lisbon earthquake with the invention and promulgation of the gaiola system of timber and masonry frame construction (Fig. 19)?

Surya Acharya, a civil engineer at the National Society for Earthquake Technology (NSET) in Nepal said: *"All the monuments [in Nepal] were built with earthquake-safe technology 400 years ago, using timber, brick, stone or mud, and lime. Those buildings survived many big earthquakes – this one was not so big. Many of the historical structures even survived the last major earthquake here, in 1934, but materials weaken due to age and poor maintenance"* (Fleeson, 2015).

The problem at this moment, just a short two weeks following the earthquake, is that the impression is that Bhaktapur and other traditional construction areas are devastated. What is missing at this early stage is an assessment of those structures which have survived without collapse. For that we will have to wait for further research. Later we will return to discuss Nepal, but first, we turn to nearby Kashmir.

2 INDIA AND PAKISTAN KASHMIR

The Vale of Kashmir in India is located in the western part of the Himalayan mountain range on the site of a prehistoric lake created by the uplift of the mountains between Indian and Pakistan Administered Kashmir. Over geological time, this lake gradually silted in, and the alluvium from the mountains became the fertile soil of the valley floor. This is responsible both for the area's rich agriculture and for its earthquake vulnerability. Srinagar lies on one of the most waterlogged soft soil sites for a capital city in the world, not unlike Mexico City.

The timber-laced masonry historic construction systems found here are mentioned in texts from the 12th century (Langenbach, 2009). Unreinforced masonry is strong in compression, but suffers both from differential settlement on soft soils and in earthquakes from a lack of tensile strength which allows for brittle failure from shear forces within the walls, or from overturning of the walls from differential settlement or out-of-plane earthquake vibrations.

Timber lacing and a strong tie between the timbers in the walls and the floors serve to restrain the walls from spreading and hold the building together while still allowing the system as a whole to be flexible. In traditional environments in developing countries, strength is not always possible, so flexibility or "give" is essential. In fact, in 1875, after spending some years in Kashmir, a British geologist, Frederick Drew, wrote *"These mixed modes of construction are said to be better against earthquakes (which in this country occur with severity) than more solid masonry, which would crack"* (Drew, 1917).

At the beginning of the 19th century the systems evolved into what are now the two main traditional construction systems: taq (timber-laced masonry bearing walls) and dhajji dewari (timber frame with masonry infill – like what in Britain is called "half-timber". Most of the traditional buildings in Srinagar and the Vale of Kashmir can be divided into these two basic systems (Fig. 4). In Pakistan, timber-laced masonry is

Figure 4. An older building in central Srinagar, Kashmir, India, that has *taq* timber-laced construction on the first two floors, and *dhajji dewari* infill frame construction above (Credits: Randolph Langenbach).

Figure 5. A small lane in central Srinagar showing typical streetscape of the historic city that is now getting rare as street widening and demolition and replacement with concrete structures have wreaked havoc with what had been one of the most remarkably well preserved historic urban environments in the world (Credits: Randolph Langenbach).

known by the Pashto word bhatar, and the timber frame with infill is simply called dhajji.

There are so many influences on the development of building construction traditions that it is not easy

Figure 6. A three and a half story building in central Srinagar, Kashmir, India, of taq timber-laced construction partially demolished for a street widening (Credits: Randolph Langenbach).

to isolate any one reason for the use of timber lacing in the masonry, but its effectiveness in holding the masonry together on soft soils undoubtedly has played a major role. It has also proven to be effective in reducing damage in earthquakes, which may help explain why variations of it can be found in the mountains, where soft soils are not a problem.

Taq (bhatar) Construction: Taq (or bhatar), consists of load-bearing masonry walls with horizontal timbers embedded in them. These timbers are tied together like horizontal ladders that are laid into the walls at each floor level and at the window lintel level. They serve to hold the masonry walls together and tie them to the floors (Fig. 6).

There is no specific name in Kashmiri to identify this timber-laced construction method itself, but the closest name used to describe it is *taq* because this is a name for the type of buildings in which it is commonly found. *Taq* refers to the modular layout of the piers and window bays, i.e. a five-*taq* house is five bays wide. Because in Srinagar this modular pier and bay design and the timber-laced load-bearing masonry pier and wall system go together, the name has come to identify the structural system as well.

The best early account of the earthquake performance of *taq* construction maybe the one by British traveler Arthur Neve, who was present in Srinagar during the earthquake of 1885 and published his observations in 1913: *"The city of Srinagar looks tumbledown and dilapidated to a degree; very many of the houses are out of the perpendicular, and others, semi-ruinous, but the general construction in the city of Srinagar is suitable for an earthquake country; wood is freely used, and well jointed; clay is employed instead*

Figure 7. Example of cribbage construction, The Khankah in Pampore, near Srinagar, ca. 1600. When photographed in 2007, the interior was being clad with plywood in order to, as they stated, to "modernize" the interior (Credits: Randolph Langenbach).

Figure 8. Example of dhajji dewari construction in Srinagar. This is an example with only rectangular panels. There is often in the present a belief that diagonals are necessary, just as they were in Lisbon in the gaiola that was invented after the 1755 earthquake, but there is increasing evidence that they are not necessary, and may even be counter-productive (Credits: Randolph Langenbach).

of mortar, and gives a somewhat elastic bonding to the bricks, which are often arranged in thick square pillars, with thinner filling in. If well built in this style the whole house, even if three or four storeys high, sways together, whereas more heavy rigid buildings would split and fall" (Neve, 1913).

An important factor in the structural integrity of *taq* is that the full weight of the masonry is allowed to bear on the timber lacing and the ends of the floor joists penetrate the exterior walls, thus holding them in place. These timbers in turn keep the masonry from spreading. Engineers now often find themselves uneasy about the absence of any vertical reinforcement, but in my own opinion, that is part of the brilliance of this system – it does not have elements which could shift this overburden weight of the masonry off and onto columns buried in the walls. It is this weight, and the resulting compression of the mud-laid masonry, that is such an essential component of what it needs in order to resist the earthquake forces.

***Cator* and Cribbage:** Several of the historic mosques in Srinagar are of "cribbage" construction, a variation of timber-laced masonry construction that can be found in the Himalayan mountains of northern India, northern Pakistan near the Chinese border, and parts of Afghanistan (Fig. 7). This has proven to be particularly robust in earthquake-prone regions, but as wood supplies became depleted it must have been found to be extravagant. This may in part explain the origins of the *taq* and *bhatar* systems, where the timber lacing is limited to a series of horizontal interlocking timber bands around the building, thus requiring significantly less wood in its construction.

A combination of cribbage at the corners with timber bands, known as "*cator* and cribbage", can be found in the Hunza region of Northern Areas of Pakistan. Examples can also be found in the Himalayan regions of northern India. This is a heavier, more timber-intensive version of timber-laced masonry than *taq* and *bhatar* that dates back some 1,000 years (Hughes, 2000). The corners consist of a cribbage of timber filled with masonry. These are connected with timber belts (*cators*) that extend across the walls just as they do in *taq* and *bhatar* construction.

There is evidence that many of these construction traditions have followed patterns of migration and cultural influence over centuries, such as the spread of Islamic culture from the Middle East across Central Asia, including Kashmir and other parts of India. In Turkey, timber ring beams in masonry, known singly as *hatıl* and plural *hatılar*, are part of a construction tradition that is believed to date back 9,000 years (Hughes, 2000). The Turkish word *hatıl* has the same meaning as *cator* does in Balti language. Also in Turkey, another common traditional construction type, *hımış*, is similar structurally to *dhajji* construction in Kashmir.

British conservator Richard Hughes has noted that *"The use of timber lacing is perhaps first described by Emperor Julius Caesar as a technique used by the Celts in the walls of their fortifications. Examples, with a lot of variations, are to be noted from archaeological excavations of Bronze and Iron Age hill forts throughout Europe."* Hughes also cites examples in the Middle East, North Africa and Central Asia (Hughes, 2000). Different variations on all of these construction types are also likely to be found in the areas outside of the regions discussed in this volume, including Nepal, Bhutan and parts of China, including Tibet.

***Dhajji dewari* Construction:** *Dhajji dewari* is a variation of a mixed timber and masonry construction type found in earthquake and non-earthquake areas around the world in different forms. While earthquakes may have contributed to its continued use in earthquake areas, timber and masonry infill frame construction probably evolved primarily because of its economic and efficient use of materials. However, its continued common use up until the present in Srinagar and elsewhere in the Vale of Kashmir most likely has been in response to the soft soils, and perhaps also to its observed good performance in past earthquakes (Fig. 8–9).

Figure 9. A cross-section of dhajji dewari construction revealed by a demolition for a road widening. Notice how thin the walls are in this form of construction. Despite this, it has proved to be remarkably resilient in earthquakes. (Credits: Randolph Langenbach).

The term *dhajji dewari* comes from the Persian and literally means “patchwork quilt wall”, which is an appropriate description for the construction to which it refers. The Persian name may provide a clue to Persian influence in the origins of this system of construction. It is also very similar to Turkish *hımış* construction, which was also common beyond the boundaries of Turkey, perhaps in part because of the widespread influence of the Ottoman Empire. *Dhajji dewari* consists of a complete timber frame that is integral with the masonry, which fills in the openings in the frame to form walls. The wall is commonly one-half brick in thickness, so that the timber and the masonry are flush on both sides. In the Vale of Kashmir, the infill is usually of brick made from fired or unfired clay. In the mountainous regions of Kashmir extending into Pakistan, the infill is commonly rubble stone.

Dhajji dewari construction has proven to be very effective in holding the walls of buildings together even when buildings have settled unevenly so as to become dramatically out of plumb. In the mountain areas, where soft soils and related settlements of buildings are not a problem, its use continued probably because timber was available locally and the judicious use of timber reduced the amount of masonry work needed, making for an economical way of building. The panel sizes and configuration of *dhajji* frames vary considerably, yet the earthquake resistance of the system is reasonably consistent unless the panel sizes are unusually large and lack overburden weight.

What many people fail to grasp is that the timber frame and the masonry are structurally integral with each other. In fact, such structures are best not considered as frames, but rather as membranes. In an earthquake, the house is dependent on the interaction of the timber and masonry together to resist collapse in the tremors. Historically, the amount of wood used, and therefore the sizes of the masonry panels, varied considerably. There is evidence that walls with many smaller panels have performed better in earthquakes than those with fewer and larger panels.

Figure 10. The debris left from the total collapse of almost all of the concrete slab and stone walled houses in Balakot, Pakistan in the 2005 earthquake (Credits: Randolph Langenbach).

There is no research that demonstrates that one *dhajji* pattern is better than another. Some patterns even lack diagonal bracing elements, relying on the masonry to provide all of the lateral resistance. The ones with random patterns probably result from the economics of using available random lengths of wood in the most efficient way possible. In fact, the quilting from which it gets the name ‘*dhajji*’ is itself produced from the reuse of scraps and small pieces of cloth.

Dhajji dewari construction was frequently used for the upper stories of buildings, with *taq* or unreinforced masonry construction on the lower floors (Fig. 4). Its use on the upper-floors is suitable for earthquakes because it is light, and it does provide an overburden weight that helps to hold the bearing wall masonry underneath it together.

3 THE 2005 KASHMIR EARTHQUAKE

The Kashmir earthquake was one of the most destructive earthquakes in world history. The death toll from this magnitude 7.6 earthquake was approximately 80,000 and over 3 million were left homeless. In a region known to be so vulnerable to earthquakes, it is reasonable to ask: Why did both the masonry and reinforced concrete buildings in the area prove so vulnerable to collapse? Why did over 80,000 people lose their lives in what is a largely rural mountainous region? Why did 6,200 schools collapse onto the children at the time of morning roll call in Pakistan alone? (Fig. 10)

This kind of scenario has played out repeatedly over recent decades in other earthquakes around the world, in cities and rural areas alike, as it has again in Nepal.

Ironically, even as the knowledge of earthquake engineering has grown and become more sophisticated, earthquakes have an increasing toll in places where steel and reinforced concrete construction have displaced traditional construction.

After the 2005 earthquake, international teams of engineers and earthquake specialists fanned out over the damage districts on both sides of the Line of Control and returned with reports on the damage to different types of structures. Most of these reports focused on the Pakistan side of the Line of Control because the epicenter of the earthquake was northwest of Muzaffarabad. In that area, which has a high population density, the death and destruction was far more extensive than on the Indian side.

None of these reports covered timber-laced traditional construction of any type. The reason for this is superficially explained by the following exchange between Marjorie Greene of the Earthquake Engineering Research Institute (EERI), an international NGO, and various local officials and technical experts in Pakistan three months after the earthquake. She asked if they were aware of any examples of traditional timber-laced construction of any type in the earthquake-affected area. The officials answered that they were *"unaware of any, but in years past there may have been"* (Langenbach, 2009).

In some ways, this lack of knowledge of the vernacular building systems in the earthquake area is not a surprise. It parallels a widespread lack of interest in such systems that exists in many countries which have recently experienced the rapid transformation from traditional materials and methods of construction to reinforced concrete. In most universities in the Middle East and South Asia, reinforced concrete frame construction remains the only system that most local engineers are trained to design.

As a consequence, after the earthquake the Government of Pakistan began to withhold reconstruction assistance funds from those people who proceeded to rebuild with *dhajji* or other timber-laced systems rather than with the government approved reinforced concrete block and slab system. For over a year after the earthquake, only those who followed the government's approved plans for reinforced concrete block and slab houses were allowed to obtain government assistance. This belief in the efficacy of reinforced concrete and concrete block continued despite its abysmal performance in that very same earthquake in Muzaffarabad, Balakot (Figure 10), and even including one middle class apartment complex in Islamabad.

However, what the experts failed to see was painfully evident to the rural villagers themselves, who, after they had climbed out of the ruins of their rubble stone houses, saw that the nearby concrete buildings were also destroyed. They could not help but notice that the only buildings still standing were of traditional *dhajji* and *bhatar* construction (Figure 11). Then on their own initiative, they revived the use of these historic technologies in the reconstruction of their own houses.

Figure 11. Country store of dhajji construction in Pakistan Kashmir near the epicenter of the 2005 earthquake. This and other buildings like it are what the local residents saw that inspired them to rebuild dhajji houses (Credits: Randolph Langenbach).

Eventually, after the architects in the disaster response and recovery NGOs could see this and brought it to the attention of the government's consulting engineers, both systems were approved by the Government of Pakistan as "compliant" for government assistance. As a result, there may be as many as a quarter of a million new houses using one of these two traditional systems, which before the earthquake had largely fallen out of use.

Returning to the Indian side of Kashmir, one of the most important of the post-earthquake reconnaissance reports was published by EERI. This report was written by Professors Durgesh C. Rai and C. V. R. Murty of the Indian Institute of Technology, Kanpur and published in December 2005 as part of the EERI "Learning from Earthquakes" report on the Kashmir earthquake. The quotations below from the authors were based on observations made during the first several weeks after the earthquake. Describing *taq* construction, which they observed in the damage district on the Indian side of the Line of Control, Professors Rai and Murty observed:

"In older construction, [a] form of timber-laced masonry, known as Taq has been practiced. In this construction large pieces of wood are used as horizontal runners embedded in the heavy masonry walls, adding to the lateral load-resisting ability of the structure... Masonry laced with timber performed satisfactorily as expected, as it arrests destructive cracking, evenly distributes the deformation which adds to the energy dissipation capacity of the system, without jeopardizing its structural integrity and vertical load-carrying capacity" (Rai and Murty, 2005).

It is interesting to compare their observation with that of Professors N. Gosain and A.S. Arya, after an inspection of the damage from the Anantnag Earthquake of 20 February 1967, where they found buildings of similar construction to Kashmiri *taq: The*

Figure 12. The owner and carpenter building a new dhajji house in Topi, near Bhag, Pakistan, to replace one of rubble stone on the right, which collapsed after seeing the survival of the building in Figure 11 and others like it (Credits: Randolph Langenbach).

Figure 13. Villager standing near his house in a remote village between Batagram and Besham, in NWFP, Pakistan with bhatar construction which survived the earthquake. This inspired the new construction in bhatar seen on right (Credits: Randolph Langenbach).

timber runners...tie the short wall to the long wall and also bind the pier and the infill to some extent. Perhaps the greatest advantage gained from such runners is that they impart ductility to an otherwise very brittle structure. An increase in ductility augments the energy absorbing capacity of the structure, thereby increasing its chances of survival during the course of an earthquake shock (Gosain and Arya, 1967).

The concept of ascribing ductility to a system composed of a brittle material – masonry – is difficult for many modern engineers to comprehend. It can be readily observed that a steel coat hanger is ductile, as demonstrated when it is bent beyond its elastic limit, but by contrast, a ceramic dinner plate is brittle. So how can masonry, which on its own is inarguably made up of brittle materials, be shown to be ductile? Rai and Murty in 2005 avoided the use of the term "ductile" probably because the materials in *taq* are not ductile and do not manifest plastic behavior. However, what makes timber-laced masonry work well in earthquakes is its ductile-like behavior as a system. This behavior results from the energy dissipation because of the friction between the masonry and the timbers and between the masonry units themselves.

This friction is only possible when the mortar used in the masonry is of low-strength mud or lime, rather than the high-strength cement-based mortar that is now considered by most engineers to be mandatory for construction in earthquake areas. Strong cement-based mortars force the cracks to pass through the bricks themselves, resulting in substantially less frictional damping and also rapidly leading to the collapse of the masonry. Arya made this difference clear when he said: *"Internal damping may be in the order of 20%, compared to 4% in uncracked modern masonry (brick with Portland cement mortar) and 6%–7% after the masonry has cracked."* His explanation for this is that "*there are many more planes of cracking... compared to the modern masonry.*" (Gosain and Arya, 1967).

In areas subject to earthquakes, engineers have often sought to specify strong cement-based mortar. However, in the larger earthquakes, the strength of the

Figure 14. Four and five story residential buildings in the Indian Kashmir city of Baramulla showing how the unreinforced masonry collapsed, leaving the dhajji dewari bridging over the gap, while a tall rubble stone building reinforced with taq timber ring beams survived the 2005 earthquake undamaged (Credits: Randolph Langenbach).

Figure 15. A grand four story bearing wall brick masonry house on the Rainawari Canal in Srinagar of timber-laced taq construction (Credits: Randolph Langenbach).

mortar ceases to be helpful once the walls begin cracking, as they inevitably do in a strong earthquake. It is then that the "plastic cushion" and other attributes described by Harley McKee become more important. More important is that the masonry units – the stones or bricks – be stronger than the mortar, so that the onset of shifting and cracking is through the mortar joints, and not through the bricks. Only then can the wall shift in response to the earthquake's overwhelming forces without losing its integrity and vertical bearing capacity. With timber-laced masonry, it is important to understand that the mortar is not designed to hold the bricks together, but rather to hold them apart. The timbers are what tie them together. The friction and cracking in the masonry walls dissipate the earthquake's energy, while the timber bands are designed to confine the masonry, and thus prevent its spreading, which would lead to collapse.

4 NEPAL AFTER TWO EARTHQUAKES

After a little more than two weeks after the April 25th Gorkha earthquake in Nepal, on the 12th of May, a second earthquake or large aftershock struck Nepal, testing the surviving buildings still further. Reports indicated that some have failed the test.

Just before that second earthquake, a colleague sent me a paper he had found on the internet by a Nepalese scholar, Dipendra Gautam, who claimed that *"The historic urban nucleus of Bhaktapur city Nepal has … unreinforced masonry buildings which have many features particularly contributing [to] better [performance] during earthquake events."* This finding, he said was *"based on detailed survey of forty two buildings."* His conclusion in light of the two recent earthquakes seemed in sharp contrast to the cascade of photographs of partially and totally collapsed brick buildings in the Kathmandu Valley city of Bhaktapur which he said is the *"culturally most preserved city of Nepal"* (Gautam, 2014).

With only the news photos to go on, in the first weeks after the earthquake, the many collapses of masonry buildings in Bhaktapur would seem to undermine his conclusions (Fig. 17). However, his observations came with the authority of thorough building-specific research. His findings also contrasted with my own more brief observations from visits to Nepal a decade earlier, on which I had written about in several papers, and in the UNESCO book *Don't Tear It Down!*" in which I had said timber bands were less common

Figure 16. A photo taken in Nepal fifteen years before the 2015 earthquake showing the progressive structural deterioration of a masonry bearing wall building which lacks the timber bands of *taq* and *bhatar* (Credits: Randolph Langenbach).

Figure 17. Unreinforced brick building collapses in Sankhu, near Kathmandu, Nepal showing the collapsed end of a row of dwellings, which lacked timber bands (Credits: Xavier Romão & Esmeralda Paupério).

than in Kashmir "*except in some of the palaces and temples*" (as, for example, in Figure 16)."

A compelling source for evidence was a book of photographs of the heavy damage inflicted by the 1934 earthquake, which had devastated large parts of Kathmandu, including some of the palaces and temples. In those photographs, there was no evidence of timber lacing that could be seen in the ruins.

Based upon a conversation with Mr. Gautam about how the 42 buildings in his study fared in the earthquake, there appeared to be evidence that those with timber lacing survived the earthquake intact. His study sample consisted of houses, rather than palaces or temples. His reply to this question – based on his initial reconnaissance in Bhaktapur after the both the first and second earthquake was "*I re-inspected [the 42 buildings and] I am really excited with their performance...The timber bands, double boxing of openings, struts, subsequent load reduction mechanism are*

Figure 18. Hanuman Dhoka Palace, Kathmandu after the earthquake showing a section with timber bands – visible as horizontal lines on the brick façades (Credits: Kai Weise, Kathmandu, Nepal).

genius. The smaller openings, building symmetry and others are also excellent... Inside many of the houses ...there were only minor diagonal cracks... Till date, I haven't found any collapsed house [with] timber bands."

His prior research and publication, together with his post-earthquake findings, ultimately leads to an important contribution towards the preservation of the historic structures that make of the context for the World Heritage Site in Bhaktapur. If the effectiveness of these aseismic features – particularly timber bands – can be shown to have kept the buildings from collapsing, the survival of particular masonry buildings thus would be determined no longer to be a matter of chance. This knowledge can then help both (1) lead to a program of reinforcement of masonry buildings, and (2) help give confidence in such systems, so as to counteract the present belief that all masonry construction is at risk of collapse in the future.

In the months following the completion of this chapter, more information will likely become available to help to answer the question of why some houses and not others were timber reinforced. However, the earthquake and its aftermath in the media have already proved that such aseismic construction was far from universal. It will be interesting to learn from further research why timber bands were not included in the construction of so many masonry buildings. Was it a result of a rise in price of timber, or some other factor, or simply that the technology was not widely known?

These are important questions to raise at a time when concrete construction, which has already displaced most of timber and masonry construction in the rest of Kathmandu outside of Bhaktapur, stands poised to be used after these earthquakes to replace the masonry buildings in the heritage areas. It is easy to see that for many people the immediate impression is that the concrete structures proved to be safer, despite the collapse of many of them spread out through the city. One Nepali heritage professional, Kai Weise, reported

Figure 19. Interior of post-1755 Lisbon, Portugal, earthquake building in Baixa, Lisbon with interior walls of *gaiola* exposed during a remodeling (Credits: Randolph Langenbach).

his experience in a Kathmandu coffee shop "*my waiter, who brought me my latte... explained that all the load-bearing houses cracked open horizontally and vertically, while the "pillar system"* [the local name for reinforced concrete frame structures] *withstood the earthquake.*

This then raises the question of what now might become evidence of a 'Seismic Culture' in Nepal after these two earthquakes. Will the collapsed masonry buildings get reconstructed with timber bands? Or will people look around and see that in these particular earthquakes that the reinforced concrete buildings for the most part remained standing and proceed to rebuild in concrete, despite the increasingly disappointing record of reinforced concrete in other earthquakes including massive collapses in Ahmedabad in 2001 and in nearby Sikkim in 2011.

In a sense, this could be reminiscent of what happened in Lisbon, after the 1755 earthquake with the 'invention' of the *Gaiola.* This was a technology that was not new – but which was derived from the traditional form of construction which could be seen to have survived the earthquake – a form of construction that can be found around the world from Elizabethan England, medieval Germany, Eastern Europe, Spain, Turkey, Kashmir, and in Lisbon itself – where medieval half-timber buildings were found to be still standing amidst the devastation of the earthquake. Their resilience was proven by their survival, and so they inspired the design and mandatory use of the *Gaiola* – a technology that became such a compelling part of Lisbon's subsequent rebirth.

REFERENCES

Bilham, Roger. (2004, April/June) "Earthquakes in India and the Himalaya: techtonics, geodesy and history," Annals of Geophysics, Vol. 47, No. 2/3.

Drew, Frederick. (1875). The Jummoo and Kashmir Territories, Edward Stanford, London, p. 184.

Fleeson, Lucinda. (2015, May 3). "How to rebuild a safer Nepal?" Philadelphia Enquirer. Retrieved from http://www.philly.com. (reprinted in Emergency Management Magazine, entitled "Whether a Rebuilt Nepal Will Be Better and Stronger Remains a Question," Retrieved from http://www.emergencymgmt.com.

Gautam, Dipendra. (2014). Earthquake Resistant Traditional Construction in Nepal: Case Study of Indigenous Housing Technology in the Historic Urban Nucleus of Bhaktapur City, Unpublished paper posted on www.researchgate.net.

Gosain, N. and Arya, A.S. (1967, September). A Report on Anantnag Earthquake of February 20, 1967. Bulletin of the Indian Society of Earthquake Technology, No. 3.

Hughes, Richard. (2000). "Cator and Cribbage Construction of Northern Pakistan," Proceedings of the International Conference on the Seismic Performance of Traditional Buildings, Istanbul, Turkey.

Langenbach, Randolph. (2009). Don't Tear it Down! Preserving the Earthquake Resistant Vernacular Architecture of Kashmir, UNESCO, New Delhi.

National Seismological Centre (NSC). (2015). Historical Earthquakes, Govt. of Nepal, Ministry of Mines and Geology, Kasthmandu, on website at: http://www.seismonepal.gov.np/index.php?linkId=56.

Neve, Arthur. (1913). Thirty Years in Kashmir. London, p. 38.

New York Times. (2015, May 5). ["Tally of Deaths " graphic]. Retrieved from http://www.nytimes.com/interactive/2015/04/25/world/asia/nepal-earthquake-maps.html

Rai, Durgesh and Murty, C. V. R. (2005). Preliminary Report On The 2005 North Kashmir Earthquake of October 8, 2005. Kanpur, India, Indian Institute of Technology Kanpur. (Available at www.EERI.org).

Seismic Retrofitting: Learning from Vernacular Architecture – Correia, Lourenço & Varum (Eds)

 ISBN 978-1-138-02892-0

Traditional construction in high seismic zones: A losing battle? The case of the 2015 Nepal earthquake

X. Romão
Faculty of Engineering, University of Porto, Porto, Portugal

E. Paupério
Construction Institute, University of Porto, Porto, Portugal

A. Menon
Department of Civil Engineering, Indian Institute of Technology Madras, Chennai, India

ABSTRACT: The 25th of April 2015's earthquake in Nepal, along with its subsequent aftershocks, left several of the country's districts with high levels of damage and significant losses. In this context, the present chapter addresses the particular pattern of the damage scenario in the housing sector. The severe impact of the earthquake in traditional constructions is discussed, and compared to that of other construction systems. Earthquake-resistant features of traditional Nepalese construction are discussed along with possible reasons for their inadequate performance in this earthquake. Potential repercussions of the damage sustained by traditional construction systems are also discussed based on the experience of past earthquakes in order to highlight possible threats to their survival during the reconstruction stage.

1 INTRODUCTION

On the 25th of April 2015, at 11:56 AM local time, an M_w 7.8 earthquake struck Nepal. Its epicentre was located in Barpak in the historical district of Gorkha; about 76 km northwest of Kathmandu, and its hypocentre was at a depth close to 15 km (USGS, 2015a). Over the next two months, more than sixty aftershocks with a moment magnitude M_w 4.0 or higher were recorded. During this time, the most important aftershock occurred seventeen days later, on May 12th at 12:50 PM local time. This event had a moment magnitude M_w 7.3, and struck Nepal hard for a second time. The epicentre of this event was located 19 km SE of Kodari, approximately 80 km to the east-northeast of Kathmandu (USGS, 2015b). Figure 1 presents a map of Nepal, illustrating the location of the epicentres of these two events.

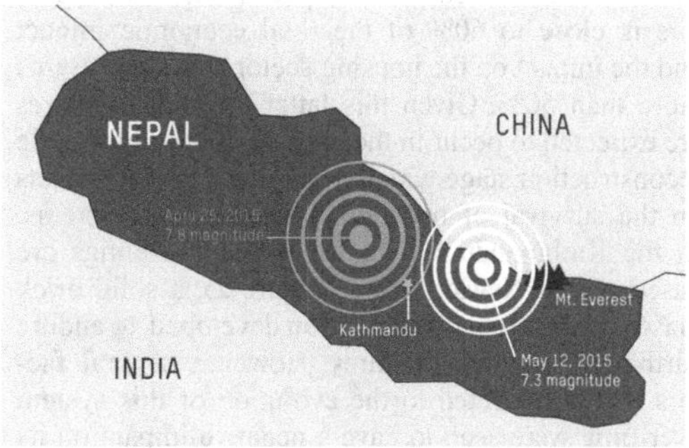

Figure 1. Map of Nepal and locations of the two major earthquake events of April 25 and May 12 2015 (credits: OXFAM, 2015).

After two months, over 8832 casualties and 22309 injuries were reported (NDRRP, 2015). It has also been estimated that the lives of eight million people, almost one-third of the population of Nepal, have been impacted by these events. Thirty-one of the country's seventy five districts have been affected, out of which fourteen were declared "crisis-hit" for the purpose of prioritising rescue and relief operations; another seventeen neighbouring districts were partially affected (NPC, 2015). With respect to the built environment, the total number of government and private houses fully damaged is 530502 and an additional 281598 were also partially damaged (NDRRP, 2015). In terms of cultural heritage, 741 buildings and sites have been severely affected (133 fully collapsed, 95 suffered partial collapse, and 513 are partially damaged), according to the Department of Archaeology (THT, 2015). The economic impact resulting from this disaster is believed to be considerable, and the recently finished Post-Disaster Needs Assessment report indicates that the total economic value of this impact is expected to be close to US$ 7000 million (NPC, 2015). In particular, the economic impact to the housing, health, education and cultural heritage sectors are expected to be US$ 3505 million, US$ 75 million, US$ 313 million

Figure 2. Example of a Newari house from Kathmandu (credits: Esmeralda Paupério & Xavier Romão).

and US$ 192 million, respectively (NPC, 2015). As it can be seen, the economic impact in these four sectors is close to 60% of the total economic impact and the impact on the housing sector alone represents more than 50%. Given this latter estimate, changes are expected to occur in the housing sector during the reconstruction stage which may have negative effects on the survival of traditional construction systems. In the Kathmandu Valley, traditional dwellings are based on the Newari house (Figure 2), a solid brick masonry and timber construction developed, to endure earthquake-resistant features. However, several factors that contributed to the evolution of this system over time were seen to have a negative impact on its seismic performance. As such, based on the damage scenario observed in several areas affected by the 2015 Nepal earthquake and also based on the reconstruction experiences after the 2005 Kashmir earthquake in Pakistan and the 2011 Sikkim earthquake in India, the present chapter addresses some concerns related to the reconstruction of the affected areas, especially those combining higher levels of damage with lower levels of development.

2 OVERVIEW OF THE DAMAGE SCENARIOS IN AREAS AFFECTED BY THE 2015 NEPAL EARTHQUAKE

2.1 *Initial remarks*

The following observations regarding the damage scenarios in areas affected by the 2015's Nepal earthquake are mostly based on data collected during two reconnaissance missions that took place within two months after the 25th of April earthquake. These missions took place from the 24th to the 31st of May and from the 16th to the 24th of June and involved researchers from the University of Oporto, Portugal, and from the Indian Institute of Technology Madras, India, which were part of a joint initiative promoted by ICCROM, ICOMOS, ICOM and the Smithsonian Institute.

Given the scope of the present chapter, only the damage pattern to the housing sector will be addressed herein. Furthermore, the general comments presented in the following are valid for the more developed cities of the Kathmandu Valley (Kathmandu, Bhaktapur and Lalitpur) as well as for the more rural villages (e.g. Sundarijal, Bungamati, Sankhu, Harsiddhi or Nikoshera).

2.2 *Overview of the earthquake damage pattern in the housing sector: Old vs new*

During the first week after the 25th of April earthquake, videos and photographs of a devastated region were broadcast around the world through TV and internet news reports. Apocalyptic scenes of historical monuments of the Kathmandu Valley reduced to rubble, of piles of bricks lying on the ground, where houses used to stand; and photographs of destroyed villages hanging on landslide-prone hills and mountains were seen all over the world. However, reality is somehow different. After stepping outside the airport of Kathmandu and driving into the city, many constructions remain standing and undamaged. According to the 2011 census, the total number of houses in the Kathmandu district is 436344, and the total number of government and private houses fully or partially damaged by the earthquake, in the district is 88088 (NDRRP, 2015), i.e. 20% of the houses, in the district. Furthermore, the damaged houses that are actually in the city of Kathmandu are only 24041 (NDRRP, 2015), i.e. 5.5% of the total number of houses in the district. A similar trend can also be found for the Bhaktapur district: the total number of houses is 68636 and the total number of government and private houses, fully or partially damaged by the earthquake, in the district is 28010 (NDRRP, 2015), i.e. 40% of the houses in the district; the damaged houses that are actually in the city of Bhaktapur are only 8078 (NDRRP, 2015), i.e. 12% of the houses in the district (Fig. 3). Given these numbers, it is seen that most of the damages to the housing sector occurred outside the main cities, especially in more rural areas of the main districts or in more rural districts such as Sindhupalchowk, Dolakha, Nuwakot, Gorkha or Dhading where the total number of fully or partially damaged houses is around 100%, 100%, 88%, 88% and 85%, respectively (KS, 2015).

Aside from analysing its geographical distribution, to fully understand the implications of the type of damages to the housing sector it is also important to observe the typology of the structures that was damaged. After walking in several affected areas, both

Figure 3. Undamaged street in Bhakapur (credits: Esmeralda Paupério & Xavier Romão).

Figure 4. Damaged traditional masonry constructions close to apparently undamaged RC buildings in Kathmandu (credits: Esmeralda Paupério & Xavier Romão).

urban and rural, the same damage pattern can be witnessed: traditional and vernacular masonry buildings performed poorly under the earthquake, while modern reinforced concrete (RC) structures performed much better. Although there are still no official statistics on the subject, this damage pattern was already highlighted by the media (Awale, 2015), and becomes evident from the on-site visits. Scenarios such as those presented in Figures 4 to 8, where almost undamaged RC buildings stand near severely damaged or collapsed traditional masonry constructions, are common in many urban and rural areas.

Figure 5. Damaged traditional masonry constructions close to apparently undamaged RC buildings in Bhaktapur (credits: Esmeralda Paupério & Xavier Romão).

Figures 4 and 5 illustrate this type of scenario for urban areas, i.e. the cities of Kathmandu and Bhaktapur. In both cases, the photos were taken in the streets surrounding the city's Durbar Square area. For more rural areas, Figures 6, 7 and 8 illustrate the situation found in the villages of Bungamati, Sankhu and Harisiddhi, respectively. For the case of Bungamati, according to the local population, of the close to 1500 houses of the village, 825 collapsed, and more than 300 suffered severe damage. In Sankhu, an ancient Newari village, which was on the Tentative List of UNESCO's World Heritage Sites, more than 90% of the houses suffered from severe damage or collapsed. In Harisiddhi, another Newari village, close to 90% of the houses were also severely damaged or collapsed. In these three villages, damaged or collapsed houses were mostly traditional masonry constructions while RC buildings stand nearby apparently without damage or with only minor damage. These observations can be sustained using preliminary data available from the Global Shelter Cluster (GSC, 2015), which shows that, for the fourteen districts declared "crisis-hit", around 70% of the traditional masonry constructions collapsed or exhibited severe damage; while 87% of RC buildings exhibited from moderate to no damage.

The reasons for the different performance of these two types of constructions are more complex than the simple fact that one is made of RC, and the other is made of masonry. Aside from aspects related to the construction details of the masonry constructions that will be addressed in the following section, the ground motion characteristics, particularly the frequency content, and the highly variable geotechnical conditions (e.g. see Paudyal *et al.*, 2012, 2013; Chamlagain and Gautam, 2015) are also key factors that influence the performance of constructions under earthquake loading. Also, in more urban areas, many traditional constructions were not constructed as isolated buildings but instead as wall-to-wall buildings, forming building, blocks with a much more complex dynamic behaviour and performance. Nevertheless, to the untrained eye, the evidence is overwhelming: new

Figure 6. Damaged traditional masonry constructions close to apparently undamaged RC buildings in Bungamati (credits: Esmeralda Paupério & Xavier Romão).

Figure 7. Damaged traditional masonry constructions close to apparently undamaged RC buildings in Sankhu (credits: Esmeralda Paupério & Xavier Romão).

Figure 8. Damaged traditional masonry constructions close to apparently undamaged RC buildings in Harisiddhi (credits: HFHI, 2015).

RC buildings are safe while old traditional masonry constructions are not. Among other aspects, this empirical statement regarding the seismic performance of the RC building stock in the Himalayan urban and rural areas raises a serious cause for concern: the quality of design and execution of this building typology is very often owner- and contractor-driven and largely afflicted by almost no structural engineering input and poor workmanship. Therefore, such constructions are expected to perform poorly as demonstrated in recent earthquakes in the region, with total collapses reported even in regions with modified-Mercalli intensities of VI and VII (Menon *et al.*, 2012). So what happened in this earthquake? Was it sheer luck (República, 2015)?

The observations made regarding the earthquake damage pattern on the housing sector can have multiple adverse implications on the upcoming reconstruction stage. Before addressing these implications, some of the reasons for the inadequate performance of traditional masonry buildings are analysed next.

3 WHAT WENT WRONG: REASONS FOR THE INADEQUATE PERFORMANCE OF TRADITIONAL MASONRY BUILDINGS

Throughout history, people around the world have been exposed to many different and recurrent hazards. It is well known that earthquakes are among those with more devastating effects, and people started to develop built structures designed to accommodate and mitigate those effects long ago. Different systems have been developed in different parts of the world, which represent a local cultural adaptation of people to earthquakes, e.g. see (Krüger *et al.*, 2015). In the earthquake-prone region of the Himalayas, in particular in India and in Nepal, traditional construction systems have also been developed to mitigate the effects of earthquakes. The Sumer house of the Yamuna Valley in India (Saklani *et al.*, 1999), the Dhajji-Dewari in Kashmir (Ali *et al.*, 2012), the Ekra house of Sikkim in India (Menon *et al.*, 2012) and the Newari house of the Kathmandu Valley in Nepal (Marahatta, 2008) are examples of such traditional construction systems.

With respect to the Newari house, several construction details have been identified as being the main features for its earthquake-resistant properties. Examples of such details are the fact that Newari houses have (Jigyasu, 2002; D'Ayala, 2006; Marahatta, 2008):

- A symmetric plan arrangement;
- A low height (usually up to three stories);
- Double leaf/Whyte masonry façades and sidewalls usually made of *ma appa* bricks in the external leaf/Whyte and *dachi appa* bricks in the internal one. These walls are continuous and connected at the corners;
- A reduction of the weight of the building over its height, by using timber partition walls, timber columns, and by reducing the thickness of the main masonry walls in the upper storeys;
- Small square timber windows with lintels extending well into the surrounding masonry. The small size of the windows allows for the development of larger masonry piers with adequate shear resistance

Figure 9. Earthquake resistant details of Newari houses involving timber bands and timber pegs (credits: Esmeralda Paupério & Xavier Romão).

Figure 10. Independent behaviour of masonry leaves/wythes due to lack of timber pegs (left); out-of-plane collapse of a wall from lack timber pegs tying the wall to the floor joists (right) (credits: Esmeralda Paupério & Xavier Romão).

between the openings. These windows are usually built with a double frame, one within the external masonry leaf/wythe, and a slightly larger one within the internal masonry leaf/wythe. The two frames are connected by timber elements embedded in the masonry;

- Ring timber bands and plates to tie the roof to the walls or placed at different heights of the walls to tie the different leaves/wythes of the wall (Fig. 9);
- Timber pegs called *chokus* to restrain floor joists from sliding over walls (Fig. 9). At the roof level, two vertical pegs are inserted through a joist, on each side of the wall. For the intermediate storeys, the common practice is for the joists to be anchored with pegs on the internal face of the external wall and in-between the two masonry leaves/wythes. These pegs are very effective in preventing relative sliding of the floor structure on the walls in the presence of lateral forces, creating a box effect. The pegs are also effective in limiting out-of-plane movement of the external walls.

Even though the effectiveness of the construction details of the traditional Newari house has been previously acknowledged, current masonry constructions do not follow this traditional model entirely. From the late eighteenth to the mid nineteenth centuries, changes in the way of life, and the need for more space to accommodate larger families, led to the need to increase the height of the houses and to modify their architectural and structural components. While new houses built in this period were now four or five storey buildings, additional storeys were also built on older houses. An increase in the size of the openings, due to changes in the internal subdivision of the houses, has led to a reduction in the width of lateral masonry piers between consecutive openings. As a result, the lateral capacity of the façade walls is now only that of the flexural capacity of the piers, which is much smaller than the shear capacity of the piers of the original houses. Further changes in the construction style were also witnessed during the late nineteenth and early twentieth centuries, when the neoclassical Rana style was introduced based on British construction practice that has no traditional earthquake-resistant features (Shrestha, 1981). Nonetheless, two of the most important factors leading to the significant reduction in the seismic performance of current Nepalese traditional masonry houses are related to economic constraints. The first is related to the increasing cost of timber that leads to a reduction of its use in the construction of houses. Since some of the more important earthquake-resistant features of traditional masonry houses depend on timber elements, one can anticipate a reduction in the seismic performance of these constructions. For example, the lack of timber elements to tie the different leaves/wythes of walls will make them to behave as independent elements, a fact then leading to the typical out-of-plane collapse of the external leave/wythe, often witnessed after earthquakes (Fig. 10). In this context, regulatory governmental policy aimed at curbing deforestation, can also play a crucial role in determining the evolution of the use of timber in these housing typologies.

The second factor is related to the maintenance of these constructions. Even for houses with adequate earthquake-resistant features, the level of maintenance is a crucial factor to their seismic performance. Material ageing and degradation, as a result of natural or anthropological factors, will undermine the expected structural behaviour and, ultimately, lead to a poor performance under earthquake loading. Moreover, binder materials such as earth mortar and lime mortar, widely used in traditional constructions, need of regular renewal for an effective performance.

As referred to, these two problems stem from economical constrains, and it is well known that such factors often lead to structural performance issues that jeopardized the overall safety of the constructions. In the aftermath of the earthquake that left a damage pattern with the previously highlighted characteristics, further economic issues might arise during the reconstruction phase. Some of these issues and their potential repercussions are discussed in the following, making reference to the reconstruction experience in other earthquake-damaged countries of similar traditional construction typologies.

4 POSSIBLE REPERCUSSIONS FOR THE RECONSTRUCTION STAGE

As previously mentioned, the damage pattern left by the 2015 Nepal earthquake is likely to yield changes on the reconstruction of the housing sector, and such changes may have negative and irreparable effects on the survival of traditional construction systems. Before addressing some concerns related to the specific case of the 2015 Nepal earthquake, reference is previously made to other relevant aspects from the reconstruction experiences of two other earthquakes: the 2005 Kashmir earthquake in Pakistan, and the 2011 Sikkim earthquake in India.

According to Mumtaz *et al.* (2008), a significant change in construction materials and building types was observed after the 2005 Kashmir earthquake, due to time and economic pressure, increasing material costs after the earthquake, and also due to the financial and material support available from the government. Due to these factors, people adopted atypical materials for the sake of a cheap and rapid reconstruction, a fact that will lead to a profound change in their cultural identity. Another interesting fact is that, after this earthquake, many people lost faith in the construction typologies that performed poorly, and many damaged buildings that could have been easily repaired or strengthened, were simply demolished (Mumtaz *et al.*, 2008). Again, for the case of traditional construction, such behaviour can severely impact on the survival of traditional construction systems.

In the case of the 2011 Sikkim earthquake, reference is made to a notification from the State Government of Sikkim, referring the traditional earthquake-resistant Ekra typology as being "informal", as opposed to formal housing. Such classification resulted in this typology not qualifying for government subsidies, which yielded an implicit, but possibly unintended State discouragement towards its use (Sheth and Thiruppugazh, 2012). As a result, most of the traditional houses in Gangtok, Sikkim's capital city have systematically been replaced by non-engineered poor quality RC buildings (e.g. see Figure 11), even though almost all buildings that suffered from serious damage in the earthquake were RC constructions. Aside from the inevitable impact of such policies on traditional construction systems and cultural identity, the drive for development which leads common man, seems to aspire for RC houses may also be responsible for the growing stock of poor quality non-engineered RC constructions. As a result, the overall seismic vulnerability of the Himalayan urban, semi-urban and rural areas is possibly growing also.

These scenarios can also occur in the case of the 2015 Nepal earthquake. Since the reconstruction stage has not yet begun, information regarding official reconstruction policies is unavailable. Still, two particular aspects need to be highlighted. First, the fact that several locals mentioned that there are recommendations towards reconstructing traditional and cultural heritage buildings, using the Façade retention

Figure 11. A traditional Ekra construction sitting beside its replica in RC – Sikkim earthquake, 2011 (credits: Esmeralda Paupério & Xavier Romão).

Figure 12. A building in Bhaktapur where the façade mimics traditional construction details while the rest of the construction system is made of RC (credits: Esmeralda Paupério & Xavier Romão).

or "Façadism" technique, where a building is constructed using modern building technology behind its retained historical façades or envelope (Fig 12). Second, given the damage scenarios left by the earthquake, one can expect that people will want to rebuild their homes in RC instead of traditional construction. Unless economic, material and technical support is made officially available by the government, such trend will be difficult to overturn. In this context, it is noted that Nepal possesses a building standard addressing the construction of masonry buildings with adequate earthquake-resistant features (NBC, 1994). As in traditional construction, some aspects related to the combination of timber elements with brick masonry, to enhance the seismic performance of these constructions are clearly mentioned in this standard. Still, adequate conditions for its practical implementation need to be put in place, namely in terms of available technical expertise, quality control of the constructions, and availability of the construction materials at affordable prices.

ACKNOWLEDGEMENTS

The authors wish to thank ICCROM, ICOMOS, ICOM, the Smithsonian Institute, the University of Porto, Portugal, and the Indian Institute of Technology Madras, India, for the financial support to carry out the two reconnaissance missions to the areas affected by the 2015 Nepal earthquake.

REFERENCES

Ali, Q., Schacher, T., Ashraf, M., Alam, B., Naeem, A., Ahmad, N. & Umar, N. (2012). *Plane behaviour of the Dhajji-Dewari structural system (wooden braced frame with masonry infill). Earthquake Spectra,* 28(3), 835–858.

Awale, S (2015) A concrete future. Nepali Times #758.

Chamlagain, D. & Gautam, D. (2015). Seismic Hazard in the Himalayan Intermontane Basins: An Example from Kathmandu Valley, Nepal. *Mountain Hazards and Disaster Risk Reduction, Springer Japan.*

D'Ayala, D. (2006). Seismic Vulnerability and Conservation Strategies for Lalitpur Minor Heritage. *Proceedings of the Getty Seismic Adobe Project.*

GSC (2015). Global Shelter Cluster. Nepal Earthquake 2015. Available at https://www.sheltercluster.org/global. Retrieved on June 2015.

HFHI (2015). Habitat for Humanity International Nepal Disaster Response. Available at http://hfhi-nepal.blogspot.pt/2015/05/the-nepal-earthquake-2015.html. Retrieved on June 2015.

Jigyasu, R. (2002). *Reducing disaster vulnerability through local knowledge and capacity – the case of earthquake prone rural communities in India and Nepal* (PhD Thesis Norwegian University of Science and Technology, Trondheim).

Krüger, F., Bankoff, G., Cannon, T., Orlowski, B. & Schipper, E. (Eds.) (2015). *Cultures and disasters: understanding cultural framings in disaster risk reduction.* UK: Routledge.

KS (2015). Karuna-Shechen – Humanitarian Projects in the Himalayan Region. Available at http://karuna-shechen.org/.

Marahatta, P. (2008). Earthquake vulnerability and Newari buildings: a study of indigenous knowledge in traditional building technology. *'VAASTU' the Annual Journal of Architecture,* 10.

Menon, A., Goswami, R., Narayanan, A., Jaiswal, A., Gandhi, S., Satyanarayana, K., Raghukanth, S., Seth, A. & Murty, C. (2012). Observations from damages sustained during 2011 (India-Nepal) Sikkim earthquake. *15th World Conference on Earthquake Engineering, Lisbon, Portugal.*

Mumtaz, H., Mughal, S., Stephenson, M. & Bothara, J. (2008). The challenges of reconstruction after the October 2005 Kashmir earthquake. Proceedings of the NZSEE Annual Conference. Wairakei, New Zealand.

NBC (1994). Nepal National Building Code: NBC 202 1994. Mandatory Rules of Thumb: Load Bearing Masonry. Government of Nepal. Ministry of Physical Planning and Works. Department of Urban Development and Building Construction

NDRRP (2015). Nepal Disaster Risk Reduction Portal. Government of Nepal, Kathmandu, Nepal. Available at: http://drrportal.gov.np/. Retrieved on June 2015.

NPC (2015) Nepal Earthquake 2015: Post Disaster Needs Assessment – Executive Summary. National Planning Commission, Government of Nepal, Kathmandu.

OXFAM (2015). Nepal earthquake. Oxford Committee for Famine Relief. Available at http://www.oxfamamerica.org/take-action/save-lives/nepal-earthquake/.

Paudyal, Y., Yatabe, R., Bhandary, N. & Dahal, R. (2012). A study of local amplification effect of soil layers on ground motion in the Kathmandu Valley using microtremor analysis. *Earthquake Engineering and Engineering Vibration,* 11(2), 257–268.

Paudyal, Y., Yatabe, R., Bhandary, N. & Dahal, R. (2013). Basement topography of the Kathmandu Basin using microtremor observation. *Journal of Asian Earth Sciences,* 62, 627–637.

República (2015). Nepal quake could have been much worse: Here's why. República, Nepal Republic Media. Available at http://www.myrepublica.com/society/item/20216-nepal-quake-could-have-been-much-worse-here-s-why.html.

Saklani, P., Nautiyal, V. & Nautiyal, K. (1999). Sumer, Earthquake Resistant Structures in the Yamuna Valley, Garhwal Himalaya, India. *South Asian Studies,* 15(1), 55–65.

Sheth, A. & Thiruppugazh, V. (2012). Seismic risk management in areas of high seismic hazard and poor accessibility. *15th World Conference on Earthquake Engineering, Lisbon, Portugal.*

Shrestha, M. (1981). Nepal's Traditional Settlement: Pattern and Architecture. *Journal of Cultural Geography,* 1(2), 26–43.

THT (2015). Rebuilding heritage monuments will take years. The Himalayan Times. Available athttp://thehimalayantimes.com/business/rebuilding-heritage-monuments-will-take-years/.

USGS (2015a). M7.8 - 34 km ESE of Lamjung, Nepal. Available at http://earthquake.usgs.gov/earthquakes/eventpage/us20002926#general_summary.

USGS (2015b). M7.3 – 19 km SE of Kodari, Nepal. Available at http://earthquake.usgs.gov/earthquakes/eventpage/us20002ejl#general_summary.

Seismic Retrofitting: Learning from Vernacular Architecture – Correia, Lourenço & Varum (Eds)
© 2015 Taylor & Francis Group, London, ISBN 978-1-138-02892-0

Case study: Local seismic culture in vernacular architecture in Algeria

A. Abdessemed
Lab ETAP, Institute of Architecture and Urban Planing, University of Blida 1, Algeria

Y. Terki & D. Benouar
CAPTERRE, Ministry of Culture, Timimoun, Algeria

ABSTRACT: This text concerns the study of the local seismic culture in Algerian vernacular architecture. The case study, explains how ancient populations built their constructions taking into account seismic risk during the last centuries and show earthquake resistant constructive techniques used in vernacular architecture to avoid seismic loads.

The seismicity of northern Algeria is light to moderate and the country is considered a moderate-seismic hazard territory (Benouar, 1994), in spite of the earthquakes from Orléansville 1954, El Asnam 1980 and Boumerdes 2003. The country has experienced in the past, several moderate seismic events that caused loss of human lives and damage to property in different regions, Algiers (1365, 1716), Cherchell (1732), Dellys (1731), Oran (1780), Blida (1825, 1857), Jijeli (1856), Constantine (1858) and Biskra (1869), and thus it is susceptible to earthquake occurrence and damage in the future. Based on this information, the fundamental questions of interest to architects and engineers are: How did the local population repair and retrofit their houses? What architectural elements and structural techniques did the local population use for repairing and retrofitting their houses to protect themselves against earthquakes? In northern Algeria, where light and moderate earthquakes are quite frequent, the specific actions of the physical environment lead to the development of architectural and structural capacities to resist these events, which was usually well incorporated into the local culture.

The structural failure of vernacular buildings during earthquakes has often led to a better understanding of their performance and improvements in their design. This was emphasized during the identification of traditional architecture seismic retrofitting techniques found in many vernacular buildings in northern Algeria historical cities, which eventually made these buildings more resistant to earthquake damage.

These techniques are the result of a continuous learning cycle of trial and error, which gradually improved along time to adapt and to resist the changing requirements of the physical seismic environment.

This has allowed the local population to develop a local seismic culture in northern Algeria cities, which can be observed in the Casbah of Algiers and Dellys (central Algeria) and in Batna (eastern Algeria) (Adjali, 1986). According to Abdessemed-Foufa (2005) and Abdessemed-Foufa and Benouar (2010), the main features of local seismic cultures are: 1) A seismic isolation technique that was used by placing wood logs of Thuya under the foundation of some buildings (Figs 1a, 1b) Reinforced masonry arches, where one or two layers of logs are superimposed in layers of bricks (Figs 1b, 1c) Reinforced load-bearing walls, where logs of wood called Thuya or Juniper are embedded along the length of the walls (consolidating the angles) and evenly distributed at every 80 to 120 cm (Fig. 1c–2–3–4). A sub-division of the walls into smaller panels makes the walls more resistant to

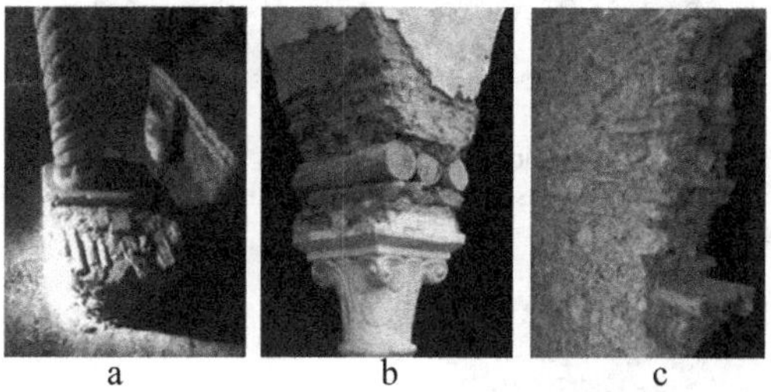

a b c

Figure 1. a) Seismic Isolation (Algiers) b) Reinforced arches Foundations (Algiers) c) Reinforced bricks masonry walls by logs of Thuya (Algiers) (credits: ©Abdessemed-Foufa, 2005, 2010).

Figures 2–3. Reinforced earthen masonry wall by logs of Juniper (Menaa-Aures) (credits: Kays Djilali © Ministry of Culture, Algeria, 2009).

Figure 4. Consolidating angles by juniper (Amenthane-Aures) (credits: Kays Djilali © Ministry of Culture, Algeria, 2009).

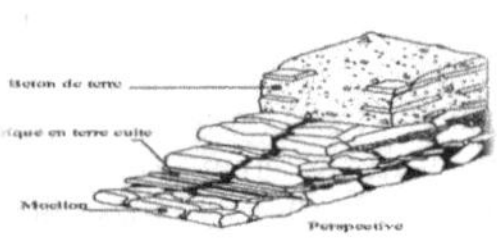

Figures 5-6. Reinforced Rammed Earth in Blida, Cherchell, Ténès and Kolea cities (credits: ©Abdessemed-Foufa, 2010).

the shear force and thus prevents large cracks and collapse. 4) The consolidation of walls is achieved by alternating the joints of Thuya wood logs. 5) Wooden framing is found around the openings to strengthen the openings. 6) Reinforced rammed earth walls with layers of bricks are used in the medieval cities, such as Blida, Ténès, Koléa and Cherchell (Abdessemed-Foufa 2010). This reinforced earth wall is called "tapia valenciana" in Spain (Cristini and Checa 2009) and is part of the "knowledge transfer" by Andalusia moors who founded these cities (Figures 5–6). 7) A number of arches, built out of stones or bricks to transfer lateral loads to the ground, were used in urban framework of Algiers as discharging arches for the structural continuum (Fig. 7).

Figure 7. Discharging arches in urban framework (Algiers) (credits: ©Abdessemed-Foufa, 2005).

REFERENCES

Abdessemed-Foufa, A. (2005). Contribution for a Catalogue of Earthquake-Resistant Traditional Techniques in Northern Africa: The Case of the Casbah of Algiers (Algeria). *European Earthquake Engineering*, 19(2), 23.

Abdessemed-Foufa, A., & Benouar, D. (n.d.). Damage survey of the old nuclei of the Casbah of Dellys (Algeria) and performance of preventive traditional measures in the wake of the Boumerdes 2003 earthquake. *European Earthquake Engineering Journal*, 3, 10.

Abdessemed-Foufa, A. A., & Benouar, D. (2010). Investigation of the 1716 Algiers (Algeria) Earthquake from historical sources: effect, damages, and vulnerability. International Journal of Architectural Heritage, 4(3), 270–293.

Abdessemed-Foufa. (2010). Typologie architectural et constructive comme valeur de la ville. *La Città Storica. The Historic City, Mediterranea*, 2, 42–45.

Benouar, D. (1994). Materials for the investigation of the seismicity of Algeria and adjacent regions during the twentieth century. *Special Issue of Anali Di Geofisica*, 37(4).

Adjali, S. (1986). Habitat traditionnel dans les Aurès: le cas de la Vallée de l'Oued Abdi. *Annuaire de l'Afrique Du Nord*, 25, 271–280.

Cristini, V., & Checa, J. (2009). A historical Spanish traditional masonry techniques: some features about "tapia valenciana" as a reinforced rammed earth wall. In *Proceedings of the 11th Canadian Masonry Symposium, Toronto, Ontario.*

Seismic Retrofitting: Learning from Vernacular Architecture – Correia, Lourenço & Varum (Eds)
© 2015 Taylor & Francis Group, London, ISBN 978-1-138-02892-0

Case study: Assessment of the seismic resilience of traditional Bhutanese buildings

T. Ilharco, A.A. Costa, J.M. Guedes, B. Quelhas & V. Lopes
NCREP – Consultancy on Rehabilitation of Built Heritage, Ltd., Porto, Portugal

J.L. Vasconcelos & G.S.C. Vasconcelos
Atelier in.vitro, Porto, Portugal

ABSTRACT: Architectural, structural and material assessment of Traditional Bhutanese Buildings (TBB) was performed by NCREP as part of the project Bhutan: Improving Resilience to Seismic Risk, namely on its part C: Improving Seismic Resilience of TBB, requested by the Division for the Conservation of Heritage Sites of the Department of Culture – Ministry of Home and Cultural Affair of Bhutan. The present paper describes some of the steps of the study of this interesting set of vernacular architecture.

1 INTRODUCTION

The project developed by NCREP integrated the study of 18 traditional rammed earth buildings of the villages Pathari, Kabesa and Tana, Zome, both in Punakha district, in Bhutan. Buildings aged up to 200 years old, as in Figure 1. The objective was to characterize the main constructive features of these valuable examples of Bhutanese vernacular architecture, in order to understand their structural behaviour and improve their resilience to seismic risk (Ilharco et al., 2015).

2 THE BUILDINGS

Most of the buildings have two floors and an accessible attic, with areas per floor varying between 50 m^2 and 180 m^2. The buildings are made of rammed earth walls and timber floors and roofs.

Figure 1. Traditional Bhutanese building (credits: NCREP).

The exterior walls' thickness varies between 58 cm and 77 cm. In some cases there are interior rammed earth cross walls although with poor connections with the façades. Almost all the walls have a stone masonry footing at the ground level. There are some timber-framed walls (*Ekra-walls*) in the interior and in the main façades of the first floors (Fig. 2). The timber floors have joists with cross-sections from 8×10 cm^2 to 16×22 cm^2, whereas in older buildings there are circular beams with a diameter up to 18 cm. The joists support the floor planks and a layer of earth and straw mix, and are spaced between 30 cm to 100 cm (Fig. 3).

The timber roofs are usually made of 4 to 6 main timber trusses supporting timber purlins. Roofs are usually single gabled, sometimes with a small gable roof over the main gable roof (*Jamtho* roof). Some are two tiered gabled (*Drangim*).

Figure 2. Interior timber-framed wall (credits: NCREP).

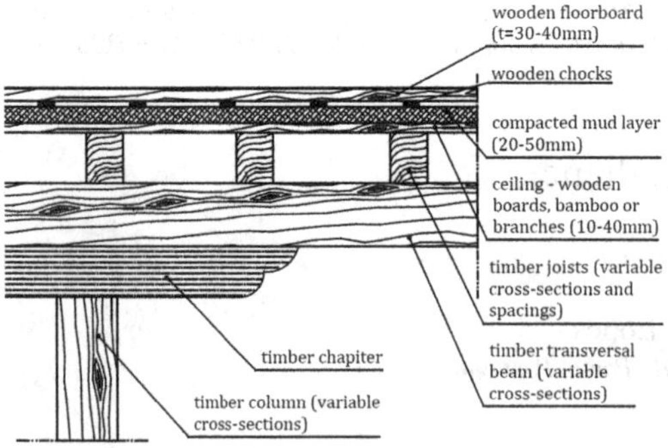

Figure 3. Cross-section of a timber floor (credits: NCREP).

3 THE SEISMIC FEATURES

No specific strengthening details or seismic resilient features were observed on the surface of most of the walls. As the walls are intact, it was neither possible to observe these features inside. However, in some cases timber pieces and flat stones were observed at the joints between lifts on both surfaces of the walls in a length of about 500/600 mm. These are apparently measures to strengthen the walls.

The houses were in a reasonable state of conservation, presenting however some structural damages and some constructive configurations that can be weaken points in the occurrence of a seismic action.

4 THE VULNERABILITY ANALYSIS

The ratio (%) of shear walls per floor area (considering only rammed earth walls) was computed to estimate the buildings behaviour under seismic excitation. Figure 4 shows that the % decreases along the height in both directions, mainly due to the existence of timber-framed walls in the main façades, not considered in the computation of shear walls.

A simple vulnerability analysis allowed estimating the level of damage in the buildings, according to the intensity level of an earthquake occurrence. The expected damage can be considerably reversed if some strengthening solutions are implemented, such as improving the connections between external and internal walls and between horizontal and vertical structures (Costa et al., 2011).

Another point addressed was the presence of openings near the edges and cantilever elements at the top, which may form local mechanisms prone to collapse during a moderate seismic action.

In order to perform a vulnerability seismic assessment of the existing buildings, the application of the macroseismic method was made resorting to the Giovanazzi and Lagomarsino (2004) proposal. Moreover, a detailed analysis of each building features was made, such as the evaluation of the regularity in plan and

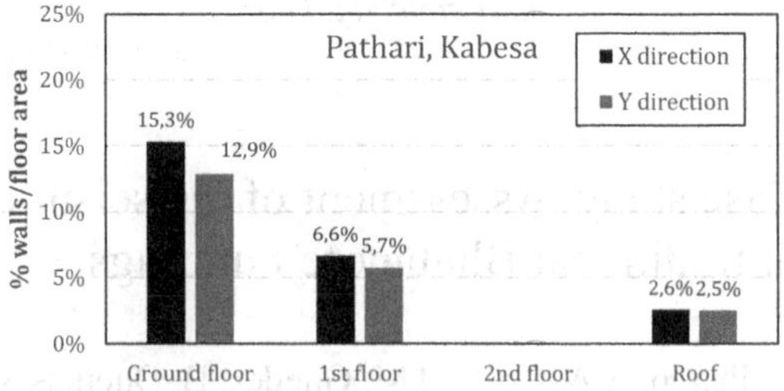

Figure 4. Average distribution of shear walls along the height for the surveyed buildings of Pathari, Kabesa (credits: NCREP).

height, of the presence of main shear walls in both direction and of the discontinuity of these walls, Bothara and Brzev (2011). The main results obtained in the analysis will be used to improve the seismic features of Bhutanese traditional buildings.

However, a thorough vulnerability analysis of Bhutanese traditional buildings is fundamental to achieve a complete knowledge of their seismic behaviour and to define the most proper strengthening solutions. Moreover, a refined calibration of the vulnerability curves used may allow spreading the vulnerability analysis to other earthquake prone areas. Detailed post-earthquake surveys of the rammed earth buildings damaged during 2009 and 2011 should be used to calibrate all the information gathered, as it possesses crucial information for the better understanding of their seismic behaviour.

ACKNOWLEDGMENTS

The authors thank the Division for the Conservation of Heritage Sites of the Department of Culture – Ministry of Home and Cultural Affair of Bhutan for the support during the development of the project.

REFERENCES

Bothara, J., and Brzev, S. (2011). A tutorial: improving the seismic performance of stone masonry buildings. Oakland, California, United States of America: Earthquake Engineering Research Institute.

Costa, A. A., Arêde, A., Costa, A., & Oliveira, C. S. (2011). In-situ cyclic tests on existing stone masonry. Earthquake Engineering and Structural Dynamics, 40(4), 449–471.

Giovanazzi, S., and Lagomarsino, S. (2004). A macroseismic method for the vulnerability assessment of buildings. 13th World conference on Earthquake Engineering, (p. peper 896). Vancouver, Canada.

Ilharco, T., Vasconcelos, J.L., Costa, A.A., Dorji, C., Vasconcelos, G., Paiva, L. (2015) Study of Typology of Bhutanese Rammed Earth Buildings. Pathari, Kabesa and Tana, Zome. Division for the Conservation of Heritage Sites of the Department of Culture – Ministry of Home and Cultural Affair of Bhutan.

Seismic Retrofitting: Learning from Vernacular Architecture – Correia, Lourenço & Varum (Eds)
© 2015 Taylor & Francis Group, London, ISBN 978-1-138-02892-0

Case study: Vernacular seismic culture in Chile

N. Jorquera
Department of Architecture, Universidad de Chile, Santiago, Chile

H. Pereira
PROTERRA Iberian-American Network and Universidad Tecnológica Metropolitana, Santiago, Chile

ABSTRACT: As Chile is one of the most seismic countries in the world; different vernacular earthquake-resistant strategies have been created adapted to each context in this vast territory. In this case study, four examples of these strategies will be presented, based in the geometrical configuration and the use of lightweight structures or wooden reinforcements.

1 CHILEAN SEISMIC CULTURES

In Chile, an earthquake with a magnitude higher than 7, occurs approximately every 10 years. There have been more than 100 of these earthquakes since 1570 to date (National Seismological Centre, Universidad de Chile). This, along with the geographic, climatic and cultural diversity of the Chilean territory, has prompted a variety of vernacular architectures and 'seismic cultures' or technical strategies to face earthquakes, where they are frequent (Pierotti & Ulivieri, 2001). In the arid north of Chile where stone, earth and cactus materials are the only building materials, earthquake-resistant strategies are based in the geometry of the buildings. In the centre and south of the country, where the temperate and cold climates allow the growth of large trees, wooden reinforcements are the most common seismic resistant solutions.

2 GEOMETRICAL STRATEGIES IN ANDEAN MASONRY ARCHITECTURE

Andean vernacular architecture is built in adobe and stone masonry, the only available resources in the highlands of the arid regions of the north of Chile (17°30°–26°05′S). There, the absence of wood obliges the use of very massive walls and the adoption of trapezoidal shape geometry, both in the entire building as in every single wall. These strategies allow lowing the centre of gravity of the building by concentrating its mass closer to the ground. In larger buildings, like churches, buttresses and side chapels are used in more stressed areas to counteract horizontal forces caused by earthquakes (Fig. 1).

Figure 1. Andean church of Cariquima, region of Tarapacá (credits: Natalia Jorquera, 2014).

3 LIGHT-WEIGHT CANE STRUCTURES IN PICA AND MATILLA

In Pica and Matilla, two little oases in the Tarapacá, region (20°5′S-69°3′W), the growth of cane and the anhydrite soil (Ca SO4) are used together to build vernacular architecture with *quincha*. *Quechua* is a term for a timbered structure with a secondary cane structure fill with soil (Fig. 2). The elastic properties of wood and the lightness of cane and anhydrite soil allow the deformability of walls during an earthquake, without reaching the breaking point. The church of Matilla, built originally with *quincha,* was inadequately intervened with modern techniques during the 80's, but in the 2007 restoration project (Fig. 3), the original technique was re-used to recover its good performance during earthquakes.

Figure 2. Cane and anhydrite *quincha* in Matilla's church, region of Tarapacá, Chile (credits: Hugo Pereira, 1992).

Figure 3. Restoration of Matilla's church re-using the original traditional *quincha* (credits: Hugo Pereira, 2007).

Figure 4. Wooden horizontal reinforcement in adobe architecture in central valley (credits: Natalia Jorquera, 2007).

Figure 5. Wooden frame fill with 'adobillo' (credits: R. Cisternas, 2014).

4 WOODEN REINFORCEMENTS IN CENTRAL VALLEY AND VALPARAISO ARCHITECTURE

In the big macro-region called the 'central valley' of Chile (32°02′– 38°30′S), adobe and wood are used together in vernacular architecture to achieve stability of the buildings. In the colonial architecture of Hispanic legacy, adobe walls are reinforced with timber tying elements called *llaves*, horizontally positioned within the adobe masonry to improve the resistance against horizontal loads (Fig. 4). In the city harbour of Valparaiso (33°03′S, 71°38′W), the large amount of wood transported on the ships from the northern hemisphere during the XIX century, allowed the creation of a timber frame structure filled with an earthen block called *adobillo*, which has notches in two extremes to fix the block into the wooden logs (Fig. 5). This efficient connection between wood structure and infill prevents the overturning of the blocks in the structure, in case of an earthquake (Jorquera, 2015). Several of these traditional seismic resistant systems can be observed on the historical centre of the UNESCO city of Valparaiso.

REFERENCES

Jorquera, N. (2015). Wooden frame fill with adobillo block. In M. Correia, L. Dipasquale, & S. Mecca (Eds.), *VERSUS: Heritage for Tomorrow* (pp. 241–242). Florence, France: Firenze University Press.

National Seismological Centre, Universidad de Chile website. Retrieved May 24, 2015, from http://www.sismologia.cl/

Pierotti, P. & Ulivieri, D. (2001). Culture sismiche locali. Pisa: *Edizioni Plus Università di Pisa*.

Seismic Retrofitting: Learning from Vernacular Architecture – Correia, Lourenço & Varum (Eds)
© 2015 Taylor & Francis Group, London, ISBN 978-1-138-02892-0

Case study: Seismic resistant typologies technology in vernacular architecture in Sichuan Province, China

J. Yao
Cultural Relics and Archaeology Research Institute of Sichuan Province, China

ABSTRACT: Sichuan Province is located in southwest China, one of the seismic areas in the country, within which there is a Longmenshan seismic belt. The vernacular architecture in Sichuan employ column and tie construction frame, with large depth rooms and wide openings in the front eaves, as an adaptation for local hot climate. The column and tie construction is a typical style of Chinese traditional architecture, with bearing structures composed of purlins, supported directly by columns. The columns are connected with tie beams, which form the overall enclosed skeleton of the structure. Though a structural system common to the south of China, the technologies of buildings vary from area to area.

1 FOUNDATIONS AND STYLOBATE

Before the construction, the soil under the foundation of the building is replaced and rammed, especially the soil beneath the column bases. The ground is laid with dressed stones based on the length and width of the buildings, as well as the layout of internal columns. The dressed stones named "Bishui Stone" or "Liansang Stone", are made from local red or green sandstones, and used to protect the wooden groundsills above them. The stones also form a relative complete frame on the ground for the overall structure, which resembles to what we call "Di Quan Liang" (ground ring beams).

Figure 1. The Stylobates of Traditional Buildings (credits: Yao Jun).

2 WOODEN STRUCTURE

In column and tie constructions, the columns are closely distributed. Crossing-ties (a long cuboid component) connect and maintain the columns longitudinally and transversely, while crossing-ties and columns are connected in mortise and tenon joints. Mortise and tenon joint is a type of flexible connection, which enables the components to have some space for structural deformation, and help greatly, to relieve the destructive stress from all directions. Unlike the construction in rigid material like stone or concrete, it is not easily damaged or destroyed by the different conductive frequency of stress. Purlins (a cylindrical component supporting rafters) are connected the columns longitudinally, and a lap joint is placed at the top of the column. There is a cambered mortise at the top of the column, which is done based on the diameter of the purlin, so its depth is about one-third of the diameter. The purlin is just overlapped on the

Figure 2. Combinations of Columns and Purlins (credits: Yao Jun).

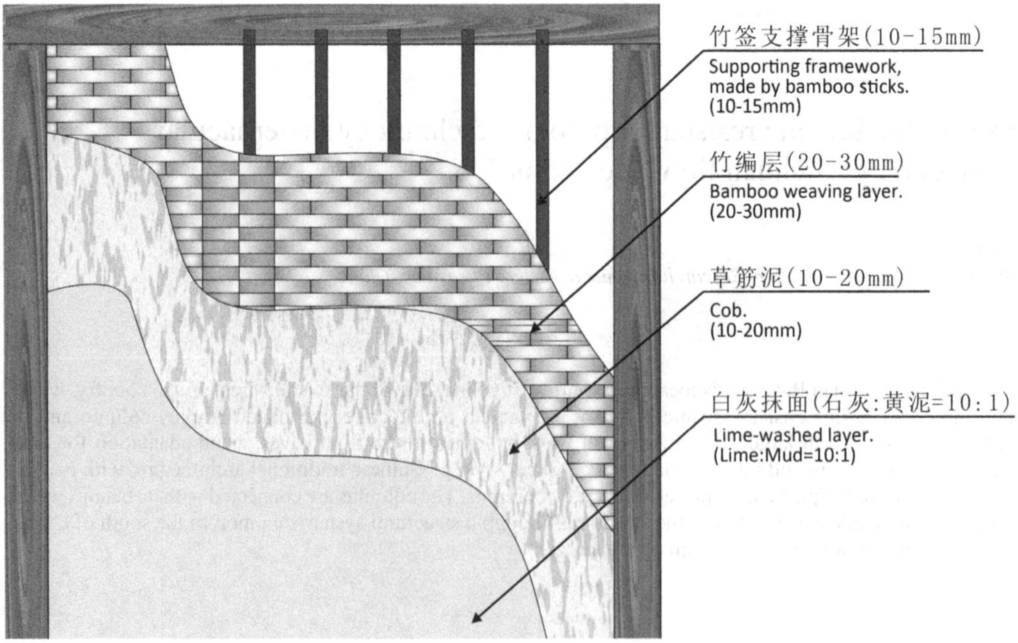

Figure 3. Bamboo Mat Wall Plastered with Mud (credits: Dai Xubin).

mortise, while the connection of purlins is also there, which can prevent rolling when they are affected by an external force, like earthquake. In buildings with a wider bay, there is another purlin or cuboid tie under the purlins, called following-purlins or following-tie.

The crossing-ties in connecting and enclosing use are generally flat. There are also short pillars supported by the crossing-ties, which were called 'stolen columns' because they never reach the ground. At the bottom of these 'stolen columns', there are some mortises. The width of the mortises is the same as the width of the crossing-tie but the height is only half of the crossing-tie, which could make the two components, match each other. To fasten crossing-ties, wooden dowels are used at the exposed ending, in order to keep columns connected with ties in a violent shaking situation, and to prevent crossing-ties' pulling out or falling off.

3 WALLS

For the traditional architecture in Sichuan Province, walls are not for load bearing, but for separation and enclosure. There are two types of walls: one type made from wood, and the other from bamboo-woven mats plastered with earth mortar. Wooden boards, about 20mm thick, are often placed to connect the lower ends of those columns, which are arranged in parallel between the front and back walls. The wooden walls are used to have both divide and enclose the rooms, and to connect the adjacent columns with mortise-tenon joints, to make the structure a complete whole.

Walls made from bamboo mats plastered with earth mortar, are constructed on the top the wooden walls. A few wooden or bamboo poles are laid in parallel between the columns, with young bamboos (slates) woven in-between to form a mat. A frame with grooves at each edge is placed to enclose the mat wall. The groove in parallel with the beam is used to fix the upper and lower ends of the poles; and the grooves next to the columns are for the fixation of the bamboo slates. The bamboo mat is plastered with earth mortar. Furs and linens on both sides reinforce the elasticity and solidity of the material. The surfaces are polished and even decorated with painting in some cases. The walls are made from lightweight materials, thus reducing the load on the beams and on the major structure. In general, the structure made of crisscrossing columns, purlins and rafters helps to disperse and descend the load from the roof towards the foundation of the building. At the same time, the columns are connected with cross beams. The walls made from lightweight, materials such as wood and bamboo, are effective both to divide and envelop the space, and to lessen the load of the structure. In such integrated means of construction, people in Sichuan Province successfully managed to improve seismic-resistant performance of the buildings.

ACKNOWLEDGEMENTS

The author would like to thank the translation of PEI Jieting and LUO Xi, and the revision done by ZHANG Peng. This case study text was just possible due to the support of SHAO Yong.

Seismic Retrofitting: Learning from Vernacular Architecture – Correia, Lourenço & Varum (Eds)

Case study: Seismic retrofitting in ancient Egyptian adobe architecture

S. Lamei
Centre for Conservation & Preservation of Islamic Architectural Heritage, Cairo, Egypt

ABSTRACT: This text concerns the study of seismic retrofitting in Egyptian architecture. The case study addresses adobe heritage in ancient Egyptians dynasties; adobe building in vernacular architecture; ancient seismic retrofitting techniques; and finally, seismic retrofitting techniques in vernacular architecture.

1 ADOBE HERITAGE IN ANCIENT EGYPTIAN DYNASTIES

Egypt is renowned by the existence of a rich architectural heritage, which was built during different Eras, since the ancient Egyptian dynasties. In the Pharaonic eras, adobe was used in profane buildings. Nowadays, it is still in use in peasant houses along the Nile River.

Adobe and mortar undergo deterioration processes. The rate and symptoms of such processes are influenced by different factors. The building materials and the structures suffer not only from the deterioration process caused by physical, chemical weathering and manmade environments, but also from structural failures caused by catastrophes, like earthquakes, floods, torrential rain and fire.

2 ADOBE IN VERNACULAR ANCIENT EGYPT

The adobe word in Ancient Egyptian is 'djebet' in Coptic is 'τωɜϵ', and in Arabic is 'ṭuba' (Capaldi, 2011).

In the course of the Early Dynastic Period (about 3100–2613 B.C.) and the Old Kingdom (about 2613–2160 B.C.) adobe remained the basic building material to build structures to live in, as palaces or vernacular houses. During the Pre-dynastic Period, local populations preferred to build in wattle and daub.

Adobe consisted of sand, silt, and clay taken from the Nile mud and mixed with chopped straw as a strengthening and binding material. The earth mix was trampled by feet, to reduce the mix to a regular consistency.

3 ANCIENT SEISMIC RETROFITTING TECHNIQUES

The word for earthquake in ancient Egypt is 'murta' and in arabic 'zilzāl'. An old building technique can be noticed at the temple of AmonRé (Karnak), where adobes have a size of 38 × 18 × 14 cm. There is a layer of mortar (or *halfa* bed[1]) between two layers of adobe assuring the cohesion of the wall. At the temple Montou wooden (acacia) pieces were placed inside the wall, as reinforcement[2]. There is also a *ḥalfa* layer, which fosters the distribution of wall vertical forces.

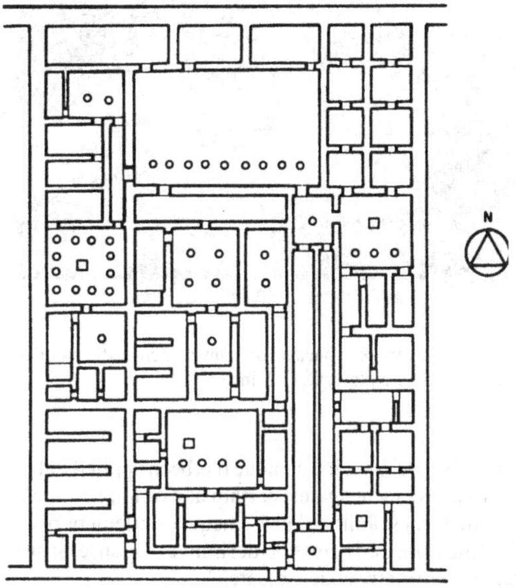

Figure 1. Egyptian mansions (credits: Badawy).[1]

Another ancient technique is the arranging of vertical joints along the total height of the wall, the use of concave/convex bed joints and inserting wooden ties between horizontal layers and wooden pieces across the adobe wall. This is noticed from the late era of

[1] Ḥalfa: plant used in general to make paper, due to its fiber.

[2] Testing explanation walls 'to Assize curves' on the great temple of Amun-Re at Karnak, Perseus (Golvin & Jaubert, 1990).

Figure 2. Trampling the earth mix by foot (credits: CIAH, Egypt).

Figure 3. Karnak, concave & convex bed with wooden pieces holes (credits: J. C. Golvin).

1000–300 B.C. Such technique has been applied either for reinforcement against earthquake or improvement of structural stability (due to unequal settlement from the Nile floods). There is no definitive response, as the topic is currently under research[3].

4 SEISMIC RETROFITTING TECHNIQUES IN VERNACULAR ARCHITECTURE

Since the earthquake of October 1992, it is recommended to undertake sufficient measures for seismic retrofitting in vernacular architecture too. Techniques commonly found are: wooden ties laid within walls, every 1.00 to 1.50 m, depending on building's height and connected at the corners in half dove tailed joints. The wooden roof joists are laid on a wooden wall plate. Both are connected into the walls, through wooden rods.

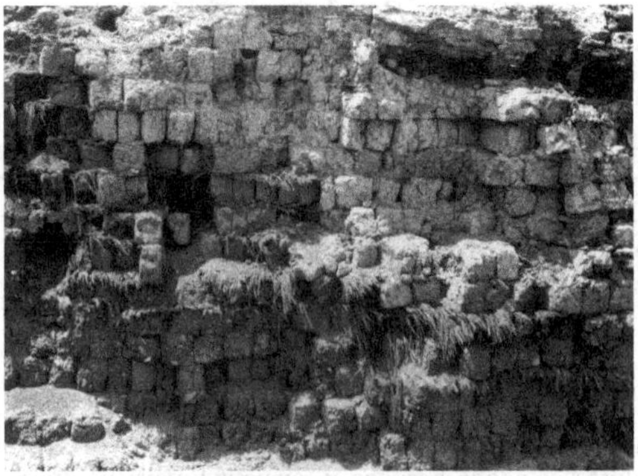

Figure 4. Ḥalfa layer between adobe bed (credits: A. Bellod).

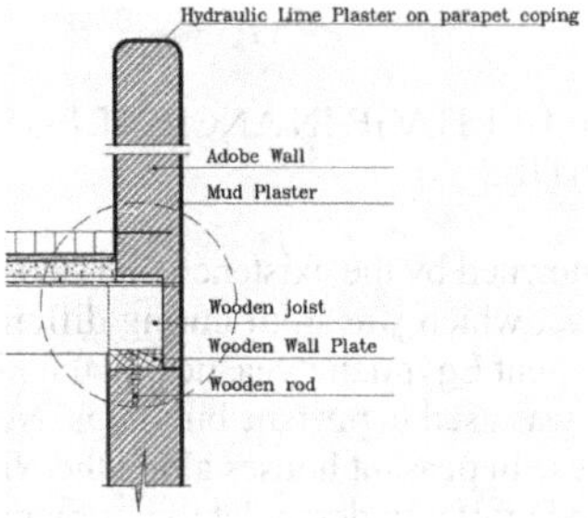

Figure 5. Details showing retrofitting of wooden roof with wall (credits: CIAH, Egypt).

Figure 6. Wooden ties, lay in adobe masonry wall corner (credits: CIAH, Egypt).

Research on seismic retrofitting techniques needs to be further developed.

REFERENCES

Capaldi, X. (2011). Ancient Egyptian Mud Brick Construction: Materials, Technology, and Implications for Modern Man. Available at: https://dataplasmid.wordpress.com/2011/04/08/ancient-egyptian-mud-brick-construction-materials-technology-and-implications-for-modern-man.

Golvin, J. & Jaubert, O. (1990). CRAIBL. *Scientific Journal of Ancient History*, 134(4), 905–946.

[3] Mr. Max Beiersdorf is writing a thesis at the University of Cottbus, Germany, and he is conducting both empirical and virtual tests on this issue.

Seismic Retrofitting: Learning from Vernacular Architecture – Correia, Lourenço & Varum (Eds)
© 2015 Taylor & Francis Group, London, ISBN 978-1-138-02892-0

Case study: Seismic resistant constructive systems in El Salvador

F. Gomes & M.R. Correia
CI-ESG, Escola Superior Gallaecia, Vila Nova de Cerveira, Portugal

R.D. Nuñez
FUNDASAL, San Salvador, El Salvador

ABSTRACT: This case study addresses the seismic resistant constructive systems, applied in housing construction in El Salvador. The traditional construction systems usually used are *adobe, bahareque, and* reinforced brick masonry *(mixto)*. The improved *bahareque* system is based on the ancient technique known locally as *Joya de Ceren bahareque*, a technique used in the past, on the world heritage site of Joya de Cerén.

1 SEISMIC CULTURE IN EL SALVADOR

El Salvador is located in a high intensive seismic region in Central America. During the 20th century, eleven destructive earthquakes took place, causing a high number of deaths and a great destruction of buildings (López et al. 2004; Martínez-Días et al, 2004). San Salvador, the capital, has suffered the impact of several of these earthquakes: in 1917 (MS 6.4); in 1919 (MS 5.9); in 1965 (MS 5.9), and in 1986 (MS 5.4) (López et al. 2004; Martínez-Días et al, 2004). The most recent and biggest event had a MS 6.6 magnitude and it happened in 2001.

2 CONSTRUCTIVE SYSTEMS IN A SEISMIC CONTEXT

In El Salvador countryside, the traditional construction systems for dwellings are *adobe*, *bahareque,* and *mixto*.

Unfired brick walls with earth mortar between layers characterize the adobe system. The adobes are made of vulcanite stone, with high rigidity, but very low strength and cohesion.

The *bahareque* consists of vertical and horizontal timber, or cane or bamboo elements, with earth mortar infill and an earth plaster finishing. The seismic resistance of the *bahareque* depends primarily on the condition of the timber and the cane elements.

The *Mixto* system introduced during the 20th century is composed of a foundation of fired clay bricks with mortar and slender elements of concrete with thin steel reinforcement. (...)

Generally, this is the method, the most used in El Salvador (López et al., 2004).

3 BAHAREQUE: PAST AND PRESENT CONSTRUCTIVE SYSTEM

The *Bahareque* is an ancient constructive technique developed by Joya de Ceren's inhabitants in El Salvador. Today its archaeological remains can still be observed at the world heritage site of Joya de Cerén (Fig. 1).

According to Lang et al. (2007), following the 1965 earthquake, *bahareque* was the technique the most frequently used in informal rural houses in El Salvador. *Bahareque* buildings are characterized by a high flexibility and elasticity, when prudently constructed and well preserved. The *bahareque* buildings originally display a good performance against dynamic earthquake loads.

The structural walls are mostly composed of vertical timber elements and horizontal struts, which are either made of timber slats, cane, bamboo or tree limb. These members are generally 2/3 inches thick and are fastened at regularly spaced intervals from the base to the ceiling height at the vertical elements. This creates

Figure 1. World Heritage Site of Joya de Cerén, El Salvador (credits: Mariana Correia, 2014).

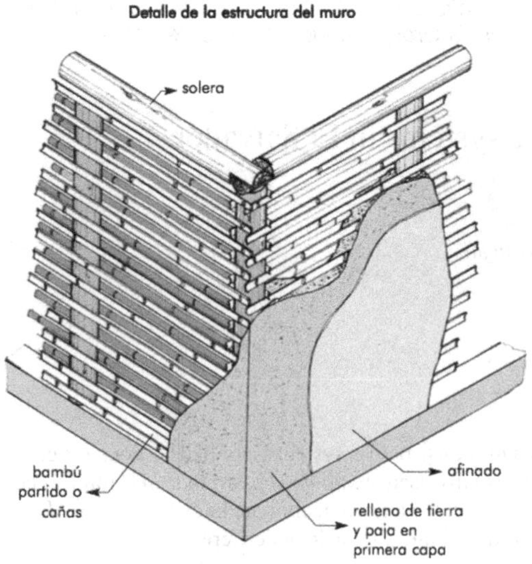

Figure 2. The adapted *bahareque* technique (credits: Wilfredo Carazas & Alba Rivero, 2002).

basketwork type skeleton (Lang et al., 2007), which is then packed with earth mortar as an in-filler, combined with chopped straw and covered with a plaster finish in some cases.

In rural areas, the walls are often left plane, without any lime plaster and whitewash, or paint, which gives them a wavy surface with an unfinished character. The *bahareque* houses in rural areas are quite different from those in urban areas, both in terms of their esthetical appearance, as well as their structural capacity (Lang et al., 2007).

The use of this improved *bahareque* system, known as *Joya de Ceren bahareque* (Carazas Aedo, 2014), is nowadays being researched by several entities, as is the case of FUNDASAL, MISEREOR and CRAterre. They seek to develop a new approach, based on the original construction forms and materials, and adapting them to contemporary requirements, according to the seismic factor. From the research findings was created a *bahareque* construction guide, based in seismic-resistant construction (FUNDASAL, 2001; Carazas Aedo & Rivero Olmos, 2002; Garnier et al. 2013).

4 VERNACULAR HOUSING WITH SEISMIC-RESISTANT CHARACTERISTICS

The ONG FUNDASAL promoted the construction of seismic-resistant houses applying the research carried out, at the structural level, in El Salvador. The entailed research concerned the adobe system, the modified mixed system with stabilized earth bricks; and the Ceren Bahareque system.

Figure 3. The adapted *bahareque* technique applied in a workshop during the 14th SIACOT (credits: Filipa Gomes, 2014).

The improvement of these traditional building systems, also contributed to the seismic response of the housing. Therefore, it improved the quality of life of Salvadoran families living in a safe house and keeping its original architectural tradition.

REFERENCES

Carazas Aedo, W. (2014). BAHAREQUE CERÉN. La vivienda nativa, una cultura construtiva ancestral en la Mesoamérica actual. El Salvador: Quijano, S. A. de C. V.

Carazas Aedos, W. & Rivero Olmos, A. (2002). Bahareque: Guia de Construccion Parasismica. France: Ediciones CRATerre.

FUNDASAL (2011). Sistemas sismo resistentes de construcción de vivienda utilizando la tierra. El Salvador: FUNDASAL.

Garnier P., Moles, O., Caimi, A., Gandreau, D., Hofman, M. (2013). Natural Hazards, Disasters and Local Development. Integrated strategies for risk management through the strengthening of local dynamics: from reconstruction towards prevention. Grenoble: CRAterre ENSAG

Lang, D., Merlos, R. Holliday, L., Lopez, M. (2007). HOUSING REPORT. Vivienda de Bahareque. World Housing Encyclopedia. Retrieved from: http://www.world-housing.net/WHEReports/wh100159.pdf

López, M., Bommer, J., Méndez, P. (2004). The seismic performance of bahareque dwellings in El Salvador.*13th World Conference on Earthquake Engineering*. Vancouver, B.C., Canada.

Martínez-Díaz, J. J., Álvarez-Gómez, J. A., Benito, B. Hernández, D. (2004). Triggering of destructive earthquakes in El Salvador. Geological Society of America, 32(1), 65–68. doi:10.1130/G20089.1

Seismic Retrofitting: Learning from Vernacular Architecture – Correia, Lourenço & Varum (Eds)
© 2015 Taylor & Francis Group, London, ISBN 978-1-138-02892-0

Case study: Seismic retrofitting of Japanese traditional wooden structures

N. Takiyama
Division of Architecture and Urban Studies, Tokyo Metropolitan University, Tokyo, Japan

ABSTRACT: This paper analyzes examples of seismic retrofitting applied to Japanese wooden structures, the structure and reinforcement of fitting-type joints, the construction and reinforcement of mud walls, and finally, the energy-absorbing mechanisms that are independent of the wall structures.

1 JAPANESE TRADITIONAL WOODEN STRUCTURES

There have been many reports on wooden structures collapsing because of large earthquakes in Japan. On the other hand, there are many traditional wooden structures that have continued to resist collapse despite experiencing many earthquakes. Some examples of traditional wooden houses are shown in Figure 1.

These traditional wooden structures were constructed using "fitting-type" joints formed without the use of nails or hardware, as this technique was deemed superior and carried a higher status. In addition, such structures were often built using locally grown wood, which may be difficult to obtain now. It is usual for every traditional area to retain a carpenter who is responsible for construction and maintenance.

2 FITTING-TYPE JOINTS

There are many kinds of fitting-type joints. A typical example would be a tenon of one member being inserted into a mortise in another member; sometimes being secured with a cotter (Fig. 2a) shows a typical example). Another version involves a crosspiece passing through a column and/or beam; called *nuki* (Fig. 2b) shows a typical example). Figure 3 shows some particularly interesting examples. Structures in Ine, Kyoto, feature oblique *nuki*, which pass through both columns and beams (Fig. 3a). Many of these oblique *nuki* are oriented at an angle of <45°. The oblique *nuki* is designed such that its upper width is larger, and it is installed from above the beam. The principle is that, when an earthquake strikes, the joint will loosen, but it will gradually re-set itself under the force of gravity. The large roof frame in Kurayoshi, Tottori, is composed of long horizontal members with short vertical struts (Fig. 3b). The long span supports a very large roof.

However, buildings that are not earthquake-resistant often have their joints reinforced as shown in Figure 4. A connection damper is mounted on a joint to absorb seismic energy (Fig. 4a). Hardware is

Figure 1. Traditional wooden structures (credits: N. Takiyama).

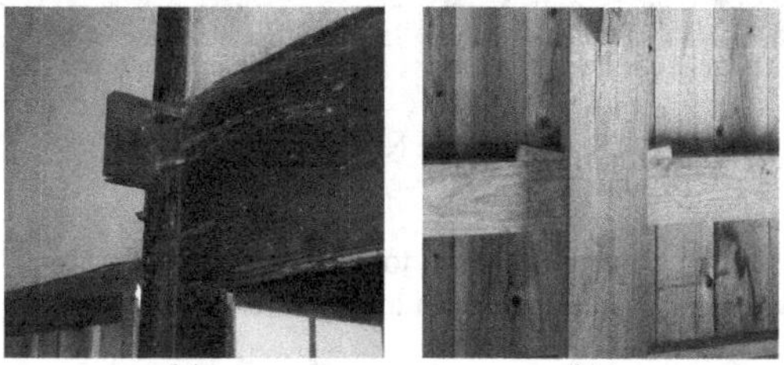

(a) (b)

Figure 2. a) Joint fastened by cotter, b) *Nuki* penetrated through column (credits: N. Takiyama).

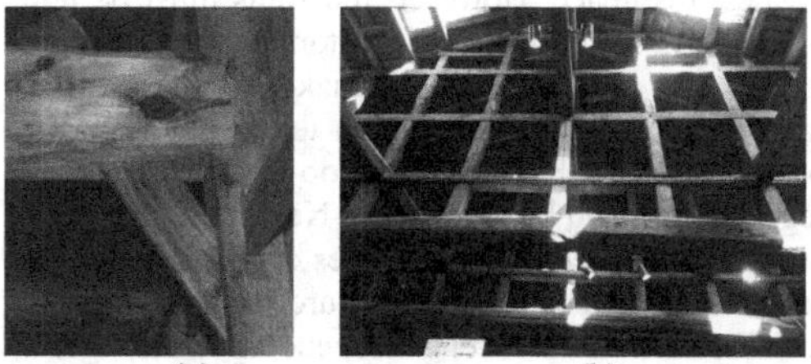

(a) (b)

Figure 3. a) Oblique crosspiece, b) Large roof frame (credits: N. Takiyama).

(a) (b)

Figure 4. a) Connection damper, b) Pulled by hardware.

(a) Bamboo-lathing (b) Magnified bamboo-lathing

(c) Daubing with mud-plaster (d) Daubing the rear face

Figure 5. Mud wall (credits: N. Takiyama).

used to connect a column to a beam so that the joint will not fail in the event of an earthquake (Fig. 4b).

3 WALLS AND SUBSTITUTES

A wooden structure resists earthquakes by virtue of its walls, such that a house with many walls offers the greatest resilience. However, the walls must be laid out evenly, so as not to cause torsion in the structure in the event of an earthquake. Mud walls are traditionally constructed using a technique based on bamboo-lathing. First, the bamboo-lathing is weaved together and held by straw rope. Next, mud-plaster is repeatedly daubed onto both sides of the lathing. The soil and the bamboo are always sourced locally, leading to differences in wall strength depending on the region. A study on the reinforcement of these mud walls examined a method whereby bamboo fiber is mixed into the mud-plaster. Provided the bamboo fibers are sufficiently long, the deformation capacity and maximum

(a) Lattice wall (b) Ladder beam

Figure 6. Reinforcement examples (credits: N. Takiyama).

(a) Column on cornerstone (b) Connected to sill

Figure 7. Column bases (credits: N. Takiyama).

shear stress intensity of the wall increase, regardless of the quantity of fiber, relative to when straw is used. To reinforce the structure, energy-absorbing mechanisms made of wood are available. Lattice walls can be installed to increase resistance without blocking light (Fig. 6a). Ladder beams, in which vertical members are placed across two horizontal members, are installed in voids without walls such as in stairwells (Fig. 6b).

4 COLUMN BASES

In Japanese traditional wooden structures, columns are often placed on cornerstones (Fig. 7a). The columns can move freely. Some columns are set on a sill (Fig. 7b).

REFERENCES

Takiyama, N., Hayashi, Y., Watanabe, C., Nambu, Y., Kobayashi, S., & Yamamoto, H., (2014). Structural Properties Evaluation of Unique Boat House Using Oblique Nuki, Part I: Structural Investigation. In *WCTE 2014, World Conference on Timber Engineering, Quebec City, Canada, August 10–14, 2014*.

Miyamoto, M., Utsunomiya, N., Takahashi, S., Yamanaka, M., Matsushima, M., & Onishi, Y. (2014). Experimental Study on Seismic Performance of Mud Wall Mixed with

Seismic Retrofitting: Learning from Vernacular Architecture – Correia, Lourenço & Varum (Eds)
 ISBN 978-1-138-02892-0

Case study: Seismic retrofitting constructive typology of vernacular Moroccan architecture (Fez)

A. Abdessemed-Foufa
Lab ETAP, Institute of Architecture and Urban Planing, University of Blida 1, Algeria

ABSTRACT: This text presents the retrofitting system used in vernacular architecture in Morocco, as well as seismic constructive typologies adopted after the great earthquakes of 1522, 1624 and 1755 in Fez, in order to avoid future seismic loads and to protect the constructions.

The seismicity in Morocco is divided into two recognised active areas: a) the Rif and Atlases chain; and b) Mesetas and Anti Atlas and its Saharan border, which are negligible to moderate.

Earthquakes of Atlantic origin with high magnitude are likely to generate substantial damage on the West Coast cities, including the interior. The major historical earthquakes in Morocco causing damage are those of Fez in 1522 and 1624, as well as that of Agadir in 1755. In the 20th century, the earthquake of Agadir in 1960 and the earthquake of El Hoceima in 1994 also spawned damage. The seismicity of Morocco is moderate with an approximate recursion of 100 years, and thus, a moderate-seismic hazard country (Cherkaoui, 2003).

The retrofitting or strengthened techniques were developed after the great historical earthquakes by empiricism and following an observation of the damage. The reconstruction held accounted for improved construction techniques that have emerged following the disaster.

Figure 1. Buttresses (credits: Abdessemed-Foufa).

On the urban framework level, it was possible to observe (Abdessemed-Foufa 2007): (1) Buttresses to maintain the high walls at the battlements and palaces (Fig. 1); (2) Wooden struts between the walls to maintain balance (Fig. 2); (3) Discharging arches built out of stones or bricks, able to transfer lateral loads to the ground and creating a structural continuum (Fig. 3).

The main earthquake resistant constructive techniques observed at Fez are: (4) Reinforced load-bearing walls, where logs of cedar called "*an'taq*" (belts) are embedded along the length of the walls, regularly distributed at every 80 cm. Such sub-division of the walls into smaller panels allow the walls to better resist the shear force, and thus to prevent large cracks and collapse (Fig. 4), links of the walls are achieved by alternate crossing with wood logs of cedar;

Figure 2. Wooden Struts (credits: Abdessemed-Foufa 2006).

Figure 3. Discharging Arches (credits: Abdessemed-Foufa, 2006).

Figure 4. Reinforced masonry walls with cedar beam (credits: Abdessemed-Foufa, 2006).

(5) An "*opus mixtum*" facing wall of bricks laid flat which overlap other inclined at 45° called "*mabsut-m'ramam*" (Figure 5). Such sub-division of the wall into smaller panels acts as the reinforced wall with wooden logs; (6) Consolidating the angles with cedar beam, regularly inserted inside the walls (150 cm) (Figure 6). In the absence of any vertical element to

Figure 5. Opus mixtum facing walls (credits: Abdessemed-Foufa, 2006).

Figure 6. Reinforced angles by cedar beam (credits: Abdessemed-Foufa, 2006).

prevent bending or spreading of the wall, this traditional technique constitutes the angle reinforcement; (7) Strengthening of pillars by inserting wooden plates arranged every 50 cm, thorough the section. These plates have the ability to absorb the chaos generated by earthquake or soil settlement (Touri et al., 1999).

REFERENCES

Abdessemed-Foufa, A.A. (2007). Contribution pour la redécou-verte des techniques constructives traditionnelles sismo-résistantes adoptées dans les grandes villes du Maghreb (Alger, Fès et Tunis) durant le 18ème siècle. Thèse de Doc-torat en Sciences. Alger: EPAU, 438p.

Cherkaoui, T.E. (2003). Le risque sismique dans le Nord du Ma-roc. Travaux de l'Institut scientifique Série géographie-physique, n° 21, pp. 225–232.

Touri et al. (1999). Le projet pilote de restauration et réhabilitation du palais "Dar 'adiyal" à Fès. Un exemple remarquable de coopération tripartite. Italie: Ed Diagonale, 158p.

Zacek, M. (1996). Construire parasismique, risque sismique, conception parasismique des bâtiments, règlementation. Marseille: Ed Parenthèsed, 340 p.

Seismic Retrofitting: Learning from Vernacular Architecture – Correia, Lourenço & Varum (Eds)
© 2015 Taylor & Francis Group, London, ISBN 978-1-138-02892-0

Case study: Local seismic culture in Romanian vernacular architecture

M. Hărmănescu
Ion Mincu University of Architecture and Urban Planning, Bucharest, Romania

E.S. Georgescu
National Institute for Research and Development in Construction, Urban Planning and Sustainable Spatial Development "URBAN – INCERC", Romania

ABSTRACT: This case study investigates the relation between vernacular architecture and the seismic hazard in Romania. This text shortly outlines the behaviour and the resistance of the vernacular constructive systems, under earthquake effects.

1 ROMANIA SEISMIC FRAMEWORK

Romanian seismicity is related to several epicentral areas, but Vrancea is the most important, as it presents a large area of micro – seismic effects along the NE-SW direction, covering almost 50% of the Romanian territory. It is worth mentioning the fact that the Romanian territory has a living continuity from more than 2000 years. The territory experienced the interference of cultures and civilisations, with differences between regions (Hărmănescu, Popa 2013), relief and climate, which demanded for various types of specific Romanian vernacular architecture.

This relation between ethnography and geography, the setting of vernacular material structures, and the seismic zoning map of the Romanian territory confirms this continuity, and adaptability of living in a seismic condition along history (Fig. 1).

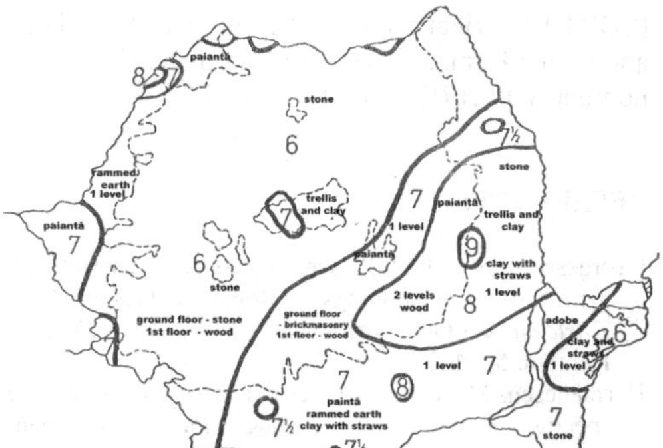

Figure 1. Correlation between vernacular architecture and seismic zoning map of Romania. Effective use of local material resources, adapting the constructive system to the seismic zones. (credits: E.S. Georgescu).

2 ROMANIAN VERNACULAR ARCHITECTURE STRUCTURE SYSTEM

In Romania, the main building material is wood, but different types related with the ethnographic area are used, either combined with stone and/or with earth. In high seismic zones (SE, V, centre and N of Romania), vernacular architecture is built on one floor level, and the constructive system consists of twigs and clay, beat earth or 'paiantă' (Fig. 2). The construction on two levels occurs on hill and mountain areas, less earthquake-prone areas but also near the epicentre. Here the constructions are also entirely of wood (Fig. 3–4). The studies of vulnerability analysis from 1838 and 1977's earthquakes release that vernacular architecture has an adequate structure, and it demonstrated that the height, stiffness, placing of the heavy and light materials are well correlated, in order to enable the inhabitants living in the seismicity zones.

a) b)

Figure 2. a) Siliştea Gumeşti, Teleorman. Vernacular building with visible patterns of the seismic culture of repair, b) Dobrogea. House in 'paiantă', stone foundation, adobe stuffing, minimal damages from 1977 earthquake (credits: M. Hărmănescu).

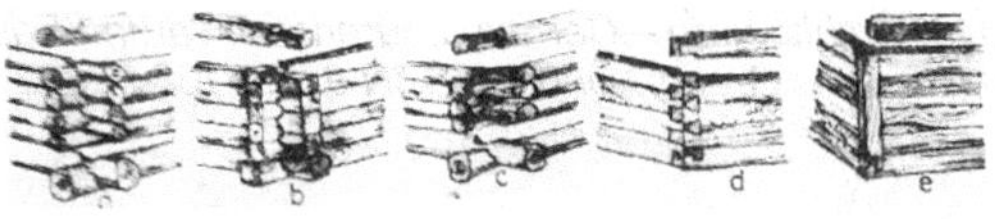

Figure 3. Details. Timber structures withstand earthquakes without collapse or major damage, only several non structural damage occur (credits: ICCPDC, 1989).

Figure 4. Năruja, Vrancea. Timber structure, house typology near to the epicentre area. The village suffered no damage in 1977. Village Museum, Bucharest (credits: E.S. Georgescu).

Figure 5. Details on the wall structure: timber frame and trills (credits: ICCPDC, 1989).

3 CONSTRUCTIVE ELEMENTS AND LOCAL SEISMIC CULTURE

There are architectural and constructive elements of the vernacular architecture that provide essential local seismic culture quality, according to Georgescu (1990): (a) reduced dimensions of Romanian rural buildings, and a relatively simple layout; (b) a relatively light roof, in 2–4 slopes, having a structure well tight with the masonry and the walls; (c) the floors are built with dense beams, tightly connected to the masonry walls; (d) the height of only 1–2 storeys, the upper floor being made of lighter materials (such as wood, trellis work) (Fig. 6); (e) the existence of a sufficient number of elements that take the horizontal loads (columns, masonry walls, wooden walls with corner joints, bracings, etc); (f) a relative symmetry of the building's layout and a symmetrical distribution

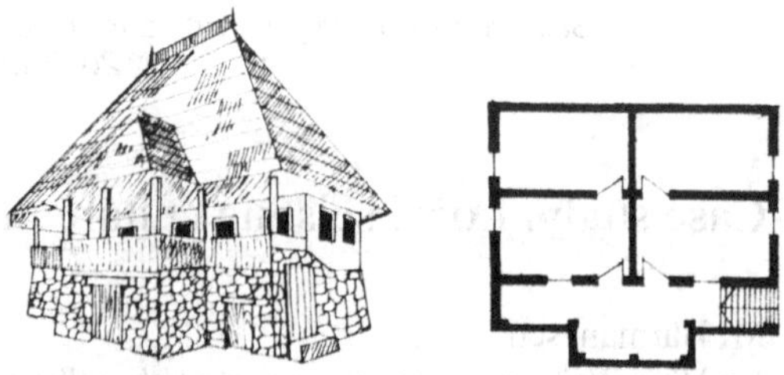

Figure 6. Stone and timber vernacular building. The light material is used for upper floor. A relative symmetry of the buildings layout, and a symmetrical distribution of the doors and windows openings (ICCPDC 1989) (credits: E.S. Georgescu).

of the doors and windows opening; (Fig. 6); (g) the presence of certain elements that provide the spatial interaction: wooden horizontal elements at corners in the earthen walls, etc) (Fig. 5); (h) the seldom use of gable walls; (i) the rigidity, specific to rural buildings made of earth and masonry, which under the earthquakes of long vibration period given by the Vrancea focus did not lead to resonance.

4 CONCLUSIONS

The main quality of Romanian vernacular architecture, due the intuitive construction methods based on the historical experience, is the safe evacuation of inhabitants, even in case of heavy damage of the earthquake (Georgescu, 1990). So, this proves a good behaviour, able to provide life safety, by using only local materials.

ACKNOWLEDGEMENTS

Special thanks are addressed to Prof. Mariana Correia, Director of CI-ESG, ESG Research Centre and her team, for all the support received for the post-doc research internship.

This paper is supported by the Sectorial Operational Programme Human Resources Development (SOP HRD), financed from the European Social Fund and by the Romanian Government under the contract number SOP HRD/159/1.5/S/136077.

REFERENCES

Georgescu, E.S. (1990). Present aspects of the aseismic protection of rural buildings in Romania. *Proceedings of Thirteenth regional seminar on earthquake engineering,* Istanbul, 5, 14–24.

Hărmănescu, M. & Popa, A. (2013). A new landscape perspective – human exercises through time in environmental perception. *Procedia – Social and Behavioral Sciences,* 92, 385–389. doi:10.1016/j.sbspro.2013.08.689

ICCPDC (1989). Village housing in Romania. Studies of traditional architecture for the conservation and valorization through typisation (in Romanian), Bucharest.

Seismic Retrofitting: Learning from Vernacular Architecture – Correia, Lourenço & Varum (Eds)
© 2015 Taylor & Francis Group, London, ISBN 978-1-138-02892-0

Case study: Local seismic culture in Taiwan vernacular architecture

Y.R. Chen
Department of Architecture, Cheng Shiu University, Kaohsiung, Taiwan

ABSTRACT: This text outlines the structures of Taiwan vernacular architecture, including Austronesian houses and Han houses. An A-shape compound column was used due to its good seismic retrofitting. However, it is rarely seen in Austronesian houses. In Han houses, the wall layout in two dimensions and the combination of wood truss and infill walls also proved to be effective seismic retrofitting elements.

1 VERNACULAR HOUSES OF AUSTRONESIAN ETHNIC GROUPS IN TAIWAN

Austronesians have lived in Taiwan for more than five thousand years. Most people of the existing sixteen Austronesian ethnic groups inhabit in modern houses. There are only very few vernacular houses standing now. These houses are one floor high, with simple wood truss supporting a thatch or slate roof.

The joints of the column and the beam adopt a rope binding or mortise. The walls, made of woven thatch, wood panels or stacked slates, are not combined with the wood truss (Fig. 1). Thus, they do not have any shape resistance against seismic force. Very few trusses combined with oblique columns form an A-shape compound column (Fig. 1) with good seismic resistance. Therefore, Austronesian vernacular houses in Taiwan do not have good capacities against seismic force in general. Fortunately, the load of the Austronesian house roof is light, so as not to cause serious damage.

2 VERNACULAR HOUSES OF HAN ETHNIC GROUP IN TAIWAN

Large amounts of Han people immigrated into Taiwan since the seventeenth century. Han people were used to build courtyard houses. The simplest house structure was built with several gable walls, covered with beams and a roof. Normal houses were built with Chuan-Dow structures; and Tai-Liang structures were adopted in temples, official buildings or luxurious houses (Fig. 2). Houses with`Chuan-Dow and Tai-Ling structure, whose joints were mortise, was always enclosed by thick wall.

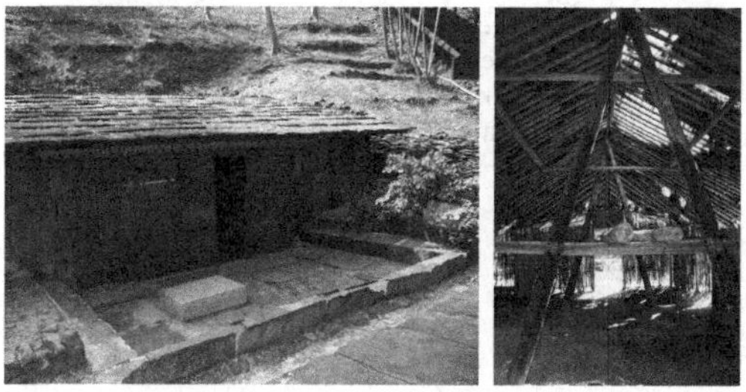

Figure 1. House of Bunun (left); House Construction of Tsou (credits: Y.R. Chen).

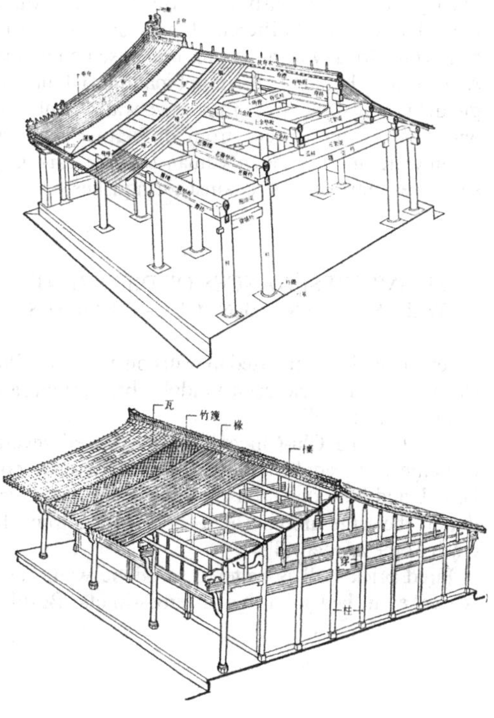

Figure 2. Tai-Liang Structural System (above), Chuang-Dow Structural System (below) (credits: H. C. Lin).

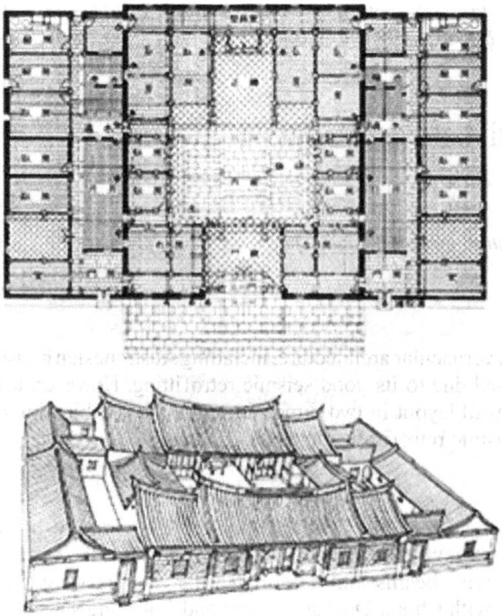

Figure 3. Han Courtyard House in Taiwan (credits: C.L. Lee).

3 WALL SYSTEMS IN HAN VERNACULAR HOUSES

The layouts of Han courtyard houses had many variations. Thick walls of different dimensions were easily and symmetrically arranged in the house compound. As shown in Figure 3, the combination of four longitudinal walls and two transverse walls is able to withstand seismic forces from different directions. Meanwhile, the trusses of Chuan-Dow or Tai-Liang structure contribute less against earthquakes.

4 SEISMIC RESISTANCES OF DIFFERENT WALLS IN HAN VERNACULAR HOUSES

Several materials were used in wall construction of the Han vernacular house, such as adobe, brick, stone, and bamboo-earth infill.

In 1999, the Chi-Chi earthquake caused several casualties. The adobe wall of the adobe houses collapsed, and bore down on the inhabitants, who died due to the large amount of dust, resulting from the crash of the adobe walls.

Fired brick wall is better than adobe wall, but it is weaker in bed joint and in cross-joints. Besides,

Figure 4. In-fill Walls with Chuang-Dow Structure, Brick Wall (above), Bamboo-Earth-Wall (below) (credits: H. C. Lin).

walls were usually made to fill the frames in Chuan-Dow structure. Brick wall and bamboo-earth-wall were often seen combined with Chuan-Dow structure, whose frame confined brick wall or bamboo-earth-wall, to form a larger unit than a single brick unit. As a result, they produce better lateral stiffness to resist seismic forces. The weak points are the openings, which comprise a lack of infill in the wall, therefore reducing the capacity of seismic force resistance.

REFERENCES

Huang, C.H. & Yang, S.H. (2012). The Collection of Measured Drawings of Taiwan Aboriginal Houses by Chijiiwa Suketaro, 1935–1944. *National Taipei University of Technology, Taiwan.*

Liu, D.J. (1982). *The History of Chinese Ancient Architecture*. Taiwan: Min-Wen Publishing.

Lee, C.L. (1980). *A History of Taiwan Architecture (1600–1945)*. Taiwan: Bei-Wu Publishing.

Lin, H.C. (1995). *Handbook of Traditional Architecture in Taiwan*. Taiwan: Artists Publishing.

Part 3: Local seismic culture in Portugal

Seismic Retrofitting: Learning from Vernacular Architecture – Correia, Lourenço & Varum (Eds)
© 2015 Taylor & Francis Group, London, ISBN 978-1-138-02892-0

Recognising local seismic culture in Portugal, the SEISMIC-V research

M.R. Correia & G.D. Carlos
CI-ESG, Escola Superior Gallaecia, Vila Nova de Cerveira, Portugal

ABSTRACT: The SEISMIC-V project was developed between 2013 and 2015. The project was funded by FCT – Foundation for Science and Technology and intended to fill the gap regarding the recognition of local seismic culture in Portugal. This paper presents the main outline of the research project, focusing essentially on its methodological approach and major outcomes. Structured under five progressive levels, the subsequent research identifies, characterises and analyses the vernacular architecture of the seismic most susceptible Portuguese regions. This paper unveils the relation between the conceptual research approach, and the expected contribution for the comprehension, strengthening and preservation of the main seismic-resistant architectural components and features of the Portuguese Vernacular Architecture.

1 INTRODUCTION

This paper presents the development and the main outcomes of the Research Project 'SEISMIC-V: Vernacular Seismic Culture in Portugal', funded by FCT *(Fundação da Ciência e Tecnologia)*, the Portuguese National Agency for R&D, (Project nºPTDC/ATP-AQI/3934/2012). The fundamental objectives were to identify the existence of a 'Local Seismic Culture' in Portugal. Also to identify and to characterise were the main seismic-resistant elements, as well as features and strategies applied to Portuguese Vernacular Heritage.

The project was integrated in the ESG/ Escola Superior Gallaecia main research lines, namely the one dedicated to Architecture and Heritage. The research was developed in accordance with other parallel research activities, such as the European Research Project 'VerSus: Lessons from Vernacular Architecture to Sustainable Development', and the International CIAV2013 annual conference. The project was submitted to the FCT program in the year 2012, under ESG's coordination, with the partnership of the Engineering Departments of the University of Aveiro and the University of Minho, and with the support of the Portuguese Ministry of Culture. SEISMIC-V had also the support of CIAV, ISCEAH and the European University Centre for Cultural Heritage (Ravello, Italy). The project counted with the scientific advices of the Professors Julio Vargas and Ferruccio Ferrigni, both SEISMIC-V consultants.

The project was approved in 2012, overall rated as excellent by the FCT evaluation panel and recommended for funding.

"The project is innovative, will bring new knowledge to the seismic response and mitigation of vernacular structures, and is carried out by a strong team." The FCT Evaluation Panel Statement and Rating report, 2012.

The project started in July 2013, and finished in September 2015, under the coordination of Professor Mariana R. Correia (PI) consolidating a set of studies and reflections, which origins dates back to the 'Taversism project', developed in 2001.

1.1 *The 'Taversism project': First Portuguese LSC overview*

The 'Local Seismic Culture' problematic was first acknowledged by Ferruccio Ferrigni at Centro Universitario Europeo per I Beni Culturali (CUEBC), located in Ravello, Italy. Ferruccio Ferrigni (project consultant) recognised, along with his team, the existence of a 'Local Seismic Culture' (LSC), consisting on the application of architectural elements with technical knowledge and comprehensible behaviour, following an efficient ensemble to reduce the impact of earthquakes. A systematic emphasis was then given to the study of historical architectural prevention and post-earthquake reaction. However, for several years it was mainly focused on the emergency reaction, and less on scientific international recognition of LSC research. The first funded project addressing this field area was the European Taversism Project (Correia, 2001).

Due to Portuguese earthquake history, the country was selected as case study. An expert on Portuguese vernacular architecture was appointed to address the research. As a result, a preliminary Report was produced as well as specific communications approaching the subject. However, a sound research on the identification of seismic resistant features in Portuguese vernacular architecture demonstrated clearly that the methodological tracing of a Portuguese Local Seismic Culture needed further development. The literature review on seismic resistant Portuguese architecture revealed that most of the studies have been focused either in the seismic resistant *Pombalino* construction, or in the architectural heritage (Mascarenhas, 1994) or urban housing (GECoRPA, 2000); but very little related to local seismic culture (Correia, 2001). The awareness and subsequent production of recommendations for reinforcement of in-use vernacular architecture can save lives, in case of an earthquake occurrence. It was obvious that this research problem needed to be methodologically addressed, in order to contribute to knowledge, and be incorporated into practice.

2 RESEARCH METHODOLOGY

The long-term goal of SEISMIC-V was to contribute to the awareness of LSC, but also to propose recommendations to reinforce existing solutions, and to avoid common errors. Thereby, it would be necessary to collect data, concerning the efforts taken by the population in the past, and to contribute to the restoration and repair of the buildings that sustained damage from the earthquake.

Regarding the general aims assessment, the project intends:

1. To identify local seismic culture and architectural seismic-resistant features, throughout the Portuguese regions exposed to frequent tremors, even with low intensity, and earthquakes with high intensity, but less frequent.
2. To identify materials and techniques to repair and restore damaged in-use vernacular buildings, related to local population reactive efforts to earthquake occurrence.
3. To identify actions addressed by the local community, on their attempt to prevent from earthquake damage.

The research project was divided into five progressive stages, corresponding to the main scientific activities:

a) Characterisation of the risk areas and main housing typologies;
b) Experimental characterization (in situ);
c) Numerical modelling and parametric studies;
d) Identification and description of the most efficient seismic resistant reinforcement solutions;
e) Recommendations for the awareness and strengthening of Local Seismic Culture.

2.1 *Characterization of the risk areas and main housing typologies*

It constitutes mainly in the definition of the areas of study, according to the seismic hazard. This activity was undertaken throughout survey missions and preliminary analysis.

This task assumed the analysis of the state of the art of Local Seismic Culture, and its adaptation to Portugal actual context, which was still an unexplored research field. The definition of the areas and their limits were executed under two premises: 1-According to the intensity and frequency of the register earthquakes, the impact of these events on the building environment, and the level of the community preparedness to the occurrence; 2-The implementation of Survey Missions to understand and to confirm the actual state of conservation and authenticity of in-use Vernacular Architecture. This was accomplished by the analyses of a uniform sample and the inclusion of isolated case studies with documental evidence for the extrapolation of solutions. The task was developed with the study of six specific regions, plus Lisbon downtown, which the preliminary analyses highlighted as striking reference for the Portuguese community.

Lisbon was assumed as an isolated reference, given the significance that the 1755's earthquake had on Portugal and the world. There is much information published on that earthquake, both from written accounts of that time, immediately following the earthquake, and all throughout the 20th century. After considering the information regarding the 6 proposed sites, the team developed a comprehensive understanding of the vernacular housing typologies, through the characterisation of their morphological and construction components, with special focus on prevention and reactive solutions.

This 1st phase of the research, coordinated by ESG, resulted in a Portuguese Vernacular Architecture Atlas of Local Seismic Culture, presented as a graphical tool for dissemination and awareness.

2.2 *Experimental characterisation, to study the materials and their application through benchmarking in paradigmatic cases*

Vernacular buildings are often complex structures, both in terms of structural behaviour, and the mechanical properties of their components and materials. Moreover, usually, specific provisions for seismic demands are not present in this type of construction. However, recent earthquakes all over the world show that, in many cases, vernacular constructions may show an adequate seismic performance. Therefore, the identification and understanding of construction solutions and the detailing that may enhance their seismic performance should be analysed, and such solutions should be studied for eventual application to existing structures (traditional masonry, rammed earth or adobe constructions), so as to reduce their vulnerability to earthquakes.

A comprehensive study of the geometry, structural solutions and constituent materials of vernacular buildings was fundamental to study how these parameters can influence their seismic performance, as well as to determine the more adequate repair and/or retrofitting solutions. Excessively intrusive assessment techniques should be avoided, in order to prevent these constructions from losing their "vernacular value".

In this task, existing vernacular buildings were selected as case-studies, analysing different situations in terms of construction materials (adobe, masonry, and timber), structural solutions (type of connections between wall-floor, wall-roof, etc.), building configurations and dimensions. The selection of the case studies considered the location and period of construction. In this way, the case studies would be representative of the existing vernacular construction in different regions of Portugal. Non-destructive tests were performed *in situ*, on the structures selected, aiming particularly at their dynamic properties characterisation. To this end, natural frequencies were measured, and modal shapes were identified.

The collected data was analysed and comparisons were established between the results obtained for the different case-studies, leading to the conclusions on the influence of the type of construction material, and the structural solutions adopted, in the dynamic behaviour of the structures. In particular, the results analysis allowed preliminary conclusions about the construction solutions in vernacular constructions that may improve their seismic performance. The results obtained within this task also contributed to a better definition and understating of the structural fragilities of the case studies assessed. Moreover, the results provided valuable information for the calibration of the numerical models and parametric studies, which were, subsequently, developed within Task 3. This task was coordinated by the University of Aveiro research team, under the supervision of Professor Humberto Varum.

2.3 *Numerical modelling and parametric studies*

As aforementioned, the project focuses on the use of vernacular architecture and its behaviour response in earthquake events. Different constructive systems are envisaged, namely buildings of the following building techniques, fired brick, adobe and stone masonry, wattle-and-daub and rammed earth. Besides, an important goal of the project is the design of retrofitting solutions for vernacular architecture buildings, in order for their mechanical behaviour to be improved under seismic actions, taking into account the particularities of each constructive system.

The sound understanding of the seismic behaviour (resisting and ductility features), regarding vernacular architecture buildings, and the proposal of retrofitting solutions should be backed by numerical modelling. Moreover, numerical modelling based on nonlinear static numerical analysis of ancient masonry buildings represents a step forward in the technical and scientific knowledge, as few results are available in literature. In this task, it is planned to use finite element modelling (FEM) for the global seismic analysis (walls, floors, spandrels, connections), by following the common macro-modelling approach, and considering masonry as an isotropic and homogeneous media. The mechanical nonlinear behaviour of masonry was based on advanced plastic constitutive models. The numerical simulation intends to: (1) calibrate the numerical models based on the *in situ* experimental testing (task 2); (2) understand, in a more thorough way, the resisting mechanisms of the different structural elements of the masonry buildings under seismic loading; (3) assess the influence of distinct factors on the seismic behaviour of masonry buildings, based on a parametric study; (4) evaluate distinct retrofitting techniques of ancient masonry buildings.

The calibration of the numerical model was possible by comparing the numerical results with the experimental results attained in Task 2, in terms of stress and strain fields, crack patterns, and nonlinear force-displacement diagrams. The numerical models were based on nonlinear static (pushover) analysis. In relation to pushover analysis, alongside with the conventional procedures in terms of applied loads (proportional to mass, and proportional to the first mode), it was also intended to follow a more advanced procedure, based on an adaptive procedure, taking into account the damage progress on the shape of the first mode of vibration. After the completion of calibration of numerical model, it was possible to carry out parametric analysis, aiming at identifying the influence of possible parameters on the seismic response masonry buildings.

Distinct parameters were considered in the parametric analysis, namely: (1) distinct loading conditions; (2) distinct geometric configurations (plan and side views, and variation of the openings size); (3) modelling of the proposed retrofitting techniques, and comparative analysis on the seismic performance of each one. The results obtained in this task were essential for Task 5, where retrofitting guidelines for ancient vernacular masonry buildings were provided. The main output of this task included: (1) experimental assessment of the global seismic behaviour of old masonry buildings; (2) valuation of the main parameters influencing the seismic response; (3) assessment of the seismic performance of distinct retrofitting techniques. Given the large expertise in the numerical analysis of masonry structures, the task was led by ISISE (UMinho), under supervision of Professors Paulo B. Lourenço and Graça Vasconcelos.

2.4 *Identification and description of the most efficient seismic resistant reinforcement solutions*

An important part of the research was dedicated to the description of the most efficient seismic-resistant strengthening solutions, as well as the most frequent faults and errors found in implemented solutions. Since Vernacular Architecture is based on empiric knowledge transference, sometimes the origin of the solution is lost, and its principles are perverted along the generations. According to Task 2 and Task 3 it was possible to identify and to confirm, which solution presented a satisfactory improvement in the resistance of the construction, which were incipient, or even which of them were deceivingly harmful to the Seismic resistance of the buildings.

This task was of an overall importance for the building culture of the considered sites, as the gap between master builders and current repair professionals is huge, and sometimes impossible to fill in. First of all, there was a dissection between the preventive or reactive seismic retrofitting nature of each selected solution. To achieve this, it was essential understanding the cause that originated the solution, the real effect of it and, most of all, the expected result.

Considering the characterisation method, the solutions were systemised by sort of reinforcement, and then classified under the respective construction system. This would be more assertive in terms of material characterisation, and before constructive techniques comparison (26). This task furnished the possibility to classify the observed solution, under determined and specific typologies (timber wall trusses, half-buried floors, vaulted ceilings, counter arches, buttresses, Tie-rods, etc.), foundation grim systems, or even to recognise original ones (regarding the Vernacular Architecture flexibility to adapt to very particular contexts). There was an attempt to correspond solution types to the identified risk areas, addressed in the VSC Atlas.

2.5 *Recommendations for the awareness and strengthening of Local Seismic Culture*

The final task consisted in the Systematisation of all the collected and produced data, concerning the analysed solutions. It also provided findings, emerging from buildings performance under earthquakes, and based on simulation variables, in order to reinforce solutions for in-use vernacular architecture, regarding the seismic resistant features.

The corresponding information is summarised in the section related to the Project's Results, which also comprises the fundamental content of the present publication. The articles of section 6 can be considered as the synthesis of this research conclusion.

3 REGION SELECTION AND CHARACTERISATION

The study areas were elected according to their history of seismic activity. The case studies were chosen, when considering the seismic resistance applied elements; the existent typological characteristics; the vernacular morphology, and the current preservation of these strategies. The fact is that population, sometimes, had a preventive response, based on their past recurrent experiences. In other cases, the population just had reactive responses to the seismic event, especially if there had not been frequent past seismic occurrences. On the other hand, local builders are the true leaders of the vernacular architecture efforts in their villages and regions. Therefore, it is the vernacular architecture, which reveals, more clearly, the existence of a Local Seismic Culture.

Conclusively, it is assumed that a portion of the Local Seismic Culture, including both, the preventive or reactive response, is influenced either by the intensity of the earthquakes, or by the frequency of earthquakes in the region; or both.

The process of region selection was based on the following criteria, considering Ferrigni's (1990) relevance to the LSC definition: 1) the impact per region, of the earthquakes of higher intensity, though less frequent; 2) the frequency of the earthquakes of low and moderate intensity; and 3) the quantity and consistency of the observed retrofitting structural solutions.

The buildings that had three or more identified reinforced techniques were considered as having reactive measures applied. When a group of a minimum of three buildings shared common features, then Local Seismic Culture was identified in the region.

Based on the materials and the techniques to repair and to retrofit damage of in-use vernacular buildings, related to local population reactive or preventive efforts to earthquake occurrence, specific case studies were selected for a deeper research.

Reference case – The Lisbon downtown

The 1755's earthquake in Portugal is an obvious reference, given the extensive damage across the country, with emphasis on the town of Lisbon, the national capital. There was significant published information on the earthquake damage, from written observation immediately following the earthquake, as the Marquis of *Pombal* requested a damage status from all the parishes in the country. Thus, the 1755 earthquake, and all the planning addressed in Lisbon, in terms of rebuilding the city's downtown, as well as the damage status requested throughout the country, emerged as a basis of reference for the development of the project. The reconstruction process of the Lisbon downtown set a new conceptual and technological seismic approach in Portugal. The applied strategies and techniques influenced many of the new coming constructions, especially the ones located at seismic prone areas. With more or less complexity, many local building cultures appropriated some of the '*pombalino*' solutions, adjusting them to their regional conditions. The '*gaioleiro*' (cage) system is perhaps the greatest manifestation of its influence in the Portuguese vernacular heritage.

Region 1 (R1) – Tagus Lezíria

The Tagus *Lezíria* region is situated in a high intensity seismic area (Fig. 1). The region is also characterised by the occurrence of relevant earthquakes, as it is located near the fault line of '*Vale Inferior do Tejo* (VIT)'.

The village of *Benavente* was selected as a case study, especially due to the fact that there had occurred three major earthquakes, 1531, 1909 and 1914, of which the epicentres were located near this village. The earthquake of 1909 had important consequences in *Benavente*. According to local bibliographic data, the earthquake left destruction, or in partial ruin, a significant part of the village buildings (Vieira, 2009). Some years later, the village of *Benavente,* was reconstructed, but in general, there was new housing. However, some of the reconstructed houses integrated seismic resistant techniques, such as, the *Pombalino* system, as well as symmetrical patterns (Vieira, 2009). Horizontal reinforcement on the houses was also observed. In regards to the preventive and reactive seismic reaction, *Benavente* is based on the use of the two approaches: a) The reactive reaction, because there was no previous concern, before the 1909's earthquake; b) The preventive reaction, as elements of seismic resistance were incorporated in some of the new construction.

Region 2 (R2) – Coastal Alentejo

The Coastal Alentejo region is a high intensity seismic area, as it can be observed in Fig. 1.

After some investigation and fieldwork, *Alcácer do Sal* was selected as a case study. The village is characterised as an area of high intensity and regular seismic activity. In the region, old buildings have been found with *pombalino* walls, (Correia & Merten, 2001). Horizontal reinforcement of the houses and the use of buttresses and tie-rods were also observed.

The earthquake of 1858 was one of the most important ones that affected the Portuguese mainland. The area was severely affected, and the village of *Melides* was partially destroyed. Following field missions, the village of *Melides* was also selected as a case study, due to historic and regular seismic activity, but also due to the identification seismic reinforcement in buildings. A constructive culture is based on the improvement of material properties of the constructive systems. As a reaction to the effects of earthquake, the population presented reactive solutions through the use of structural seismic resistant features, such as horizontal reinforced walls, buttresses, tie-rods, and reinforced plinth course. embasamento saliente? N é este o termo, é: reinforced plinth course.

Region 3 (R3) – Central Alentejo

The Central Alentejo region never suffered from the consequences of a severe earthquake, as it can be observed in Table 5. However, the region has been exposed to numerous earthquakes of medium intensity that produced minor damage to buildings, and created memories of fear and panic in the population.

During the initial fieldwork, it was possible to identify evidence of seismic prevention, through the placement of seismic resistant elements, such as, counter arches, in specific buildings. Therefore, the *Évora Historical Center* was also selected as a case study, due to several evidences, especially, the use of seismic resistant features, as a reaction to the various events. The historical centre has implicit evidence of strengthened buildings, through the use of counter arches, buttresses and wall reinforcements.

Region 4 (R4) – Lower Alentejo

The historical seismicity on the Lower Alentejo region can be characterised as frequent, but of medium intensity, as it can be seen in Table 6, with the intensities of earthquakes growing during the twentieth century. This region is characterised by the use of preventive response. There was a probable reaction to various events, and an implicit need for strengthening of the buildings. Along the lower Alentejo, it was observed, during several field missions, the use, in several houses, of isolated elements of reinforcement, such as stiffeners, buttresses and reinforced plinth courses. The villages of S.Brissos, Baleizão e Trindade, near the municipality of Beja, constitute good examples of such approach.

Region 5 (R5) – The Algarve

The Algarve is illustrated by a strong historical seismicity with earthquakes that caused major damage. The 1719's earthquake in the *Portimão* area had a maximum intensity of IX; the 1722's earthquake on the coast of *Tavira* had a maximum intensity of X; and the 1856's earthquake in *Loulé* had a maximum intensity of VIII.

In the survey missions carried out in the Algarve region, several reinforcement techniques were identified on the vernacular architecture, such as buttresses, tie-rods and *pombalino* walls. *Lagos*, *Tavira* and *Vila Real de Santo António* emerged as relevant case studies, due to consistent architectural evidence, although they represent completely different approaches to the earthquake hazards.

Region 6 (R6) – The Azores

Based on the literature review (historic seismic occurrences and damages), and a mission to the Azores Islands, the islands of *Terceira* and Faial emerged as case studies. The islands are located in a complex area with an intense volcanic activity, near the boundary of 3 tectonic plates, referred to as the *Azores* triple junction (Nunes et al. 2004), *Terceira* is in a high intensity seismic area with numerous earthquakes (Table 2). The 1980's earthquake, according to the collected data, had extensive destructive consequences to communities throughout the island, but particularly in *Angra do Heroismo*.

4 OUTCOMES

For a better dissemination of the results, based on past research projects experience, as TERRA (IN)CÓGNITA II (Funded by the European Commission-Bando 2009) or CADIVAFOR (funded by FEDER: INTERREG III-A), the final outcomes were presented through nine milestones, divided into 4 different dissemination categories:

4.1 *Conferences*

a) International Conference CIAV 2013 | 7°ATP | VerSus – One of the themes of the conference was directly related with the literature review, and the data collection of the project
b) Closure conference of the SEISMIC-V project, presenting the final results, providing lectures of the research team, the project consultants and involved institutions representatives.

4.2 *Scientific dissemination*

a) *Scientific Publication* – CRC/ Balkema/Taylor & Francis publisher
b) *Papers in international conferences* – 8 papers: CIAV2013, 14th SIACOT, VerSus 2014, TERRA2016, among several others.
c) *Papers in national conferences* – 3 papers.
d) *Posters in Conferences* – Several posters were developed throughout the project, to reach different disciplinary areas audiences. Posters were created to: CIAV2013 in Vila Nova de Cerveira, 14th SIACOT in El Salvador, 9th IMS – 9th International Masonry Conference in Guimarães, and for the Seminar 'Paredes de Alvenaria 2015: Reabilitação e Inovação' in Lisbon.

4.3 *Reaching local technical agents*

Publication of the booklet 'Local Seismic Culture in Portugal' – The publication was issued in Portuguese and in English. It is a graphic booklet that was distributed among the 308 Portuguese municipalities. Two copies will be offered to each municipality: one for the Municipal Library, and another for the Technical agents that work at the Municipal Architecture and Engineering offices.

Laboratory seminars – the partners of the project organised these seminars in Aveiro and Guimarães. All the technical agents from the Municipalities were invited to attend the seminar, in order to receive a specific training on the subject area.

4.4 *Reaching the general public*

a) *Website* – The website, in English and Portuguese, presents the main results of the SEISMIC-V project and will enable the free download of the bilingual booklet: '*Local Seismic Culture in Portugal*'.
b) *Flyers* – These simple leaflets were developed at the beginning of the project, and were offered to the 300 participants of CIAC2013 conference. The flyers were also distributed at the Order of the Architects in Porto, and in Lisbon. The flyer is available online, for free download.

5 FINAL COMMENTS

Seismic-V was a challenging project that demanded a robust teamwork among the partners, aiming for high-standards concerning the findings and the project results. One of the main aims was to reach different publics, nationally and internationally.

All the project aims were accomplished and the dissemination of results went beyond the expected. This was possible due to the booklet dissemination, among all the Portuguese municipalities; due to free Internet access to download the booklet in Internet; but also due to this publication that reached different disciplinary publics.

ACKNOWLEDGMENT

The authors gratefully acknowledge the support by the Portuguese Science and Technology Foundation (FCT) to the research project 'SEISMIC-V – Vernacular Seismic Culture in Portugal' (PTDC/ATP-AQI/ 3934/2012).

REFERENCES

Carlos, G. D., Correia, M., Viana, D., & Gomes, F. (2014). Vernacular morphology as a preventive solution of local seismic culture. In C. Mileto, F. Vegas, L. García Soriano, & V. Cristini (Eds). *Vernacular Architecture: Towards a Sustainable Future* (pp. 267–272). VerSus 2014 | 2° MEDITERRA | 2° ResTAPIA. London, UK: CRC Press/Balkema/Taylor & Francis Group.

Correia, M., Carlos, G., Rocha, S., Lourenço, P.B., Vasconçelos, G., & Varum, H. (2014) Vernacular Seismic Culture in Portugal. In M. Correia, G. Carlos, & S. Sousa, (Eds). Vernacular Heritage and Earthen Architecture: Contribution to Sustainable Development. Proceedings of CIAV 2013 | 7°ATP | VerSus. London (UK): CRC Press/Balkema/Taylor & Francis Group.

Correia, M., Carlos, G. D., Viana, D., & Gomes, F. (2014). Vernacular seismic culture in Portugal: Ongoing research. In C. Mileto, F. Vegas, L. García Soriano, & V. Cristini (Eds). Vernacular Architecture: Towards a Sustainable Future (pp. 217–224). VerSus 2014 | 2° MEDITERRA | 2° ResTAPIA. London, UK: CRC Press/Balkema/Taylor & Francis Group.

Correia, M., Gomes, F., & Carlos, G. D. (2014). Projeto de Investigação Seismic-V: Reconhecimento da cultura sísmica local em Portugal [Seismic-V Research Project: Recognition of the local seismic culture in Portugal]. In M. Correia, C. Neves, & R. D. Núñez (Eds). Arquitectura de Tierra: Patrimonio y sustentabilidad en regiones sísmicas (pp. 230–238). 14° SIACOT – 14° Seminário Iberoamericano de Arquitectura y Construcción con Tierra. San Salvador, El Salvador: FUNDASAL/PROTERRA.

Correia, M. (2005). Metodología Desarrollada para la Identificación en Portugal de Arquitectura Local Sismo Resistente. In SismoAdobe2005: Seminário Internacional de Arquitectura, Construcción y Conservación de Edificaciones de Tierra

Correia, M. (2001). Report of the Local Seismic Culture in Portugal. In Taversism Project—Atlas of Local

Seismic Cultures. Ravello: (Italy): EUCCH— European University Centre for Cultural Heritage.

Felix, L., Correia, M., Vasconcelos, G. & Feio, A. (2014). 'Preservation of Vernacular Housing: Importance and Valorization'. In REHAB 2014, International Conference on Preservation, Maintenance and Rehabilitation of Historical Buildings and Structures. Green Lines Institute for Sustainable Development, Tomar, Portugal, 19th – 21h March 2014.

GECoRPA (2000). Sismos e Património Arquitectónico – Quando a terra voltar a tremer. In Revista Pedra & Cal; n°8; Out.-Dez. 2000. Lisboa: GECoRPA.

Gomes, F., Correia, M., Carlos, G. D., & Viana, D. (2014). Local Seismic Culture in Portugal: Melides dwellings, a reactive approach case study. In C. Mileto, F. Vegas, L. García Soriano, & V. Cristini (Eds). *Vernacular Architecture: Towards a Sustainable Future* (pp. 339–342). VerSus 2014 | 2° MEDITERRA | 2° ResTAPIA. London, UK: CRC Press/Balkema/Taylor & Francis Group.

Lourenço, P. B., Mendes, N., Ramos, L. F. & Oliveira, D. V., (2011). Analysis of masonry structures without box behavior, In International Journal of Architectural Heritage, 5 (1), 369–382.

Mascarenhas, J. (1994). Baixa Pombalina, Algumas Inovações Técnicas. In 2° Encontro Sobre Conservação e Reabilitação de Edifícios. Vol. II. Lisboa: Laboratório Nacional de Engenharia Civil.

Ortega, J.; Vasconcelos, G. & Correia, M. (2014). 'An overview of seismic strengthening techniques traditionally applied in vernacular architecture'. In 9th IMS – 9th International Masonry Conference. University of Minho and IMS-International Masonry Society, Guimarães, Portugal, 7th – 9th July 2014.

Sadeghi, N., Oliveira, D., Correia, M., Bondarabadi, H., Orduña, A. (2014) "Seismic Assessment of Adobe Vaulted Architectures in Iran". In SEE7 Conference. Risk Management Research Center at the International Institute of Earthquake Engineering and Seismology (IIEES), Tehran, Iran.

Vasconcelos, G., & Lourenço, P. B., (2009). In-plane experimental behaviour of stone masonry walls under cyclic loading, Journal of Structural Engineering (ASCE), 135(10): 1269–1277.

Seismic Retrofitting: Learning from Vernacular Architecture – Correia, Lourenço & Varum (Eds)
© 2015 Taylor & Francis Group, London, ISBN 978-1-138-02892-0

Seismic hazard analysis: An overview

J.F.B.D. Fonseca & S.P. Vilanova
CERENA, Instituto Superior Técnico, Lisbon, Portugal

ABSTRACT: Seismic hazard assessment is a pre-requisite to the mitigation of seismic risk through earthquake-resistant building techniques. While such assessment may be qualitative – i.e., part of the seismic culture of a community – modern building codes have at their core a thorough quantitative analysis of seismic hazard. Mapping the hazard is also critical for sound land-use planning. In this paper, the concepts and terminology used in seismic hazard analysis are introduced, distinguishing between deterministic and probabilistic hazard. Recent results at the European scale are presented, and a detailed analysis for Portugal is also discussed. The Plate Tectonics rationale behind the global distribution of seismicity and seismic hazard is introduced.

1 INTRODUCTION

Exposure to natural hazards is inherent to the human condition: irrespective of their degree of technical sophistication, humans interact with Nature to harvest useful resources, and in the process they become exposed to adverse natural phenomena for which complex sets of coping adjustments are introduced (Kates, 1971). Earthquakes rank high among natural disasters in terms of social aversion, because *terra firma* has been perceived throughout the times as a safe haven for the risks of the seas (and, more recently, air travel), and ground shaking has the symbolic value of the ultimate subversion.

Earthquakes cannot – at least at the current level of scientific development – be predicted or averted, and therefore the coping strategies for earthquake risk are focused on the reduction of vulnerability and on the increase of community resilience. While in traditional societies the supernatural played an important role in providing mechanisms of absorption well represented in the mythologies of ancient or contemporary cultures, modern responses favour instead the mutualisation of loss (insurance), effective emergency response based on early-warning systems, land-use planning and, above all, good building practices. The latter component is of paramount importance since it is the collapse of built structures, not ground shaking, that kills people during earthquakes. At the basis of any sound implementation of earthquake-resistant building techniques is a quantitative assessment of seismic hazard (Reiter, 1991).

2 BASIC CONCEPTS AND TERMINOLOGY

The term "hazard" can be used to refer to the adverse phenomenon itself: in the context of natural risks, a flood, the lava flow following a volcanic eruption, or rock falls from a nearby steep slope can be described as hazards. More technically, **hazard** is a measure of the likely intensity of an adverse phenomenon at a particular location. In what concerns earthquakes, hazard is often communicated as the intensity of ground motion that has a given probability of being exceeded during a specified time interval.

The key choice in describing the hazard concerns the physical measure of ground motion intensity. Usual measures of ground motion are the peak ground acceleration (PGA), peak ground velocity (PGV), or the spectral accelerations at different frequencies, SA(f). In older hazard studies it is also common to find the macroseismic intensity, often expressed in the Modified Mercalli Intensity (MMI) scale.

Spectral accelerations are particularly convenient to quantify the hazard, because they are frequency dependent, and a built structure responds differently to different frequencies of ground motion. Closely related to the notion of spectral acceleration is the concept of response spectrum. While ground motion can be suitably described through a times series of accelerations, velocities or displacements – which are measurable physical quantities – the response spectrum tries to portray the way in which a generalized structure responds to the ground motion. The simplest structure that can be considered is a harmonic oscillator (a mass attached to an ideal spring, see Figure 1A), which has a single natural (or resonance) frequency at which the amplitude of vibration reaches infinity. If the harmonic oscillator is damped (in a real structure, this is the result of internal friction) the amplitude of vibration remains finite, but the transfer of energy from the ground to the oscillator is maximum at the natural frequency (Figure 1B).

Because a structure has in general multiple natural frequencies, the response spectrum (for a particular

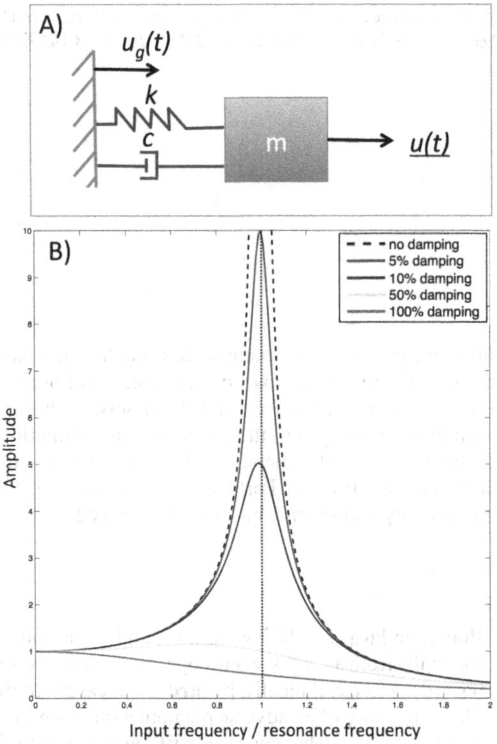

Figure 1. A) Schematic representation of the driven damped harmonic oscillator where *m* represents the mass, *k* the elastic constant, *c* the damping constant, $u_g(t)$ the ground motion, and $u(t)$ the motion of the mass relative to the ground. B) Amplitude of motion versus the input frequency of the driven damped harmonic oscillator. The resonant effect increases with decreasing damping values.

ground motion time series) is computed analysing the response of a suite of damped harmonic oscillators with different natural frequencies. This is done in the following way: at each frequency, the motion of the corresponding harmonic oscillator (typically with 5% damping) in response to the ground motion is estimated and the maximum displacement is noted; scanning through the frequencies of interest and repeating the procedure, a sequence of maximum displacements is obtained; spectral (i.e., frequency dependent) accelerations SA(f) may be approximated mathematically from the maximum displacements if damping values are low. When derived in this indirect way, spectral accelerations are called pseudo-spectral accelerations, PSA(f). The ensemble of the spectral accelerations as a function of frequency is the **response spectrum**. Since it portrays the response of a continuous suite of damped harmonic oscillators, it helps characterizing the response of a complex structure with multiple natural frequencies. Figure 2 depicts the acceleration response spectrum with 5% damping for a seismographic record (the 28 February 1969 M7.8 St. Vincent earthquake).

Besides the ground motion descriptor, a probability of exceedance and a time interval of exposure need

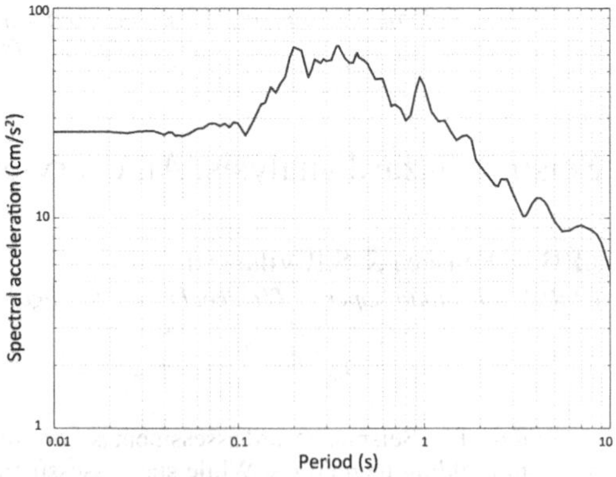

Figure 2. Acceleration response spectrum with 5% damping for the averaged horizontal components of the 28 February 1969 M7.8 St. Vincent earthquake, recorded in Lisbon, 290 km away from the epicentre.

also to be specified. Typical values are 10% probability and 50 years of exposure. If the Poissonian statistical model of earthquake occurrence is assumed (a common practice in hazard assessment), the pair probability of exceedance/exposure interval can be replaced by a single value: the **return period**. For 10% and 50 years, the corresponding return period is 475 years. It must be stressed that the return period is only a statistical concept, and does not imply any periodicity of occurrence. A site can be said to have a seismic hazard of 2.5 ms^{-2} peak ground acceleration at the return period of 475 years, meaning that there is a probability of 10% that the peak acceleration of the ground motion exceeds 2.5 ms^{-2} at least once on any interval of 50 years. Implicit in the latter formulation is the assumption that seismic hazard is stationary in time, which results from the adoption of the Poissonian model of occurrence.

Because the nature of the soil cover strongly influences the ground motion observed at the surface, most studies of seismic hazard are conducted for rock conditions, and a correction factor is applied to take into account the characteristics of the upper sedimentary layers. In general, loose sediments tend to amplify the ground motion, increasing the hazard.

3 SEISMIC HAZARD ANAYSIS

Two basic approaches can be followed when estimating the seismic hazard at a site: i) the **probabilistic analysis**, whereby all possible seismic occurrences (within a given radius from the site) are considered; and ii) the **deterministic analysis**, which considers only one scenario of occurrence, usually the worst-case scenario. The main difference is that the probabilistic approach treats the different occurrences as possible and assigns them probabilities, whereas the deterministic analysis considers the event as certain (hence the name) and computes the resulting ground motion at the site. It should be noted that, given the uncertainties

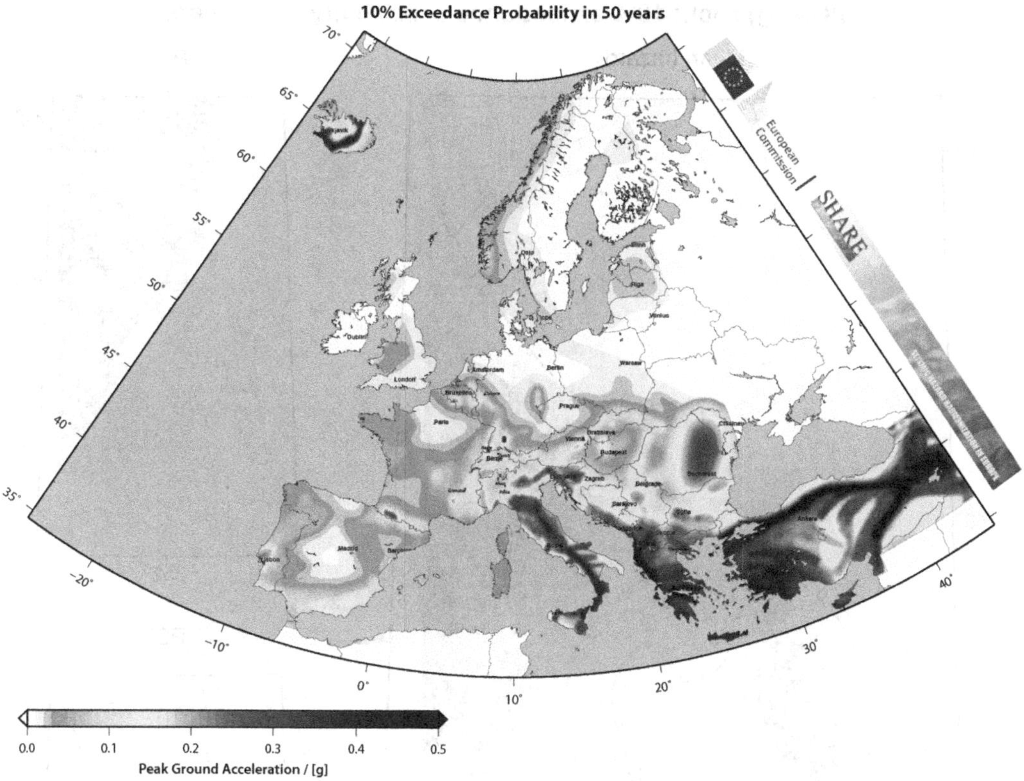

Figure 3. Mean seismic hazard map of Europe for 10% exceedance probability in 50 years (return period of 475 years), expressed in PGA (credits: Giardini et al., 2014).

in the computation of ground motion, the result of a deterministic hazard assessment is also a distribution of probabilities for different ground motion values. Deterministic hazard analysis is typically conducted for very sensitive infrastructures such as nuclear power plants or dams, and the scenario may be the full rupture of a nearby geological fault. In short, probabilistic hazard analysis integrates all possible scenarios at a given return period, while deterministic hazard analysis considers a single scenario, at an infinite return period. After a probabilistic analysis is completed, the resulting hazard can be "disaggregated", identifying the leading scenarios and estimating the partial contribution of each one to the overall hazard at the site.

The standard technique for probabilistic seismic hazard analysis (PSHA) was proposed by A. Cornell in 1968, and has remained in use until our days through different computational implementations. To compute PSHA at a site, the required inputs are:

- A seismic source model, which specifies the location, size and occurrence rate of relevant earthquakes (i.e., earthquakes that produce relevant ground motion at the site under analysis) Both seismicity catalogues and geological data are usually used to define the seismic source model.
- A ground motion prediction equation (GMPE), characterizing the attenuation of ground motion with distance for the different magnitudes;

The GMPE, which "predicts" the ground motion resulting at the site for each value of magnitude and distance to the source, is associated with a significant (aleatory) uncertainty that must be taken into account during the calculations. In addition, it is common to adopt a "logic tree" approach, combining multiple seismic source models, different GMPE's, etc., thus producing a collection of different hazard values for a given site, whose dispersion captures the (epistemic) uncertainty inherent to the analysis. The adopted result is usually the mean or the median of the distribution of values. A thorough discussion on methodological procedures, documentations, and the use of experts, can be found in the SSHAC (1997) and NUREG (2012) reports.

Seismic hazard can be conveniently mapped by specifying a return period (or exceedance probability and exposure time interval) and computing on a grid of locations the corresponding ground motion (expressed as PGA, PGV, SA at a given frequency, etc.). Figure 3 shows the seismic hazard map of Europe for a return period of 475 years, expressed in PGA according to Giardini et al. (2014). It is a common practice

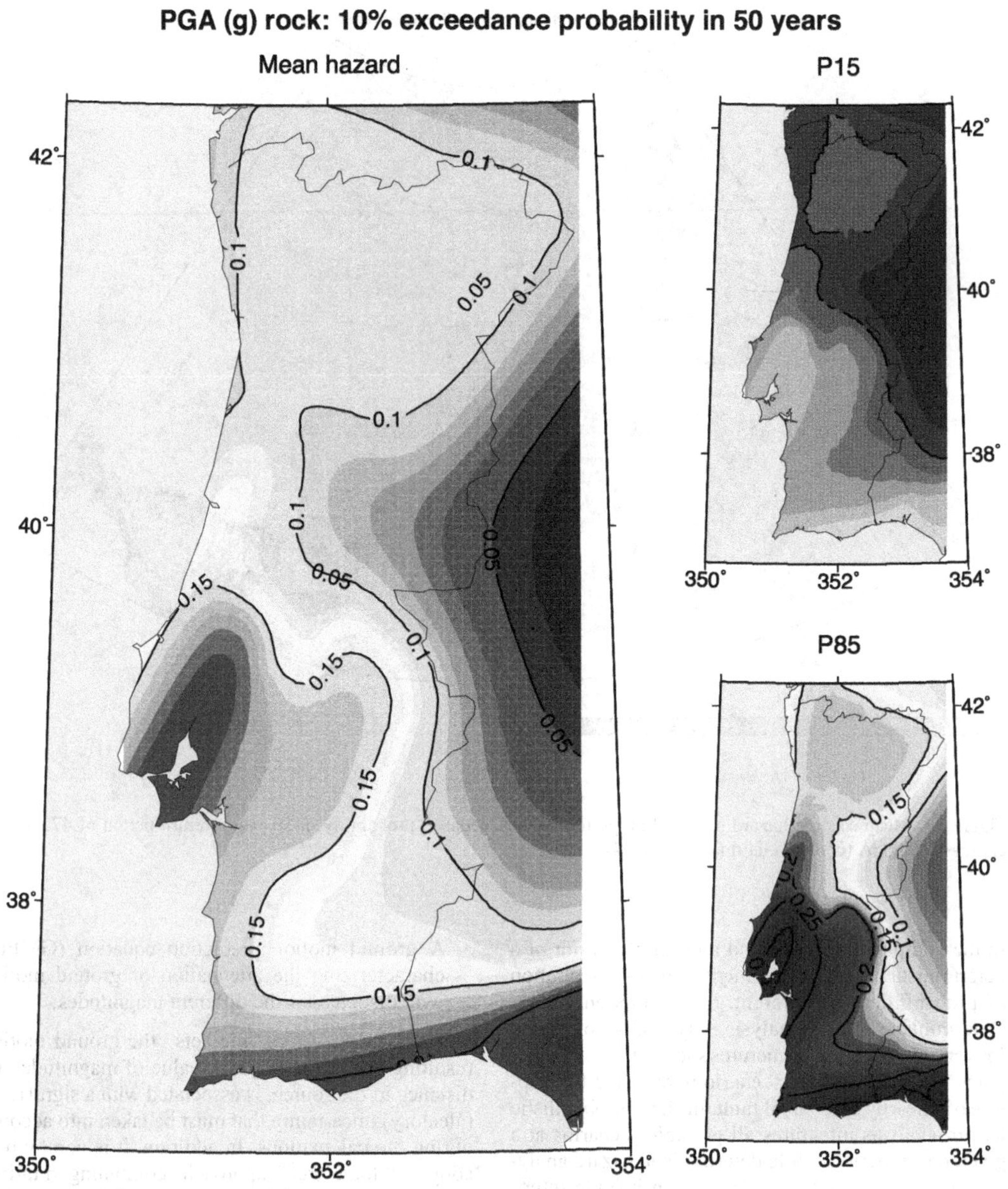

Figure 4. Mean seismic hazard map for Portugal together with the 15th and 85th percentiles for 10% exceedance probability in 50 years (return period of 475 years) expressed in PGA (Vilanova and Fonseca, 2007).

to give ground accelerations in units of g, whereby $1\ g = 9.8\ ms^{-2}$, the acceleration of gravity.

4 GLOBAL DISTRIBUTION OF SEISMIC HAZARD

Earthquakes are not randomly distributed in space. The Earth' crust is composed of large tectonic plates that move slowly on top of the molten rocks that lay underneath, and the earthquakes concentrate in the boundaries were the plates collide, spread or slide against each other. Plate tectonics, derived in the 1960's and now observed directly through satellite geodesy, provides a rationale for most of the planet's seismicity. The strongest earthquakes (magnitudes 8.5 and above) occur at subduction zones, where oceanic plates dive underneath less dense continental crust. This process is taking place around the Pacific Ocean, leading to strong earthquakes in Japan, Chile and Alaska, for instance. In the Indian Ocean, Java and Sumatra (Indonesia) are also the locus of strong subduction earthquakes. Regions of continent-continent collision are also prone to large earthquakes, typically in the magnitude range 7–8.5. These plate boundaries are broader, with the deformation spreading over a belt that can reach ~1000 km in width. Examples are the continental collision of India with the Eurasian plate. Transcurrent plate boundaries (transform faults) where tectonic plates slide past each other, such as the San Andreas Fault or the North Anatolian Fault, generate maximum earthquakes in the range 7–8.

All plate boundary types display high seismic activity rates that are associated with high levels of seismic hazard. The interiors of the plates, which have significantly lower seismicity rates, display relatively lower levels of seismic hazard.

5 SEISMIC HAZARD OF PORTUGAL

Quantitative seismic hazard analysis for Portugal was initiated in the 1980's, in association with the evolution of the building code (LNEC, 1983). These early attempts were strongly scenario-based approaches (although not formally deterministic), stipulating two ground motions for each site: a "distant-source" ground motion, somewhat related to the scenario of repetition of the 1755 earthquake, and a "near-source" ground motion, which can be traced to the scenario of repetition of the Benavente earthquake of 1909 (Choffat and Bensaude, 1912). The country was divided in zones of uniform hazard, and for each zone two values of PGA were estimated, one for the distant source and the other for the near source. At each site, both scenarios were considered and the strongest ground motion was adopted for design purposes. A response spectrum "shape" was specified in the building code, which was scaled up or down to fit the PGA value (which, by definition, must coincide with the spectral acceleration in the high-frequency limit). With small adjustments (e.g., two different zonations for the different "near-source" and "distant-source" scenarios) have remained in force in the Portuguese building code until the present version, the Portuguese National Annex to Eurocode 8 (IPQ, 2010).

In parallel with the seismic zonation for regulatory purposes, more PSHA assessments have been conducted in the last decades. Vilanova and Fonseca (2007) mapped the distribution of PGA for Portugal at 475 years of return period using a logic tree approach to address the epistemic uncertainties. Their work includes a thorough revision of the seismic catalogue and a new homogenized magnitude parameterization for most earthquakes. On the other hand Sousa and Costa (2009) presented PSHA maps for different return periods using macroseismic intensity as the ground motion parameter. Besides using different ground motion descriptors, the studies diverge also on the approach regarding the choice of GMPE's. While Vilanova and Fonseca (2007) used a combination of published GMPE's developed from robust datasets for regions within similar tectonics environments, Sousa and Costa (2009) derived GMPE's from the limited dataset of macroseismic intensities available for Portugal. The merits and demerits of each approach are discussed in Fonseca and Vilanova (2011) and Costa and Sousa (2011).

Figure 4 depicts the results of Vilanova and Fonseca (2007) for PGA for a return period of 475 years. The map produced by Costa and Sousa (2009) shows maximum ground-motion values at the extreme southwest of Portugal, a result that is in agreement with the zonation scenarios adopted in the building code. The results of Vilanova and Fonseca (2007) are however consistent with other recent studies, developed both at national (Silva et al., 2015) and European scales (Giardini et al., 2014).

6 SYNTESIS AND CONCLUSIONS

Earthquakes are highly disruptive phenomena that societies have to cope with. Although earthquakes cannot be deterministically predicted neither in time nor size, their devastating effects may be significantly minimized through the adoption of suitable land-use planning and adequate building techniques. Seismic hazard analysis is the discipline that bridges the seismological and engineering communities. It faces the important challenge of conveying the ultimate research in earthquake science to the society. According to the best international practices, the level of earthquake resistance required by building codes should be based on a robust and suitably documented PSHA assessment, which properly addresses the uncertainties involved.

ACKNOWLEDGEMENTS

The authors acknowledge funding support from the Portuguese Foundation for Science and Technology (FCT), through Excellence Research Line project SEICHE (EXCL/GEO-FIQ/0411/2012) and research position Investigador FCT IF/01561/2014 (S.P.V.). CERENA research unit is funded by FCT through strategic project UID/ECI/04028/2013.

REFERENCES

Cornell C.A. (1968). Engineering Seismic Risk Analysis, *Bulletin of the Seismological Society of America*, 58 (5), 1583–1606.

Choffat, P., and A. Bensaude (1912). *Estudos sobre o sismo do Ribatejo de 23 de Abril de 1909.* Lisboa: Comissão do Servico Geologico de Portugal, Ed. Imprensa Nacional.

IPQ (2010) NP EN 1998-1. *Eurocódigo 8: Projecto de estruturas para resistência aos sismos. Parte 1: Regras gerais, acções sísmicas, e regras para edifícios.* Caparica, Portugal: Instituto Português da Qualidade.

Fonseca, J.F.B.D. & Vilanova, S.P. (2011). Comment on Sousa, M. L. and Costa, A. C., "Ground motion scenarios consistent with probabilistic seismic hazard disaggregation analysis. Application to Mainland Portugal", *Bulletin of Earthquake Engineering*, 9 (4), 1289–1295. Doi: 10.1007/s10518-011-9275-1

Giardini, D., Woessner, J., Danciu, L., Valensise, G., Grünthal, G., Cotton, F., Akkar, S., Basili, R., Stucchi, M.,

Rovida, A., Stromeyer, D., Arvidsson, R., Meletti, F., Musson, R., Sesetyan, K., Demircioglu, M. B., Crowley, H., Pinho, R., Pitilakis, K., Douglas, J., Fonseca, J., Erdik, M., Campos-Costa, A., Glavatovic, B., Makropoulos, K., Lindholm, C., & Cameelbeeck, T. (2014). Seismic Hazard Harmonization in Europe (SHARE): Online Data Resource, doi: 10.12686/SED- 00000001-SHARE.

Kates, R.W. (1971). Natural hazard in human ecological perspective: hypotheses and models. *Economic Geography*, 47 (3), 438–451.

LNEC (1983). *Regulamento de Segurança e Acções para Estruturas de Edifícios e Pontes*, Lisboa: Laboratório Nacional de Engenharia Civil.

Reiter, L. (1991). *Earthquake Hazard Analysis: Issues and Insights*. New York: Columbia University Press.

Silva, V., Crowley, H., Varum, H., & Pinho, R. (2014). Seismic risk assessment for mainland Portugal. *Bulletin of Earthquake Engineering*, 13 (2) 429–457. DOI 10.1007/s10518-014-9630-0

Sousa, M.L. & Costa, A.C. (2009). Ground motion scenarios consistent with probabilistic seismic hazard disaggregation analysis. Application to Mainland Portugal, *Bulletin of Earthquake Engineering*, 7 (1), 127–147. Doi: 10.1007/s10518-008-9088-z

Sousa, M.L. & Costa, A.C. (2011). Reply to "Comment on Sousa, M.L. and Costa, A.C., 'ground-motion scenarios consistent with probabilistic seismic hazard disaggregation analysis. Application to Mainland Portugal' " by João F. B. D. Fonseca and Susana P. Vilanova. *Bulletin of Earthquake Engineering*, 9(4), 1297–1307. Doi: 10.1007/s10518-011-9279-x

SSHAC – Senior Seismic Hazard Analysis Committee (1997). *Recommendations for probabilistic seismic hazard analysis: Guidance on uncertainty and the use of experts*. Washington, D.C.: U.S. Nuclear Regulatory Commission Report NUREG/CR/6372.

USNRC – United States Nuclear Regulatory Commission (2012). *Practical Implementation Guidelines for SSHAC Level 3 and level 4 Hazard Studies*. Washington, D.C.: U.S. Nuclear Regulatory Commission Report NUREG-2117.

Vilanova, S.P. & Fonseca, J.F.B.D. (2007). Probabilistic seismic hazard assessment for Portugal, *Bulletin of the Seismological Society of America*, 97 (5), 1702–1717.

Seismic Retrofitting: Learning from Vernacular Architecture – Correia, Lourenço & Varum (Eds)
© 2015 Taylor & Francis Group, London, ISBN 978-1-138-02892-0

A brief paleoseismology literature review: Contribution for the local seismic culture study in Portugal

M.R. Correia
CI-ESG, Escola Superior Gallaecia, Vila Nova de Cerveira, Portugal

M. Worth
University of Auckland, New Zealand

S. Vilanova
Instituto Superior Técnico de Lisboa, Portugal

ABSTRACT: On the framework of the FCT research project 'SEISMIC-V: Vernacular Seismic Culture in Portugal', the team was recommended, by the Evaluation Panel, to perform a review on the field of Paleoseismology and its application to the Portuguese territory. This paper intends to fulfil this request, by presenting an introduction to the history of Paleoseismology, its application and main techniques. The paper also includes a literature review regarding Portuguese geology, and a brief summary on the paleoseismic studies in Portugal, focusing on the following regions: the Gulf of Cadiz, and the Lower Tagus Valley. Finally, it presents an overview with general conclusions.

1 BACKGROUND: PALEOSEISMOLOGY

Paleoseismology is a branch of geology that is dedicated to studying the geologic and geomorphic evidence of past earthquakes. Because paleoseismic studies extend back the seismic history of a fault, they may significantly increase the knowledge on the long-term seismic behaviour of active faults, and are therefore essential to seismic hazard analysis assessments at long return periods.

Paleoseismology focuses on the most recent geological period, the Quaternary, which spans from approximately 1.8/2.6 million years ago to present, according to different definitions (e.g., Yeats et al. 1997). It combines geology with seismology, archaeology and history (Michetti et al. 2005).

1.1 *History*

The beginning of Paleoseismology date to the late nineteenth century. One of its forefathers was Grove Karl Gilbert, who first identified the representative features of the 1872 Owens Valley earthquake along the Wasatch fault (e.g., Yeats & Prentice, 1996; Ran & Deng, 1999). G. K. Gilbert, a pioneering geologist, also identified the relationship between faults and earthquakes, leading to the idea that faults generate earthquakes (Hough & Bilham, 2006). This concept only began to be commonly accepted by the geological community about 1908, including the idea that a fault will break again and again in the same place (Wallace, 1987).

Figure 1. Fault trench in Bam, Iran (credits: Mariana Correia, 2011).

Originally research had been focused on analysing the surface rupture area and fault scarp (Ran & Deng, 1999). However, in the late sixties of the 20th century, paleoseismic data started to be collected using the trenching technique. Some of the first trench studies were performed by Jay Smith of Converse Davis and Associates in 1968, at San Bernardino (Wallace, 1987), Malde (1971) at Idaho, and Clark et al. (1972) at Borrego Mountain California. Figure 1 shows an example of a fault trench.

1.2 *Paleoseismic techniques*

Paleoseismic analysis uses many different research techniques, including geomorphological techniques that aim at identifying and mapping paleoseismic landforms, and stratigraphic techniques that focus on mapping subsurface deformation exposed by trenching (McCalpin, 1996).

The analysis of landforms allows locating the surface paleoseismic deformations (fault scarps, offset river channels, sag ponds, etc.) using large-scale remote sensing imagery, and it is the first-step of a paleoseismic research.

Fault-trenching investigations are essential, because they can provide a direct assessment of the deformation associated with fault movements. The timing of paleoearthquakes can be determined using Late Quaternary dating methods, such as radiocarbon dating, and optically stimulated luminescence dating (McCalpin & Nelson, 1996).

The magnitude, timing and recurrence rates of a paleoearthquakes can also be determined using off-fault deformation. Examples of off-fault evidence include seismites, i.e. sedimentary structures produced by earthquake shaking, liquefaction evidence, earthquake-related landslides, broken cave formations, tsunami deposits, etc. (Michetti et al., 2005).

Paleoseismic research in marine or lacustrine environments is a relatively new field of research, and requires different techniques from those used by traditional seismology (see Gràcia et al., 2013, for a review). These include acoustic mapping techniques, seismic reflection profiling, and the analysis of sedimentary cores, amongst others.

It is important to note that the paleoseismic record is incomplete, since only large magnitude earthquakes are likely to generate enduring geological evidence at the surface. In addition, the presence of active sedimentation and fine stratigraphy is required to record the deformation associated with individual paleoearthquakes. A comprehensive paleoseismic study involves, therefore, the analysis of multiple fault trenches along different segments of a fault.

For an overall understanding of fault behaviour and the current tectonic activity, additional analysis techniques need to be combined with paleoseismic studies. These include remote sensing data, GPS and interferometry data, strain rate measurements, amongst others (Michetti et al., 2005).

1.3 *Application of Paleoseismology*

The period of historical recordings of seismic events is relatively short, in comparison to the seismic history of a fault and the mean recurrence interval for large earthquakes. This is particularly true for uninhabited regions, regions in which civilizational characteristics prevented the existence of reliable historical records, or slow deformation regions. Paleoseismic studies are able to extend back the seismic record of a fault for several thousands of years, allowing the analysis of multiple earthquake cycles (Wallace, 1987).

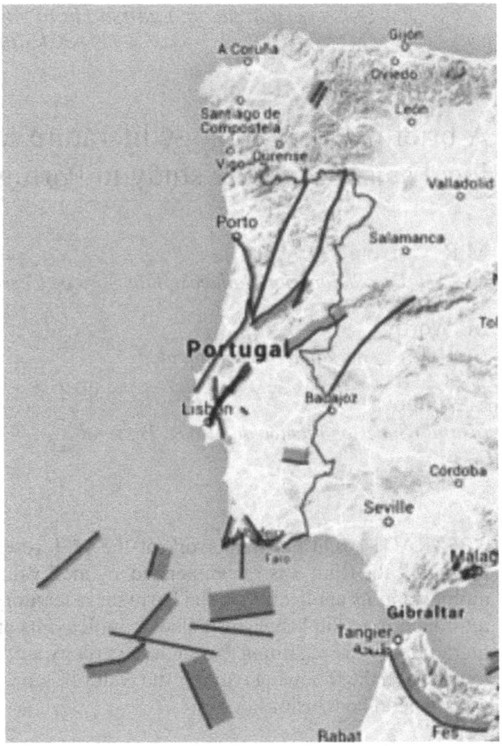

Figure 2. Map of potentially active faults in western Iberia according to the European Database of Seismogenic Faults according to Basili et al., (2013).

Magnitude and timing of paleoearthquakes and slip-rate of a fault are most critical information that can be retrieved from a paleoseismic study. This information is generally included in the models used for seismic hazard analysis. This framework is of major importance for paleoseismic investigations, which otherwise may have become a purely academic exercise (Michetti et al., 2005).

2 PALEOSEISMIC LITERATURE REVIEW

2.1 *Tectonic setting of Portugal*

Portugal is located in the vicinity of the Eurasian-Nubian plate boundary, which runs from the Azores Islands and through the Gulf of Cadiz. The Azores Islands sit at the triple junction formed by the Eurasian plate, the Nubia plate and the North American plate (e.g., Madeira & Brum da Silveira, 2003). The region exhibits trans-extensional tectonics. In its central section, the Azores-Gibraltar plate boundary has a clear geomorphological signature (the Gloria fault) and exhibits trans-current tectonics. However, to the east of the Gloria Fault, the deformation becomes transpressional and diffuses, spreading over a wide area (e.g., Buforn et al., 1988).

The intraplate region of western Iberia includes several faults that are considered potentially active

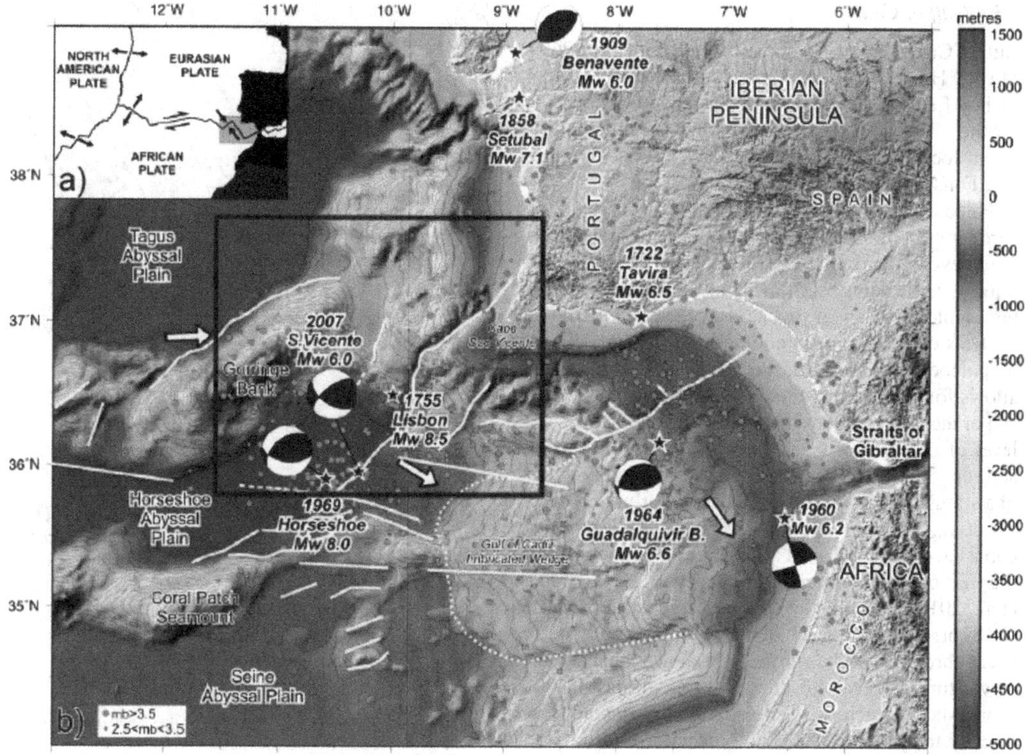

Figure 3. Moderate to large magnitude seismicity in the Gulf of Cadiz (credits: Gràcia et al., 2010).

in the current tectonic regime (Basili et al., 2013, Vilanova et al., 2014). The best studied of which are the Lower Tagus Valley Fault Zone, near Lisbon, and the Manteigas-Vilariça-Bragança Fault, in the North East of Portugal (Cabral, 2012). Figure 2 shows the potentially active faults present in the European Database of Seismogenic Faults (Basili et al., 2013).

2.2 *Brief history of paleoseismology in Portugal*

The first paleoseismic trench was excavated in Portugal, in 1979. The site was located in Ferrel, at the west coast of Portugal, and the study was sponsored by the Portuguese Electricity Company, and the Portuguese nuclear regulatory entity (Cabral, 2012). However, until the late nineties of the 20th century, most geological studies in Portugal focused at providing regional coverage of faults using neo-tectonic approaches. The lack of paleoseismic data is recognised to be a major drawback for seismic hazard evaluations, as there is limited knowledge on the earthquake potential of many faults (Vilanova & Fonseca, 2007, Cabral, 2012, Vilanova et al., 2014).

At the turn of the 20th century, an increasing number of research projects with a significant paleoseismic component have been funded by the Portuguese Science Foundation. The researches focused primarily on the Lower Tagus Valley Fault Zone, and on the Manteigas-Vilariça-Bragança Fault.

While the first paleoseismic results in the Lower Tagus Valley have been controversial (Fonseca et al., 2000, Cabral and Marques, 2001), recent studies showed increasingly compelling results (Besana-Ostman et al., 2012a, Besana-Ostman et al., 2012b, Canora et al., accepted). Research conducted in the Manteigas-Vilariça-Bragança Fault, in the context of a seismic hazard analysis assessment for a hydroelectric dam, also produced conclusive results (Rockwell et al, 2009). Paleoseismic trenches have also been excavated in the Azores archipelago, including Faial, Pico, São Jorge and São Miguel Islands (Madeira & Brum da Silveira, 2003, Carmo et al., 2013).

Since the early 1990s, numerous geological and geophysical surveys have been carried out in the Gulf of Cadiz, in order to identify sources of large magnitude earthquakes. The nature of the basin is in constant debate, due to its complex geodynamic setting (e.g., Zitellini et al. (2009), Cunha et al. (2012), Rosas et al. (2012). In the last decade, paleoseismic studies have been undertaken in the region. These consist mostly on core sampling of the ocean floor sediment, in order to identify turbidite deposits produced by seismic events (e.g., Garcia-Orellana et al., 2006, Gràcia et al 2010).

The following sections summarise the paleoseismic research that has been conducted within the Gulf of Cadiz and the Lower Tagus Valley. The reason for focusing on these regions is that they enclose the main sources of earthquakes of large magnitude that affected mainland Portugal in historical times.

2.3 *Gulf of Cadiz*

Gulf of Cadiz region is associated with both, the 1969's Mw7.8 Horseshoe earthquake, and the 1755's M8.4 Lisbon Earthquake (Figure 2). The epicentre location for the 1755's earthquake is unknown, and many different locations have been proposed in the literature (see Fonseca, 2005, for an overview).

Paleoseismic studies in this region consist of taking core samples of the ocean floor, in order to analyse turbidite events. Turbidite events occur when sediments liquefy, while being transported in ocean current (i.e. a turbulent flow), causing dense slurry to be formed. The analysis of core samples containing turbidites allows for paleoearthquakes to be identified, while core dating allows for determining the event's timing.

For the past 11,000 years (Holocene period), the sea level of the Gulf of Cadiz has been relatively stable and, therefore, earthquakes are the most likely cause of a simultaneous wide distribution of turbidites (Gràcia et al., 2010). In the late 2000s, several sediment cores have been collected at different locations within the Gulf of Cadiz (Garcia-Orellana et al., 2006, Gràcia et al., 2010). Turbidite layers observed at each sample were correlated with each other, in order to produce an event chronology. The chronology showed that at least eleven turbidite events occurred during the past 11,000 years, six of which were considered widespread seismically triggered events. Three events were found to correlate well with available instrumental and historical seismic records, and tsunami deposits. However, the thickness of the turbidite deposits does not correlate with earthquake magnitude, and therefore the size of the source earthquake cannot be determined with such approach. Based on this analysis Gràcia et al. (2010) proposed a mean recurrence interval of 1800yr for great earthquakes in the area.

2.4 *Lower Tagus Valley*

At least two damaging historical earthquakes are clearly associated with the Lower Tagus Valley Fault Zone: the Mw6.0 1909's earthquake and 1531's Lisbon earthquakes (e.g. Vilanova & Fonseca, 2004; Besana-Ostman et al., 2012a; 2012b).

The first paleoseismic study in the area revealed the presence of active thrust fault in the vicinity of Vila Chã de Ourique (Fonseca et al., 2000). This interpretation was, however, contested by Cabral & Marques (2001), who proposed that the observed deformation had a gravitational origin. A subsequent road cut showed that part of the deformation observed by Fonseca et al. (2000) was indeed landslide-related, but it also revealed the presence of a sub-vertical fault, structurally connected to the landslide. Fonseca et al. (2007) suggested that the observed deformation is consistent with a flower structure, affected by an earthquake-triggered landslide.

Besana-Ostman et al. (2012a) identified the active traces of the Lower Tagus Valley Fault Zone within the western margin of the Tagus River, over a length of

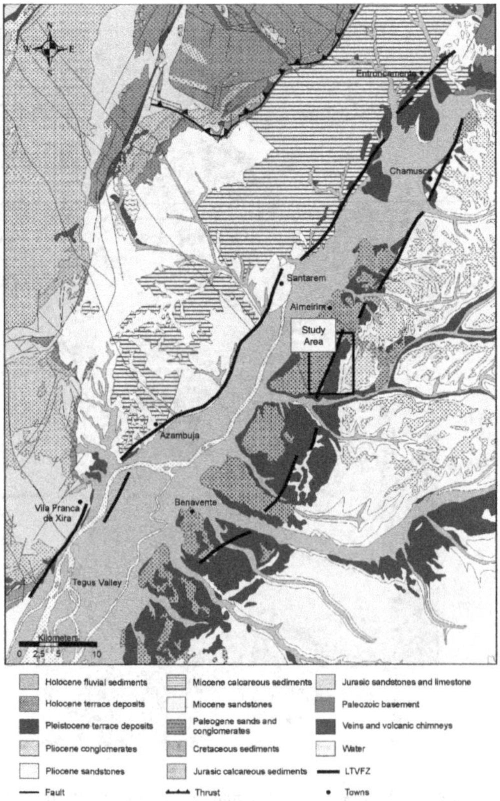

Figure 4. The Lower Tagus Fault traces according to Canora et al. (2014).

80 km (from Atalaia to Alhandra), using stereographic analysis of aerial photographs. The paleoseismic landforms identified included fault scarps, shutter ridges, transacted rivers and diverted drainages consistently exhibiting left-lateral displacements. The excavation of paleoseismic trenches at *Pombalinho*, where a fault scarp transacted a Holocene terrace, exposed horizontally stratified unconsolidated sand and clay, which showed deformation by both faulting and liquefaction (Besana-Ostman et al., 2012b). Organic samples from the trench indicated Holocene age for the observed deformation.

Recent developments include the identification of a sub-parallel strand of the fault zone within the eastern margin of the Tagus Valley (Canora et al., under review). The eastern strand of the fault is 70 km long, and also exhibits left-lateral deformation. The excavation of two trenches in the vicinity of Almeirim (the Alorna trenches) showed that the fault produced, at least, four surface ruptures during the last 8000yr, while the offsets displacing the drainage system indicate a minimum slip-rate in the range of 0.1–0.2 mm/yr. Figure 5 shows the north wall of the Alorna trench, and the corresponding interpreted log according to Canora et al. (2014).

Undergoing work using high-resolution LiDAR, based on topographic data led to the identification of

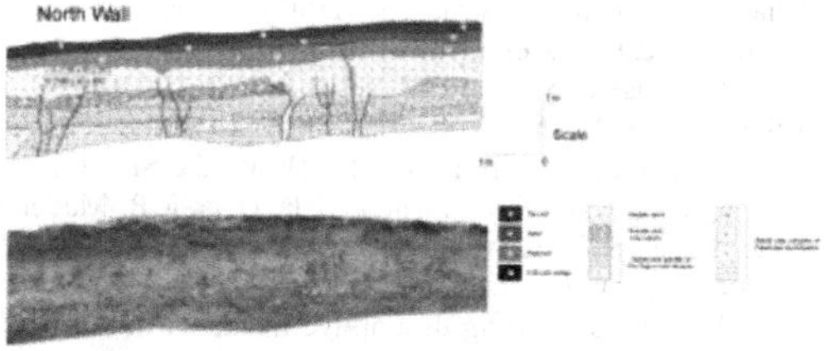

Figure 5. Alorna north trench wall and corresponding interpreted log according to Canora et al. (2014).

small displacement features (river offsets), which may be related to single earthquake ruptures. This information will be critical to estimate the size of the last earthquakes that affected the different sections of the fault.

3 CONCLUSIONS

Paleoseismic researches have the potential for extending back the seismic record for several thousands of years, and are therefore essential for understanding the long-term seismic behaviour of faults.

The paleoseismic techniques have been applied in Portugal for around 20 years with very promising results. However, most identified potentially active faults (Basili et al., 2013) still lack paleoseismic studies. The acquisition of paleoseismic data, in a regional sense, is, therefore, critical for the accurate representation of the seismic hazard in Portugal.

ACKNOWLEDGEMENTS

The authors acknowledge funding support from the Portuguese Foundation for Science and Technology (FCT), through the research project Seismic V - Vernacular Seismic Culture in Portugal (PTDC/ATP-AQI/3934/2012) and the Excellence Research Line project SEICHE (EXCL/GEO-FIQ/0411/2012) and research position Investigator FCT IF/01561/2014 (S.P.V.). CERENA research unit is funded by FCT through strategic project UID/ECI/04028/2013.

REFERENCES

Besana-Ostman, G.M., Vilanova, S.P., Falcao-Flor, A., Heleno, S., Ferreira, H., Nemser, E., Narciso, J., Domingues, A., Pinheiro, P. & Fonseca, J. (2012b). The Lower Tagus Valley Fault Zone: Overview, New Insights and Future Challenges. 15 WCEE – 15th World Conference on Earthquake Engineering. Lisbon, Portugal: International Association of Earthquake Engineering & Portuguese Society of Seismic Engineering.

Besana-Ostman, G., Vilanova, S.P., Nemser, E.S., Falcao-Flor, A., Heleno, S., Ferreira, H., & Fonseca, J.F.D.B. (2012a). Large Holocene Earthquakes in the Lower Tagus Valley Fault Zone, Central Portugal. Seismological Research Letters, 83, 1, 67–76.

Buforn, E., Udias, A. & Colombas, M. (1988). Seismicity, source mechanisms and tectonics of the Azores-Gibraltar plate boundary, Tectonophysics 152, 89–118.

Basili, R., Kastelic, V., Demircioglu, M. B., Garcia Moreno, D., Nemser, E. S., Petricca, P., Sboras, S. P., Besana-Ostman, G. M., Cabral, J., Camelbeeck, T., Caputo, R., Danciu, L., Domac, H., Fonseca, J., García-Mayordomo, J., Giardini, D., Glavatovic, B., Gulen, L., Ince, Y., Pavlides, S., Sesetyan, K., Tarabusi, G., Tiberti, M. M., Utkucu, M., Valensise, G., Vanneste, K., Vilanova, S., Wössner, J. (2013). The European Database of Seismogenic Faults (EDSF) compiled in the framework of the Project SHARE. http://diss.rm.ingv.it/share-edsf/

Cabral, J. (2012). Neotectonics of Mainland Portugal: state of the art and future perspectives. Journal of Iberian Geology, 1(38), 71–84.

Cabral, J. & Marques, F.M.S.F. (2001). Paleoseismological studies in Portugal: Holocene thrusting or landslide activity? *Eos, Transactions, American Geophysical Union* **82**, 351–352.

Canora, C., Vilanova, S., Besana-Ostman, G., Heleno, S., Fonseca, J., Domingues, A., Pinheiro, P., & Pinto, L. (2014). Paleoseismic results of the east strand of the Lower Tagus Valley Fault Zone, Central Portugal. In *EGU General Assembly Conference Abstracts* (Vol. 16, p. 10149).

Canora, C., Vilanova, S.P., Besana-Ostman, G.M., Carvalho, J., Heleno, S., & Fonseca, J.F.B.D., (accepted). The Eastern Lower Tagus Valley Fault Zone in central Portugal: Active faulting in a low-deformation region within a major river environment. Accepted for publication in Tectonophysics.

Carmo, R., Madeira, J., Hipólito, A., & Ferreira, T. (2013). Paleoseismological evidence for historical surface faulting in São Miguel island (Azores). Annals of Geophysics, 56 (6), S0671

Clark, M., Grantz, A. & Rubin, M. (1972). Holocene activity of the Coyote Creek fault as recorded in sediments of Lake Cahuilla. US Geol. Surv. Prof. Paper, 112–130.

Fonseca, J. F. B. D., Bosi, V., Vilanova, S.P., & Meghraoui, M. (2000). Investigations unveil Holocene thrusting for onshore Portugal, Eos Trans. AGU, 81(36), 412–413.

Fonseca, J.F.B.D. (2005). The source of the Lisbon earthquake (letter), Science 308, 5–6.

Fonseca, J., S. Vilanova, P. Costa, S. Heleno, & V. Bosi (2007). Vila Cha de Ourique, Cartaxo: A case of complex Holocene deformation, of probable tectonic origin. In Sismica 2007, Proceedings of the Cogresso de Sismologia e Engenharia Sismica, 1–9. Porto, Portugal: University of Porto.

Garcia-Orellana, J., Gràcia, E., Vizcaino, A., Masqueé, P., Olid, C., Martínez-Ruiz, F., Piñero, E., Sanchez-Cabeza, J.A., & Dañobeitia, J.J. (2006). Identifying instrumental and historical earthquake records in the SW Iberian Margin using 210Pb turbidite chronology. Geophysical Research Letters 33, 24.

Gràcia, E., Vizcaino, A., Escutia, C., Asioli, A., Rodés, A., Pallàs, R., Garcia-Orellana, J., Lebreiro, S., Goldfinger, C. (2010). Holocene earthquake record offshore Portugal (SW Iberia): Testing turbidite paleoseismology in a slow-convergence margin. Quaternary Science Reviews, doi:10.1016/j.quascirev.2010.01.010.

Gràcia, E., Lamarche, G., Nelson, H. & Pantosti, D. (2013) Preface: Marine and Lake Paleoseismology, Natural Hazards and Earth System Science, 13 (12), 3469–3478.

Hough, S. E. & Bilham, R.G. (2006). After the Earth Quakes: Elastic rebound on an Urban Planet. Oxford, UK: Oxford University Press.

Madeira, J. & Brum da Silveira, A. (2003). Active tectonics and first paleoseismological results in Faial, Pico and S. Jorge islands (Azores, Portugal). ANNALS OF GEOPHYSICS, 46(5), 733–761.

Malde, H. (1971). Geologic investigations of faulting near the National Reactor Testing Station, Idaho. U.S. Geological Survey Open-File Report, 267 p.

McCalpin, J.P. (1996), Field Techniques in Paleoseismology, in, McCalpin, J.P. (editor), Paleoseismology, New York, USA: Academic Press, p. 33–84.

McCalpin, J.P. & Nelson, A.R. (1996). Introduction to Paleoseismology. In, McCalpin, J.P. (editor), Paleoseismology, New York, USA: Academic Press, p. 1–32.

Michetti, A. M., Audemard M., F. A. & Marco, S. (2005). Future trends in paleoseismology: Integrated study of the seismic landscape as a vital tool in seismic hazard analyses. Tectonophysics, 408, 3–21.

Ran, Y. & Deng, Q. (1999). History, status and trend about the research of paleoseismology. Chinese Science Bulletin, 44(10), 880–889.

Rockwell, T, Fonseca, J.F.B.D., Madden, C., Dawson, T., Owen, L.A., Vilanova, S.P., & Figueiredo, P. (2009). Palaeoseismology of the Vilariça Segment of the Manteigas-Bragança Fault in northeastern Portugal. Geol. Soc. London. Spec. Publ, 316, 237–258.

Rosas, F.M., Duarte, J.C., Neves, M.C., Terrinha, P., Silva, S., Matias, L., Gràcia, E., & Bartolome, R. (2012). Thrust-wrench interference between major active faults in the Gulf of Cadiz (Africa-Eurasia plate boundary, offshore SW Iberia): tectonic implications from coupled analogue and numerical modelling. Tectonophysics, 548–549:1–21.

Vilanova, S., & Fonseca, J.F.D.B. (2004). Seismic hazard impact of the Lower Tagus Valley fault zone (SW Iberia), Journal of Seismology 8, 331–345.

Vilanova, S. P. & Fonseca, J. F. (2007). Probabilistic Seismic Hazard Assessment for Portugal. Bulletin of the Seismological Society of America, 97(5), 1702–1717.

Vilanova, S.P., Nemser, E., Besana-Ostman, G.M., Bezzeghoud, M., Borges, J.F., Brum da Silveira, A., Cabral, J., Carvalho, J., Cunha, P.P., Dias, R.P., Madeira, J., Lopes, F.C., Oliveira, C.S., Perea, H., García Mayodormo, J., Wong, I., Arvidsson, R., & Fonseca, J.F.D.B. (2014). Incorporating descriptive metadata into seismic source zone models for seismic hazard assessment: a case study of the Azores–West Iberian region. Bulletin of the Seismological Society of America 104 (3), 1212–1229.

Wallace, R. (1970). Earthquake recurrence intervals on the San Andreas fault. The Geological Society of America Bulletin, 81(10), 2875–2890.

Wallace, R. (1987). A Perspective of Paleoseismology. Directions in Paleoseismology (pp. 7–17). Albuquerque, New Mexico: U.S. Geological Survey.

Yeats, R.S., Sieh, K., & Allen, C.R. (1997). The geology of earthquakes. New York, USA: Oxford University Press.

Yeats, R.S., & Prentice, C.S. (1996). Introduction to special section: paleoseismology. Journal of Geophysical Research, 101, 5847–5853.

Zitellini, N., Gràcia, E., Matias, L., Terrinha, P., Abreu, M.A., DeAlteriis, G., Henriet, J.P., Dañobeitia, J.J., Masson, D.G., Mulder, T., Ramella, R., Somoza, L. & Diez, S. (2009). The quest for the Africa–Eurasia plate boundary west of the Strait of Gibraltar. Earth and Planetary Science Letters 280 (1–4), 13–50.

Seismic Retrofitting: Learning from Vernacular Architecture – Correia, Lourenço & Varum (Eds)
© 2015 Taylor & Francis Group, London, ISBN 978-1-138-02892-0

Portuguese historical seismicity

G. Sousa
CI-ESG, Escola Superior Gallaecia, Vila Nova de Cerveira, Portugal

ABSTRACT: All through its history Portugal has been tattered by earthquakes, some of them of great intensity and dramatic consequences. Although the occurrences of these natural phenomena, sometimes, leaving traumatic marks in the collective memory of an entire community, tend to be registered; the information is scarce and often vague. With the exception of the 1755 Lisbon's earthquake, there is an almost total void regarding the study of the impact of these events on local building culture. This paper addresses a general approach to historical earthquakes in Portugal, focusing on a more detailed analysis of three cases: the earthquakes of 1755, 1909 and 1969.

1 INTRODUCTION

1.1 *Historical seismicity*

Earthquakes are natural phenomena manifested with great violence and destruction, often bringing with them great panic and social turmoil.

> *"By an earthquake is then understood a violent agitation, shaking or trembling of some part of the earth's surface, generally attended with a terrible noise like thunder, and sometimes with an eruption of fire, smoke, water or wind."*
>
> (Owen, 1756, p. 5)

The dramatic consequences of these quakes for the population justifies, from very early, that different authors show a keen interest in studying its origin and, particularly, in finding a method to predict them. However this method still, these days, impossible to discover, despite the great development of modern seismography, allowing already a considered reasonable warning interval. Devoid of the technical and scientific means available today, the historical communities faced major disasters that lastingly marked the collective memory.

The Portuguese territory was no exception and, throughout its history, experienced numerous earthquakes. Some of them leaving a deep scar in local communities, as it the case of the earthquakes, which will be addressed in this article.

When thinking of disasters associated with earthquakes, it's intensity, more than in scale values and magnitude, is measured in two very linear ways: the number of victims, and the material damage caused.

The latter are the central aspects of this article that focus on how earthquake disasters affect the constructive culture of past communities. These influences can be seen in many ways, from the introduction of new building systems, as happened in Lisbon after the 1755's earthquake, by promoting constructive and urban policies, where seismic reinforcement systems are a priority, as in *Benavente* (1909), or simply by the awareness that a traumatic experience of this kind may awaken, as it seems to be the case of the 1969's earthquake.

1.2 *Methodology*

This article results from a documental research integrated in the SEISMIC-V project, aiming at understanding, from an historical perspective, the great impact on Portuguese regional building culture. The research concentrated on the regions identified as case studies, but by its interest and historical significance, it also included the case of the 1755 Lisbon's earthquake.

The methodology used was based on an archive documentation research for different sources, but eventually first-person accounts of earthquakes episodes are also included. The sources found can be described by three groups: graphics, key documents, where changes in certain urban areas or buildings were already identified, as probably resulting from the application of seismic reinforcement systems; and memories and official reports about the consequences of certain seismic episodes. Also journals that have reported these events and their main consequences are considered.

The structure of this article results from the analysis of the last two sets of documents, in particular: the *Memórias Paroquiais* of the study area, and other Memoirs and Reports published after the earthquake of 1755; and national or regional newspapers published on that time concerning the major earthquakes occurred in Portugal in the twentieth century.

2 HISTORICAL EARTHQUAKES IN PORTUGAL

Of course, in the case of a phenomenon of such significance, the various problems associated with

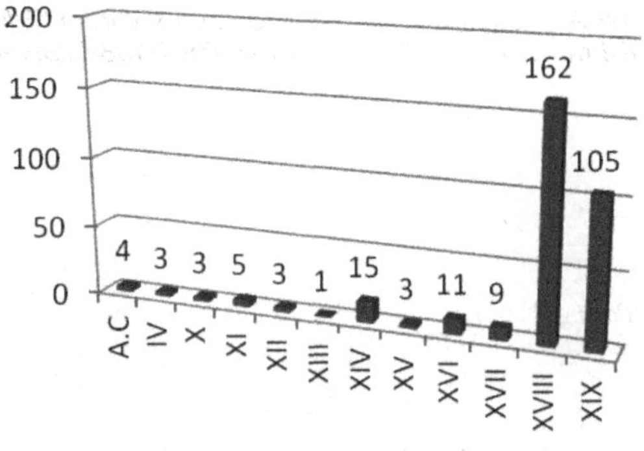

Figure 1. Number of earthquakes by century (credits: Goreti Sousa based on Martins & Víctor, 1990).

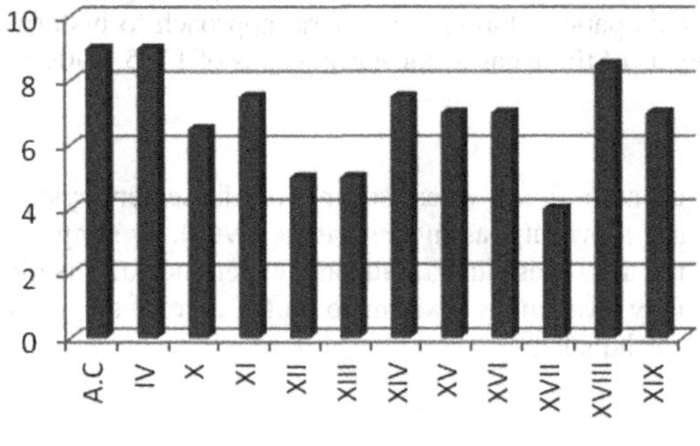

Figure 2. Maximum magnitude of earthquakes by century (credits: Goreti Sousa based on Martins & Víctor, 1990).

earthquakes such as: identification of areas with seismic risk, origin and explanation of the seismic process, etc., are subjects addressed by several authors.

The same can be said regarding the historical seismicity in Portugal. Therefore, it is not the aim of this article to make an extensive report of all earthquakes occurred in Portugal, or to give a lecture about magnitudes and epicentres; even though, it was considered essential to briefly present a systematisation of the main seismic shocks registered in the country.

As shown in the chart presented above, the XVIII and XIX centuries were those most disturbed by the occurrence of earthquakes. These are followed, with a significant margin of difference, by the XVI and XVI, which also stand out in relation to the other, concerning the number of earthquakes registered.

The great disparity in numbers between the XVIII and XIX centuries and the remaining may, however, is related with a greater availability of register sources, which, before this period, were extremely scarce. On the other hand the occurrence of earthquakes particularly violent in the XVIII century would naturally arise fear and curiosity about these phenomena which would expedite the registration of smaller quakes that would, otherwise, would never be recorded.

With regard to the intensity of these earthquakes, and analysing the maximum intensities, it is perceived that the most violent earthquakes were recorded in the fourth century AD, or other periods Before Christ,

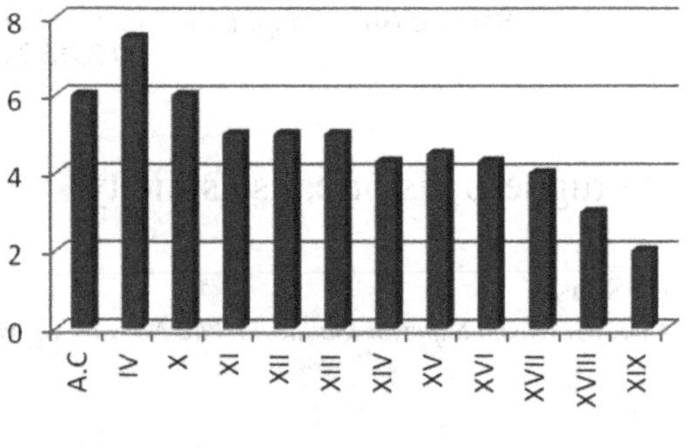

Figure 3. Minimum magnitude of earthquakes by century (credits: Goreti Sousa based on Martins & Víctor, 1990).

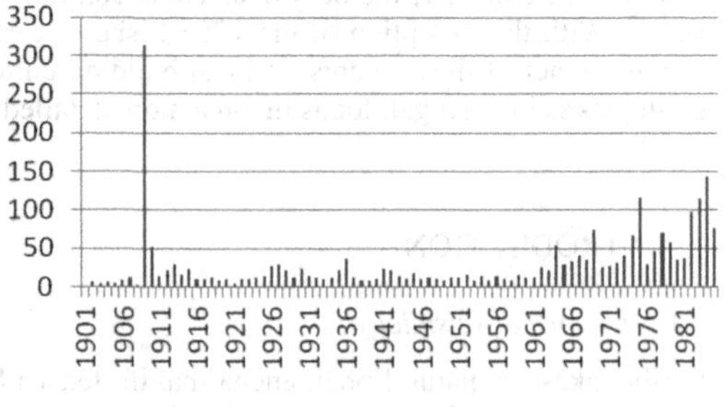

Figure 4. Number of earthquakes of the XX century by year (credits: Goreti Sousa based on Martins & Víctor, 1990).

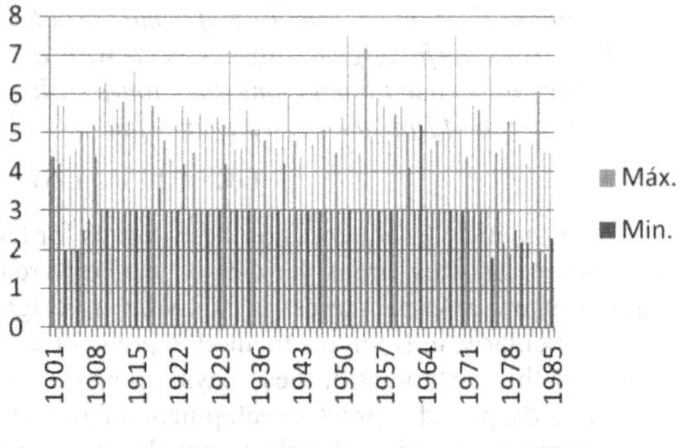

Figure 5. Magnitude of earthquakes in the XX century (credits: Goreti Sousa based on Martins & Víctor, 1990).

with magnitudes around 9°. These are followed by the eighteenth century earthquakes, in which the 1755's earthquake recorded 8°.

Discussing, now, the minimal intensities, the XIX century is the one with the lower minimum intensity register, quite possibly because of the advances in seismography monitoring systems, which allowed the record of small shocks that would otherwise be imperceptible. The IV century stands out by registering the highest value (7.5) recorded among the lowest magnitude impacts.

In the twentieth century, with the advances of seismography, this evolution can be traced annually. For the purposes of this article, only the data from the first three quarters of the century, meaning the period between 1900 and 1985, was considered.

The chart showed above presents the number of earthquakes per year recorded over this period. The year 1909 stands out with great evidence, but equally significant values are registered in the years 1983 and 1984, with 115 and 143 earthquakes respectively, as well as in 1975 (116). A little behind is the year 1969, with 75 recorded earthquakes; and 1978 with 71.

In 1931, 1951, 1954, 1964 and 1969 there are records of earthquakes with a maximum intensity level above 7°. The Earthquake of *Benavente*, despite its dramatic consequences, recorded a much lower intensity (6.2).

The information available for each of these earthquakes is extremely scarce, and even scarcer in topics related to the main focus of this research: the consequences that these phenomena may have on the local building culture. The records that mention these phenomena, when addressing its consequences focus mainly on the human casualties registered.

Only for the most recent events is available more abundant and diversified information on the issue.

On the XVIII century the first journals start to appear, and with them the first records of earthquakes in Portugal were reported nationally.

As an example, it can be mentioned the news published on the 26th October 1724, in the *Gazeta de Lisboa Occidental*. This text reported an earthquake felt in Lisbon, *Porto*, *Elvas*, *Cantanhede*, *Santarém* and *Portimão*, the night of 12th to 13th of that month. This was the most ancient news on this subject found during this study.

The availability of information multiplies exponentially in case of major disasters with great national and international impact, as it is the case of the Lisbon's earthquake in 1755, and it starts to grow with daily press.

For the second part of this article three earthquakes were selected to be presented in a more detailed manner: the 1755's Lisbon earthquake; the 1909's *Benavente* earthquake, and the 1969's Algarve earthquake.

For each case, after a brief description of the earthquake and the phenomena associated with it, it will be discussed: its consequences in terms of human and material losses; the initial response of the authorities and population; and the impact on the reflection they may have had on the local seismic reinforcement of vernacular buildings.

3 NOVEMBER, 1ST, 1755

3.1 *The Lisbon earthquake*

On the 1st day of November, 1755, the population of Lisbon, mostly concentrated in the city churches to celebrate Day of All Saints, was surprised by a huge noise, followed by a strong seismic shock (with an estimated magnitude of IX on the Richter scale). The quake lasted about seven to eight minutes, followed by several aftershocks of lesser intensity, but in very short intervals. Several witnesses testify that the aftershocks continued for the next six months.

Anthony Pereira, an eyewitness to these events, describes how "*The first earthquake was immediately followed by an extraordinary and almost incredible rising of the waters*" (1756, p. 13), reporting, the formation of a tsunami. Recent studies (Andrade, 2011) estimate that the waves that followed the earthquake could have reached about 20 m high.

But the worst was yet to come. Several fires broke out over the following days, increasing considerably the destruction that had befallen the city. These fires were probably, according to many of the analysed reports, caused by the falling of structures, and wooden furniture on burning fireplaces.

In *Setúbal*, there are reports of the sudden appearance of gushes of water that could be interpreted as geysers (Cross, 1968). In *Lagos*, different sources state a strong smell of sulphur that was felt before the quake. On mid-morning the population heard a huge noise, followed by the earthquake, and the appearance of huge fissures on the ground. Similar to what was described in Lisbon, the quake was followed by three huge waves, with about 11m high, which devastated the beach of Lagos (Trindade, 2014).

Although the most available sources report the consequences that this quake had in the capital, it also caused damage in nearly every city in the country, some of which were recorded in the 1758 inquiry that was later named *Memórias Paroquiais*.

Following these events, and already having the perception that not only the capital had suffered from the damage caused by the 1755's earthquake, the government launches an inquiry to collect a viable report of all the damaged suffered throughout the kingdom, and what had been done, or needed to be done, in order to remediate them. Despite the lacking information in some regions, this compendium is a privileged source to assess the impact of this disaster in the regions where it was completed.

A brief analysis of the available information for the regions selected as a case study under the SEISMIC-V project enabled the gathering of some reports on how the events of the 1st of November affected the buildings.

However, among the different issues listed in the inquiry, the one that seems to have roused little interest was precisely the one relating to the damages caused by the earthquake, as many of them choose not to answer this question. Many of those responsible for the answers were mainly religious authorities.

On the other hand, even those who answered this question, focus on the description of damage caused on important buildings, particularly the religious ones.

This evidence does not however diminish the importance of this documentary compendium allowing, for example, to understand that, in the district of *Beja*, in the parish of S. Luis "*The church in this village underwent some ruin on the corners, and roof, and the same happened to a few farmers in their houses, hall of which is already rigged*" (Magro, 1758).

Also in *Évora,* the parish responsible reports only the opening of some cracks on the walls and domes of churches and chapels. Only in the case of S. Manços Parish, the priest mentions "[...]*the houses of an estate of this parish called Hospital, collapsed so that no one could live there, but they are already repaired*" (Gaspar, 1758).

From the analysis carried out, it appears that these inquiries may also serve as a useful source to study other issues related to vernacular architecture, even if not directly related to the consequences of the earthquake.

In *Vila Nova de Milfontes*, the Prior Bento Dias Barreto (1758) states:

"(...) but it was God will that the houses of this village were unharmed, and none of them was severely damaged, being as they are, almost all made of earth what they call taypa, nor even some who by the rigor of time where already ruined, had more ruin caused by said the quake. (...) As also some houses, outside the village, near the river, but that belonged to its inhabitants, which only where used for storage, suffered with the flood caused by said earthquake. The water completely destroyed them leaving only a signal of where they used to be. This however was not a considerable loss; and only where destroyed three one storey houses."

In *Moura*, a parish of *Póvoa de S. Miguel*, heavy losses were also reported, "(...) *and there were not more damages (supposedly the hand of God) because they were all one storey houses (...)*" (Gracia, 1758).

3.2 *Consequences*

The number of fatal victims of this catastrophe varies greatly from source to source. Nowadays, it is estimated that it may have amounted to about 20,000 (Andrade, 2011) in a population of about 100,000 inhabitants. The vast majority of victims were buried under the debris, caused by the collapse of religious buildings (Pereira, 1756).

About 75% of the building was severely damaged. "*Most of the public buildings and the private houses have been levelled to the ground*" (Pereira, 1756, p. 9).

According to several eyewitnesses, the collapse of buildings covered the city of a black smoke, which helped to increase the panic of the fleeing population.

As reported by an eyewitness of the catastrophe: "*The floors and ceilings began to crack, the roofs to fall, the arches to give way with a horrid noise, at length some of the steeples and walls to chink, and others to tumble down. (...) In other places the tiles were seen to fly off the tops of houses*"

(Pereira, 1756, pp. 4–5).

The tsunami that followed the earthquake increased death and destruction. "The violent waves dilapidated *anchors, broke the moorings and overrun the streets, and neighbouring squares*" (Nova e fiel relação do terremoto que experimentou Lisboa e todo Portugal no 1 de Novembro de 1755, 1756, p. 4).

In *Setúbal*, it is estimated that half the homes of the parish of *Santa Maria da Graça* and almost all of the houses of *S. Julião* parish have been destroyed (Cruz, 1968). Anthony Pereira (1756) states that, as a result of the tsunami, there were also deaths by drowning in *Setúbal*, *Cascais*, *Peniche* and the Algarve.

In *Lagos*, in a population of about 3000 inhabitants, there were 400 fatalities, and around 90% of the buildings of the village were damaged. The population of Lagos could not depend upon official assistance, and the reconstruction was done by the residents themselves, using materials found in the rubble, with which they built a neighbourhood on the north part of the city walls (Trindade, 2014).

In Oporto there are also reports of panic on the 1st of November 1755. The population heard the noise and felt the earthquake, although with less intensity, but nevertheless enough to lead to the collapse of some chimneys, stones and crucifixes. Some cracked walls are also reported. The population fled the city in panic. Those who sought shelter by the river, watched as the waters receded several meters and returned with a roar (Owen, 1756).

According to the same source, the city *of Angra do Heroismo* and *Vila da Praia* have also been severely affected.

3.3 *Initial response*

In a memory published in 1758, Amador Patricio Lisboa describes with great detail the actions and measures that were planned in the days that followed the disaster, presenting the documents associated to each one of them. The most urgent measures focussed on different aspects of the disaster, and sought to mitigate the most violent consequences of it:

1. Burial of the fatal victims to prevent the propropagation of diseases;
2. Supply of groceries, as the city's warehouses were destroyed, additionally the scurry of most of the population had left no one to take charge of transportation;
3. Looting prevention;
4. Care and rescue of the wounded. Eye witnesses report the finding of survivors after 9 days buried in the rubble (Pereira, 1756);
5. Organisation of the inhabitants, who had fled the city;
6. Relocation of the survivors. The first night the population wandered out in the open, but as soon as they recovered from the initial shock, the survivors built a camp, using sheets and blankets, and later constructed wooden huts (Pereira, 1756). According to Amador Patricio Lisboa (1758), the authorities built about 9000 wooden shelters;
7. City rebuilding;
8. Help the other affected villages, in particular the region of *Setúbal* and the Algarve.

3.4 *Impact on the constructive culture*

After these events, Lisbon became a pioneer on the implementation of earthquake engineering methods. The authorities in charged with the rebuilding of the city developed a system that would become known as *Gaiola Pombalina*. This system, according to Andrade (2011), could be a derivative of shipbuilding know-how.

The system that will be discussed in other articles of this publication was designed to maintain the structure intact, even when the masonry smashes during an earthquake, thus allowing the inhabitants to exit the building safely.

However, with the disappearance of those who experienced the earthquake on first hand, worry gives way to oblivion, and the system was soon adulterated (Moita, 2013).

4 APRIL 23RD, 1909

On the 23rd April, 1909 the town of *Benavente* was destroyed by an earthquake of intensity IX, which was felt in a 380 km^2 radius. There were about 224 replicas of the main shock (Antunes, 1956).

The quake caused cracks, from which sand with water gushed (Antunes, 1956). The population was, at the time of the earthquake, busy in the fields; therefore the number of deaths was not very high. Children, who at that time were playing in the village, were, however, the main victims.

4.1 *Consequences*

Benavente and *Samora Correia* are two villages on the *Ribatejo* region particularly damaged by the earthquake. In *Samora Correia* the collapse of the bridge aggravated the problem, by delaying the arrival of assistance to populations.

The first news report that there were 14 dead, and much more injured in *Benavente* (Illustração Portugueza, No. 168, May 10), but in a more recent study Vieira (2009), it was reported that there were 30 fatalities in a total number of 3551 inhabitants.

At a meeting of the Lisbon City Council on the 28th of April Filippe da Matta, on his own behalf, and on behalf of his colleagues, who had accompanied him, in representation of the Council, in a visit to the *Ribatejo* region, states his shock with the scenario of destruction he had witnessed.

His description, detailed in the records of the city council meeting, gives an account of the devastation that fell upon the villages of *Salvaterra*, *Benavente*, *Samora Correia* and *Santo Estevão*, where few houses stood, none though in conditions to be inhabited.

At the same meeting Agostinho Fortes restated that:

> "*the farmers lacked the means to acquire furniture and clothes they lost and that the rebuilding of settlements should attend the need to build special quarters for them. It's vital to first undertake a study on soil conditions, so that the new buildings are appropriate*" (Meeting minutes of the Municipality of Lisbon Session of the 28 of April 1909, p. 249)

Choffat and Bensaúde (1912) estimate that 40% of the dwellings of *Benavente* village needed to be demolished after the earthquake; 40% would have been seriously damaged, in need of serious interventions; and the remaining 20% would require small repairs. Only one house was habitable, with no need of cares.

"*The houses shake off their walls. The rooftops collapse. The walls fall down. A cloud of dust and rubble fills the space, similar to the dense smoke of a fire.*" (Através dos escombros do Ribatejo. O Terramoto de 23 de Abril, 1909, p. 590)

The main damages registered were the opening of vertical cracks in the walls, and the collapse of eaves, corners and other structures, but rarely of the rooftops (Vieira, 2009).

4.2 *Initial response*

On the 28th of April 1909, is entitled, by order of the Ministry of public works, a workgroup charged with the study of the *Benavente* earthquake (Antunes, 1956). The next day a new decree is published, this time to create another workgroup, in order to draw regulations on building policies on seismic prone areas (Vieira, 2009).

The survivors were housed in wooden shelters. Many were still there when, in the summer of 1911, an order was issued to demolish all huts within one year time (Vieira, 2009).

The events of the 23rd of April, in *Ribatejo* region, generated a wave of national solidarity. But the turbulent political developments of the end of the monarchy, and the first years of the republic, originated that, although it was originally planned that the Commission of Official Aid would insure the reconstruction works within the stipulations of the new urban plans, most of the works were taken over by proprietors, without taking into account the requirements of the official plans (Vieira, 2009).

4.3 *Impact on the constructive culture*

The creation of a committee to draw a building regulation in seismic prone areas is a noteworthy product of the awareness for the need for seismic reinforcement of vernacular buildings.

It was a responsibility of this committee, as explained by Vieira (2009), to study the wreckage, and to establish guiding principles for reconstruction. Furthermore, this workgroup should also study the building typology, materials and construction processes employed, among other studies directly related to the topography of the territory, and the study of the seismic phenomenon itself.

The theme of seismic reinforced construction was, in 1909, on the agenda. After the earthquake of

Figure 6. Building in Alhandra damaged in the 1909 earthquake, In: Ilustração Portugueza, n.º168, p. 607 (credits: Joshua Benoliel).

Figure 7. The king visiting the affected villages, In: Ilustração Portugueza, n.º 168, p. 607 (credits: Joshua Benoliel).

Figure 8. In: The camp built by the homeless in Benavente, In: Ilustração Portugueza, n.º 168, p. 608 (credits: Joshua Benoliel)

Messina (December 1908), several international meetings gathered experts to discussed this topic.

The opinions concerned the behaviour of concrete during earthquakes. But the study of the *Benavente* wreckage lead to the dissemination of the idea that the masonry or brick dwellings were better prepared to resist an earthquake, than those constructed with rammed earth or adobe (Vieira, 2009).

Despite all these discussions, as mentioned above, most of the reconstruction was made outside the official plan, with in-site available materials, repeating the same typology, and using the techniques the population knew for centuries.

Finally, following these events, in the next years the national seismographic network was restructured and updated (Miranda & Carrilho, 2014).

5 FEBRUARY, 28TH, 1969

Again, on the 28th of February 1969 a strong earthquake was felt across the whole country. This was the stronger one felt since 1755's Lisbon earthquake. The international seismographic stations recorded the shock of 7.9 magnitudes on the Richter scale. The records from national stations were completely saturated.

The quake was felt at dawn, around 3h41a.m., and it was followed by a replica of lesser intensity at 5h28a.m. In the following hours several aftershocks where felt, some of them of considerable intensity (Miranda & Carrilho, 2014).

The population of the capital panicked, fearing a repetition of the dramatic scenery of 1755. However, the most affected region was the Algarve and the *Costa Vicentina.*

As it had happened two hundred years before, this quake was also followed by a small tsunami. Crews from several ships stationed near the coast report abnormal agitation of the water, with unusually high tides, which lasted for several hours.

5.1 *Consequences*

However, the consequences were better confined, registering only 13 deaths, only two of which as a direct result of the earthquake (Miranda & Carrilho, 2014). They also recorded dozens of injuries, 58 of them in Lisbon. The injured were essentially slightly, mainly emotional excitement, falls and trauma, caused by shedding of debris.

In buildings the same characteristic damages were recorded: opening of splits in the walls and roofs, falling of chimneys, breaking of glass and displacement of tiles. To these is now added the new inconvenienced of the cutting off in communications, and failures in the supply of energy.

Another sign of new times was the damage, caused by the fall of structure, over the vehicles parked in the street.

All over the country material damages were recorded, as a result of the earthquake. However, as mentioned above, the most affected region was the Algarve and the *Costa Vicentina*.

In Lagos there was one fatality, buried under the rubble of a house.

The material damage was also considerable. A total of 400 houses were torn down. The most affected village was *Bensafrim* (about 40 houses ruined), but the case that arose more commotion was *Fontes de Louzeiro*, 13km from *Silves*, where fifteen of the sixteen families that lived there lost their homes.

5.2 *Initial response*

In the 1969's earthquake the relief effort was more efficient than in previous cases, as it was already established a national fire service. They searched the damaged houses for possible wounded, who were then, transported to hospitals.

Most injuries were considered mild and were, later, sent home.

The municipality of Lisbon hastened to allow the reconstruction of chimneys without prior licensing, as reported in the edition of the 2nd of May 1969, from a municipal daily journal.

In the most affected localities, it was necessary to relocate the families that lost their homes. Specifically, the same newspaper, reports the case of *Fontes de Louzeiro* where only a family kept its house. The remaining families were re-housed in a warehouse.

But on the 20th of May, the government, through a law-decree (Decreto-Lei n.º 49010), approved the acquisition and improvement of installations intended for temporary accommodation, in case of an emergency relocation.

5.3 *Impact on the constructive culture*

Contrary to what was usually perceived on the vulnerability of vernacular architecture during this kind of seismic event, Braga and Estevão (2010, p. 8), refer the following concerning rammed earth buildings in the Algarve, "*(...) we can suppose the best performance of the construction in rammed earth compared to the construction with stone masonry, not having been perceived significant collapses of rammed earth buildings during the earthquake.*"

However, with regard to reconstruction, the government of Marcelo Caetano ordered (Decreto-Lei 49010) the reconstruction and necessary repairs to public buildings, or the buildings with public interest. And the houses were not forgotten either. For those who can prove not having economic means to take care of the repairs, the government would finance the works needed; and for those who actually are able to take charge of repairs personally it stipulates a fund to provide loans.

These interventions should obey a decree the government intend to issue. It was also planned the construction of new housing. None of the latter was, however, finished.

Jorge Miguel Miranda, President of the Portuguese Institute of Sea and Atmosphere, on the weekly newspaper Sol reports that the 1969 earthquake "(...) *created in the Portuguese population an awareness of the need to study, monitor and better understand earthquakes.*" Also Miranda e Carrilho, (2014) on a report prepared for the same agency, agrees that the panic caused by the seismic shock of 1969 contributed to the strengthening of national seismographic network in the following years.

6 FINAL CONSIDERATIONS

Despite the different nature of the sources available for each analysed case, the regions and the societies affected are themselves different, though some similar characteristics can be perceived for the three cases.

With regard to the perception of the earthquake by the people, it seems to be recurrent a reference to the noise that precedes the earthquake, and also a frequently mentioned strong smell associated with it. The panic of the population during the earthquake seems also to abide by similar standards. It is possible to identify, in the different descriptions, a first moment of shock and confusion, followed by moments of panic and running away; and often the recollection of disorientation, when the main shock is followed by replicas, or other associated phenomena such as a tsunami.

The most serious consequences are undoubtedly the loss of human lives. In the case of *Benavente* and the earthquake of 1969, the numbers of indirect victims was higher than those of direct victims. The emotional shock of the quake was deadlier than its physical consequences. The injuries produced by shock and trauma, caused by the fall of structures or landslides are also referred to in all the news of earthquakes.

As for the material damage, in all cases there is record of falling chimneys, the cracking of walls and ceilings, and the fall of eaves and other structures. However an inversion of priorities is detected. In the XVIII century most reports focus on the description of damage in religious or public buildings, while in the twentieth century the great consternation seems to be with the loss of housing.

It should also be highlighted the rising in the 1969 earthquake of new problems: the cut in the power supply and communications; as well as the problems in traffic.

The first measures that are taken in all cases relate to the rescue of victims, the relocation of displaced and the repair of damage to buildings. On a second phase the reconstruction is considered.

Finally, with regard to the consequences of earthquakes in the local building culture, two distinct levels of intervention are identified: the official and the private.

At an official level, all earthquakes studied, as noted above, were followed by the authorities with great concern, and on each case, an attempt to regulate reconstruction is seen, possibly to ensure that effective seismic reinforcement measures are added. It happened in the *Pombal* reform in 1755, in the *Benavente* reconstruction plans in 1909, and in the regulation of reconstruction in 1969.

However, from these three cases, only the first plan appears to have been effectively implemented. In the two subsequent cases the delay of official aid led to private initiative on most reconstructions and, in these cases, as illustrated in the situation of *Benavente*, the new building tends to reproduce the models, techniques and construction systems, practiced before the quake.

Finally, even in the case of the reconstruction of Lisbon after the earthquake of 1755, which was pioneer in the introduction of seismic reinforcement systems, with time, and the fading of memories of the disaster, the constructors tend to overlook the initial care, changing and weakening the buildings resistance to a possible earthquake in the near future.

ACKNOWLEDGMENTS

This paper is supported by FEDER Founding through the Operational Programme Competitivity Factors – COMPETE and by Nacional Funding through the FCT – Foundation for Science and Technology within the framework of the Research Project 'Seismic V – Vernacular Seismic Culture in Portugal' (PTDC/ATP-AQI/3934/2012).

REFERENCES

Andrade, H. M. C. (2011). Caracterização de edifícios antigos. Edifícios "Gaioleiros". Dissertação de Mestrado em Engenharia Civil. Lisboa: FCT-UL.

Antunes, M. T. (1956). Aspectos geofísicos do terramoto de Benavente de 23 de Abril de 1909. In Separata de Revista da Faculdade de Ciências de Lisboa, 2.ª Série B, Fasc. 6, Lisboa: FCL. pp. 5–28.

Barreto, B. D. (1758). Memória Paroquial da freguesia de Vila Nova de Milfontes, comarca de Ourique [ANTT, Memoìrias Paroquiais, Vol. 23, no 142, pp. 895 a 925].

Braga, A.M. & Estevão, J.M.C. (2010). Os sismos e a construção em taipa no Algarve, In: 8° Congresso Nacional de Sismologia e Engenharia Sísmica. Aveiro: Universidade de Aveiro.

Choffat, P. & Bensaúde, A. (1912). Estudos sobre o sismo do Ribatejo de 23 de Abril de 1909. Lisboa: Imprensa Nacional.

Cruz, M. A. (1968). Documentos para o ensino: a cidade de Setúbal. In Finisterra. Revista Portuguesa de Geografia. Lisboa: CEG. pp. 300–310.

Gaspar, A. F. (1758). Memória Paroquial de S. Manços, Évora. [ANTT, Memoìrias Paroquiais, vol. 22 no 45, pp. 285 a 288].

Gracia, M.F. (1758). Memoìria Paroquial da freguesia de Poìvoa de São Miguel, comarca de Beja [ANTT, Memoìrias Paroquiais, Vol. 30, no 238, pp. 1821 a 1824].

Portugal. (1724). Gazeta de Lisboa Occidental, N.° 43 (26 de Outubro de 1724). Officina de Pascoal da Sylva, p. 344.

A nova Messina. O Terramoto de 23 de Abril. (1909). Ilustração Portugueza, n.° 170. Lisboa: Jornal o Século. pp. 649–652.

A terra treme no terramoto do dia 23. (1909). Ilustração Portugueza, n.° 167. Lisboa: Jornal o Século. pp. 545–550.

Através dos escombros do Ribatejo. O Terramoto de 23 de Abril. (1909). Ilustração Portugueza, n.° 168. Lisboa: Jornal o Século. pp. 585–608.

Lisboa, A. P. (1758). Memorias das principais providencias que se derão no terremoto que padeceo a corte de Lisboa no anno de 1755. LOCAL E EDITOR

Nova e fiel relação do terremoto que experimentou Lisboa e todo Portugal no 1 de Novembro de 1755 (1756). Lisboa: Officina de Maniel Soares.

Magro, M. R. (1758). Memória Paroquial da freguesia de São Luís, comarca de Beja [ANTT, Memórias Paroquiais, Vol. 21, no 157, pp. 1339 a 1342]

Martins, I. & Víctor, C.A.M. (1990). *Contribuição para o estudo da sismicidade de Portugal continental.* Lisboa: IGIDL-UL.

Miranda, J.M. & Carrilho, F. [2014]. Relatório: 45 anos do sismo de 28 de Fevereiro de 1969. Lisboa: IOMA. Recuperado de: https://www.ipma.pt/export/sites/ipma/bin/docs/relatorios/geofisica/rel_sismo-1969.pdf

Moita, B. M. S. (2013) Reforço e Reabilitação Sísmica de Construções da Baixa Pombalina. Dissertação de Mestrado em Construção Civil. Setúbal: ESTB-IPS.

Moreira, V. J. S. (1991). Sismicidade Histórica em Portugal Continental. In Revista do Instituto Nacional de Meteorologia e Geofísica. Lisboa: IPMG. pp. 29–63.

Owen, W. (1756). The general theory and phenomena of earthquakes and volcanoes. London: W. Owen.

Pereira, A. (1756). A Narrative of the earthquake and fire of Lisbon. London: G. Hawkins.

Pereira, J. (1758). Memória Paroquial de Fátima

Trindade, A. D. S. (2014). Risco de tsunami da cidade de Lagos: avaliação da vulnerabilidade e modelação de rotas de evacuação numa abordagem SIG. Dissertação de Mestrado em Sistemas De Informação Geográfica – Tecnologias e Aplicações, Lisboa: FCUL.

Vieira, R. S. (2009). Do terramoto de 23 de Abril de 1909 à reconstrução da Vila de Benavente. Um processo de Reformulação e expansão urbana. Benavente: CMB.

Seismic Retrofitting: Learning from Vernacular Architecture – Correia, Lourenço & Varum (Eds)
© 2015 Taylor & Francis Group, London, ISBN 978-1-138-02892-0

Seismic behaviour of vernacular architecture

H. Varum
CONSTRUCT-LESE, Faculty of Engineering, University of Porto, Portugal

H. Rodrigues
RISCO – School of Technology and Management, Polytechnic Institute of Leiria, Portugal

P.B. Lourenço & G. Vasconcelos
ISISE, Faculty of Engineering, University of Minho, Guimarães, Portugal

ABSTRACT: Vernacular buildings are in several seismic prone regions a relevant part of the building stock. They represent an important part of the of the cultural legacy, and of the technological and heritage context of the society. The present paper presents a revision of the effect of recent earthquake disasters in old buildings, and discusses the influence of the main structural elements in the vulnerability of these buildings, as well as the more common failure mechanisms. Finally, it is presented a review on the elements that more contribute to the seismic behaviour and performance of vernacular buildings.

1 INTRODUCTION

Vernacular constructions are widely present in the built environment of many regions. Being non-engineered constructions, they result from the use of traditional materials and techniques, continuously improved based on the observations of their performance and response to different requirements and demands imposed by the social and physical environment.

It is commonly accepted that vernacular architecture has been continuously adapting to the local climate, extreme loadings, type of use, to the environment in which is inserted, etc. Thus, during centuries, Man has developed some constructive techniques to reach the comfort and performance desired, function of the local climatic conditions, the available materials and other conditions relating to their culture (Cañas & Martín, 2004).

In earthquakes, vernacular construction may suffer considerable damage, and eventually collapse, with the consequent economic losses, injuries and deaths. Some vernacular buildings are particularly vulnerable to earthquakes, in most cases due to the overall layout of the construction, to the connection between structural elements, poor constructive details and poor maintenance. Recent earthquake events have shown the vulnerability and limited capacity of some vernacular solutions, which may induce a poor seismic performance.

2 VERNACULAR CONSTRUCTION

2.1 *Concepts*

It has been estimated that 90% of the construction around the world is of vernacular type, meaning that it is for daily use (Oliver, 1997).

A vernacular construction reflects the period of time, the local environmental, cultural, technological and historical context of the society, representing a cost-effective construction, based on the local available materials.

The vernacular constructions were usually constructed to meet the needs of the residents, considering the local climate, available materials, and natural hazards that may suffer.

2.2 *Local available materials*

The structural behaviour of a vernacular construction is clearly influenced by the building technology, the construction details and the used materials.

Earth, stone or brick masonry elements and timber are the most frequently used structural materials, and their combination was currently adopted in vernacular constructions.

Earth construction solutions are among the most important materials/ techniques used in vernacular construction. In fact, it is estimated that, even nowadays, nearly 50% of the world's population lives in

Figure 1. Timber construction in Bhutan (credits: NCREP 2015).

Figure 2. Timber elements within walls in Greece (credits: Hugo Rodrigues).

Figure 3. Ties in a traditional masonry building, at L'Áquila, Italy (credits: CONSTRUCT-LESE).

earth-based constructions (Guillaud, 2008). Besides the evidence in less developed countries, where this type of construction is very popular, it can be largely found in some developed regions, due to the sustainable construction concerns and practices. The number of earth-based buildings tends to increase (Pacheco-Torgal & Jalali, 2012). Earth has been used in vernacular constructions, with several different techniques, such as rammed-earth, adobe, wattle and daub, cob, among others.

The success of a certain type of vernacular constructions may be related with the available raw materials, as it is the case of the adobe construction in Aveiro district, Portugal. The principal raw materials used in the production of the adobes in the region were coarse sand, argillaceous earth and lime. Worldwide, in the vernacular architecture, for the production of adobes, to natural earth can be added clay or sand, and it was also common the addition of natural fibres (as straw or sisal, for example) to control cracking during adobes drying in the sun (Silveira et al., 2012).

But the use of raw earth was associated to other construction techniques, adapted to the local construction practices, materials and needs. For example, in Portugal, "tabique" elements were used in different constructions all over the country. In fact, it is one of the most used traditional building techniques using raw materials, particularly for interior partition walls. A tabique constructive element can be simply described as a timber structure, more or less complex and robust, filled and plastered on both sides by a composite earth based material (Silveira et al., 2012).

2.3 *Seismic resistant constructive elements*

One of the important characteristics of the built vernacular heritage is related with its adaptation to the environment constraints, and in many situations it is evident the relation with the natural hazards incidence (Ortega, Vasconcelos, & Correia, 2014).

The occurrence of a strong natural extreme event, or the repetition of small events, like earthquakes, may push the community to react and to transpose this to act in the construction practices, developing a local risk culture (Ferrigni et al., 2005). These actions may include the production of earthquake safer features added in the current construction practice following a seismic event, which has caused damage. Methods and solutions that revealed to be vulnerable are either abandoned, or modified/improved. In reconstruction works, the details and solutions, which have withstood the event, are adopted again. Eventually, these aseismic features can take root in the building culture of the region (Jigyasu, 2002).

D'Ayala (2004) observed that, in regions of moderate seismicity, corner returns and quoins, connection with partition walls, regular masonry fabric (stone or brickwork), floor and wall ties, alternate orientation of floor structures, can typically be found. In regions of higher seismicity, these features are accompanied by timber ring-beams, monolithic lintels and stone frames around openings, framing and bracing of masonry with timber post and struts.

For example, the use of timber elements within walls, found in *Pombalino* buildings, in Portugal, and in traditional constructions in Greece, Turkey and India, has been used to improve the earthquake performance of these buildings (Pacheco-Torgal & Jalali, 2012).

In seismic prone regions it is common to find elements for the improvement of the connection between structural elements (ties and reinforcing rings); and construction elements that counteract the horizontal forces in earthquakes (buttresses and reinforcement arches).

Metal and wooden ties have been applied for a long time as a reinforcement measure in highly seismic regions, such as Italy (Figure 3). This type of strengthening measure has also been adopted in Portugal, in order to improve structural stability.

Buttresses can be found in ancient constructions, built at the same time as the building, as a deliberated feature, or they can be added to older masonry, as a consolidation, reinforcement or stability improvement measure. Buttresses can counter the rotation of the façade, thanks to their sheer mass. They can be found again, in many of the highly seismic regions, where masonry is the main construction system. Buttresses are also common in vernacular architecture, being introduced to counteract the lateral forces imposed by roof systems, arches and vaults (Ortega et al., 2014). They may also have a seismic concern origin, and they can still be found in many vernacular buildings.

In urban environments, other elements that may play a similar reinforcement role are external stairs, counter arches between buildings, vaulted passages and loggias.

3 VERNACULAR BUILDINGS AND EARTHQUAKES

Past and recent destructive earthquakes proved that most of the severe damage and collapses may occur on non-engineered buildings. As stated before, vernacular constructions result from traditions, improved with time as a response to the requirements of their social and physical environment. As so, in seismic regions, where small or moderate earthquakes are frequent, these events influence the construction practice, thus, influencing the requirements for the structural capacity of these buildings. Consequently, several elements are incorporated in the traditional construction for their seismic behaviour improvement.

However, in many situations, the poor economic condition of habitants, the state of conservation of the buildings, and the large recurrence period of this type of extreme events, are responsible for neglecting the importance of these elements.

Even with good construction practices, and incorporating seismic improvement construction practices, for strong earthquakes, it may be difficult to assure an adequate behaviour for certain type of construction, as for the ones with high mass, like load-bearing masonry structures.

In L'Áquila 2009 earthquake, several vernacular buildings have showed a deficient behaviour under the seismic demands they suffered from. The substantial re-adaptation and changes of their structural system, in the last 50 years, to account for a different use of

Figure 4. Village affected by the L'Áquila earthquake (credits: RISCO).

Figure 5. Out-of-plane collapse of a façade masonry wall (credits: CONSTRUCT-LESE).

the spaces, the consideration of modern residential requirements for comfort, may have influenced their seismic performance, contributing to several damage mechanisms identified (D'Ayala & Paganoni, 2011). The original structures include both, masonry vaults and timber floors and roofs, were replaced, in recently retrofitted buildings, by lightweight jack arches or steel beams (D'Ayala & Paganoni, 2011).

In L'Áquila, several collapses of the building façades, due to poor connections between walls, were found; and the quality of masonry dramatically influenced the level of damage (Fig. 5).

The seismic performance of traditional masonry constructions retrofitted with ties may be influenced by their location, by the material deterioration of the anchor element, or by the lack of integrity and shear capacity of the masonry to withstand the thrust generated at the anchoring plate/element, by the relative movement of the two orthogonal walls (Rodrigues et al., 2010).

Vernacular buildings are a relevant part of the building stock affected by the Emilia 2012's earthquake (Paupério et al., 2012), mainly due to the overall layout, constructive details and poor maintenance. Many of the building in the affected area have shown a very

Figure 6. Collapsed façade of a building with ties, during L'Aquila earthquake (credits: CONSTRUCT-LESE).

Figure 7. Masonry buildings collapsed during the Emilia-Romagna earthquake (credits: RISCO).

poor seismic performance. Approximately two hundred rural buildings, suffering from partial to total collapse, have been counted in an area of about 500 km^2 (Sorrentino, Liberatore, Liberatore, & Masiani, 2014).

The main cause of damage in these buildings is associated with the higher length/thickness and height/thickness ratios of the walls, and with the use of slender brick columns, connected to the roof just by friction (Sorrentino et al., 2014).

Figure 8. Slender brick columns Emilia-Romagna region (credits: RISCO).

Figure 9. Collapsed houses of an entire community near Barpak during the Gorkha earthquake, 2015 (credits: Dipendra Gautam, 2015).

The 7.8 magnitude earthquake that hit Nepal on the 25th of April 2015, and the series of aftershocks that followed it, associated with a high number of human losses, severely affected numerous vernacular buildings. In remote areas like, Gorkha, Dolakha, Sindhupalchowk, Nuwakot, Solukhumbu, Kavre, Rasuwa and Dhading, the majority of houses were vernacular constructions. Locally available stone with earth mortar or, in many cases, without binding material, are common construction solutions in those areas.

In the worst affected area, traditional single-storey masonry houses without, or with poor, binding material, was the most common building typology. In Barpak of Gorkha (the worst affected area, beside the epicentre region), the majority of the adobe houses

Figure 10. Out-of-plane façade failure during the Gorkha earthquake, 2015 (credits: Dipendra Gautam, 2015).

Figure 11. Collapsed house with a heavy roof during the Gorkha earthquake, 2015 (credits: Dipendra Gautam, 2015).

Figure 12. Damage associated to the interaction be-tween adjacent buildings – L'Áquila earthquake (credits: CONSTRUCT-LESE).

collapsed. The out-of-plane failure mechanism was commonly observed in many houses.

In many houses, the roofing system was found to be heavier than in other typologies. Many of those observed houses had roofing systems with stone covering (Fig. X). Due to climatic condition, people were adopting the use of heavy stone roofs and avoiding timber elements, and other lighter solutions in adobe masonry houses, so probably damage was further intensified by this constructive option.

4 ELEMENTS AFFECTING SEISMIC BEHAVIOUR

From the previous section it is clear the influence of the material properties and constructive solutions and details. These are critical for the structural behaviour of vernacular buildings. Based on the observations from damages and collapses in previous earthquakes, some parameters were identified that can influence significantly the structural behaviour of these buildings (Ortega et al., 2014).

One of them is the interaction between buildings, which has a significant influence in the seismic performance of structures. The interaction between adjacent buildings with different dynamic behaviour may have implications on the seismic behaviour and performance of the structures.

The shape of the building plays an important role if there is regularity in plan and in elevation. Lacking of shape regularity in plan may induce an irregular distribution of the seismic demands, due to the eccentricity between the centre of mass, relatively to the stiffness centre, enforcing the torsional demands due to earthquake actions. Regarding the typologies of walls, they can differ in the constituent material, in the quality of the material, in the arrangement of the elements/units, having a significant influence in the capacity of the masonry, and so of the building to support vertical forces and to withstand the horizontal demands resulting from seismic actions.

The diaphragms, at the floor level, may be very flexible, when wooden floors are used, which are the most common solution. Proper connections between walls and floors are important, so that the vertical structural elements do not behave independently, improving the coupled response of the building. Excessive deformability of the floors may have a negative effect on the behaviour of the façade walls. To assure the proper "box-behaviour", assuring that all structural elements contribute to the structural resistance of the building, the connections between structural elements should be carefully considered. In many cases, the walls work separately, having to bear, by themselves, the portion of load that acts on them (Ortega et al., 2014).

Strengthening devices, like ties, buttresses and other reinforcement elements, can improve the structural behaviour of the buildings and, in particular, enhance the connection between structural elements (ties and reinforcing rings). Other construction elements can counteract part of the horizontal demands due to earthquakes, such as buttresses and reinforcement arches.

Another important aspect that influences significantly, the seismic behaviour of vernacular buildings is associated with the alterations and/or inadequate interventions. Addition of floors and new elements or spaces, and removal of existing strengthening elements, may change significantly the structural behaviour of the original building.

5 FINAL REMARKS

Vernacular construction is, in many cases, the primary construction system in certain rural areas, but that is the situation also in many urban areas, where an important heritage is present.

In vernacular construction, the materials and construction techniques are very different around the world, and in many cases elements to improve the seismic behaviour of the building are included.

Seismic deficiencies of vernacular constructions are, in many cases, associated with the heavy weight of the structures, their low strength, brittle behaviour, and deficiencies in the connections between structural elements. The primary factors affecting the structural performance of a vernacular construction, in particular due to earthquake actions, is the position of the building and its relation with adjacent buildings; the presence of irregularities, the construction solutions; the quality of the materials, the lack of maintenance; eventual alterations and damage due to previous extreme events. Also, the adequate connection between the walls and floor/roof system is essential to prevent collapse.

ACKNOWLEDGEMENTS

The authors gratefully acknowledge the partial support by the research project 'SEISMIC-V – Vernacular Seismic Culture in Portugal' (PTDC/ATP-AQI/3934/2012), from the Portuguese Science and Technology Foundation (FCT).

REFERENCES

Cañas, I. & Martín, S. (2004). Recovery of Spanish vernacular construction as a model of bioclimatic architecture. *Building and Environment, 39*(12), 1477–1495. Doi: http://dx.doi.org/10.1016/j.buildenv.2004.04.007

D'Ayala, D. (2004). *Correlation of fragility curves for vernacular building types: houses in lalitpur, nepal and in istanbul, Turkey*. Paper presented at the 13th World Conference on Earthquake Engineering Vancouver, B.C., Canada

D'Ayala, D., & Paganoni, S. (2011). Assessment and analysis of damage in L'Aquila historic city centre after 6th April 2009. *Bulletin of Earthquake Engineering, 9*(1), 81–104. Doi: 10.1007/s10518-010-9224-4

Ferrigni, F., Helly, B., Mauro, A., Victor, L. M., Pierotti, P., Rideaud, A. & Costa, P. T. (2005). *Ancient Buildings and Earthquakes. The Local Seismic Culture approach: principles, methods, potentialities*.

Guillaud, H. (2008). Characterization of earthen materials. In H. G. E. Avrami, M. Hardy (Ed.), *Terra literature review—an overview of research in earthen architecture conservation* (pp. 21–31). Los Angeles (United States) The Getty Conservation Institute.

Jigyasu, R. (2002). *Reducing Disaster Vulnerability through Local Knowledge and Capacity. The case of Earthquake Prone Rural Communities in India and Nepal.* (phd Thesis), Norwegian University of Science and Technology, Trodheim.

Oliver, P. (1997). *Encyclopedia of Vernacular Architecture of the World.* Cambridge: Cambridge University Press.

Ortega, J., Vasconcelos, G., & Correia, M. (2014). *An overview of seismic strengthening techniques traditionally applied in vernacular architecture.* Paper presented at the 9th International Masonry Conference, Guimarães.

Pacheco-Torgal, F. & Jalali, S. (2012). Earth construction: Lessons from the past for future eco-efficient construction. *Construction and Building Materials, 29*(0), 512–519. Doi: http://dx.doi.org/10.1016/j.conbuildmat.2011.10.054

Paupério, E., Romão, X., Tavares, A., Vicente, R., Guedes, J., Rodrigues, H., Varum, H. & Costa, A. (2012). Survey of churches damaged by the May 2012 Emilia-Romagna eartquake sequence (pp. 27): Faculty of Engineering of the University of Porto and University of Aveiro.

Rodrigues, H., Romão, X., Costa, A. G., Arêde, A., Varum, H., Guedes, J. & Paupério, E. (2010). *Sismo de L'Aquila de 6 de Abril de 2009. Ensinamentos para Portugal"*. Paper presented at the 8° Congresso Nacional de Sismologia e Engenharia Sísmica – SISMICA 2010, Aveiro, Portugal.

Silveira, D., Varum, H., Costa, A., Martins, T., Pereira, H. & Almeida, J. (2012). Mechanical properties of adobe bricks in ancient constructions. *Construction and Building Materials, 28*(1), 36–44. Doi: http://dx.doi.org/10.1016/j.conbuildmat.2011.08.046

Sorrentino, L., Liberatore, L., Liberatore, D., & Masiani, R. (2014). The behaviour of vernacular buildings in the 2012 Emilia earthquakes. *Bulletin of Earthquake Engineering, 12*(5), 2367–2382. Doi: 10.1007/s10518-013-9455-2

Seismic Retrofitting: Learning from Vernacular Architecture – Correia, Lourenço & Varum (Eds)
© 2015 Taylor & Francis Group, London, ISBN 978-1-138-02892-0

The design of 1758's master plan and the construction of Lisbon 'downtown': A humanistic concept?

V. Lopes dos Santos
C.I.A.U.D., University of Lisbon, Lisbon, Portugal
Faculty of Architecture, University of Lisbon, Lisbon, Portugal

ABSTRACT: This paper addresses an approach to the seismic resistant main construction system and some humanistic concepts, observed for the last 20 years of research in the downtown *(Baixa)* of Lisbon. After the 1755's earthquake, Lisbon downtown was rebuilt. It took more than one century for the works to be concluded. During this period of time, different governments tried to subvert the primitive humanistic ideas of Sebastião José de Carvalho e Mello (*Pombal* Marquis, the prime minister). This paper presents an overview of the Portuguese pombalino system and its humanistic approach.

1 PORTUGUESE POMBALINO SYSTEM

The main objective of the wood frame system (*gaiola pombalina*) was to create a self-resistant structure. Regarding its elastic properties, the wood frame system could support the seismic effect, whose shockwaves were transmitted in a homogeneous way throughout the ground level, where walls were built with regular stone and ceilings with brick vaults. If the masonry walls fell down and the inhabitants were caught inside the house, the wood elements would not be affected and they would protect people from major physical injuries. The massive vaults also allow a better protection from fire spreading to the wooden frame, which only started at the second floor.

The vertical elements of this wooden structure would be, necessarily, near the inner face of the resistant walls. However, to prevent the fatigue and the natural continued deformation of its elements, and to avoid the visual effects on the plaster of the walls regarding humidity and fungi on this structure, it was decided to increase the depth of the recovering plaster, and to get some special elements (*mãos* /hands) fixing the wooden frame to the masonry emergent of the main bearings and the mid beams, and the wall corners of rock and/or equipped paving-tile, for eyebolts in plate iron in the continuation of the main wooden beams.

It was not considered acceptable to leave the bearing and crosspieces of this wooden structure in plain sight. Therefore, they were covered by plaster and mortar. Occasionally, they were covered with wooden planks.

Due to the irregularity regarding the dimensions of the stone materials of the walls, and to the dimensions of the wooden parts, cracks could appear in the plaster

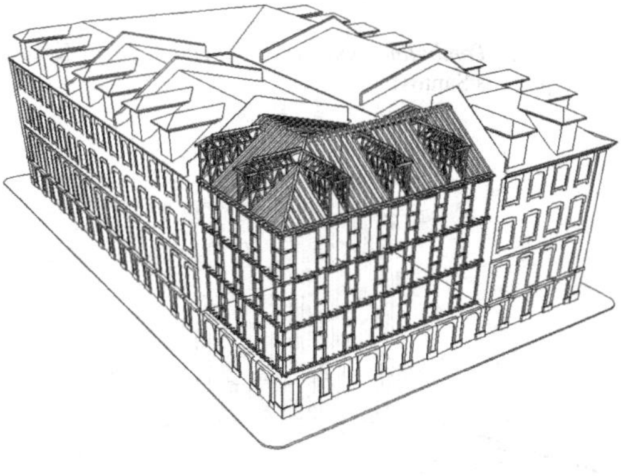

Figure 1. *Pombalino* Block (credits: V. Lopes dos Santos).

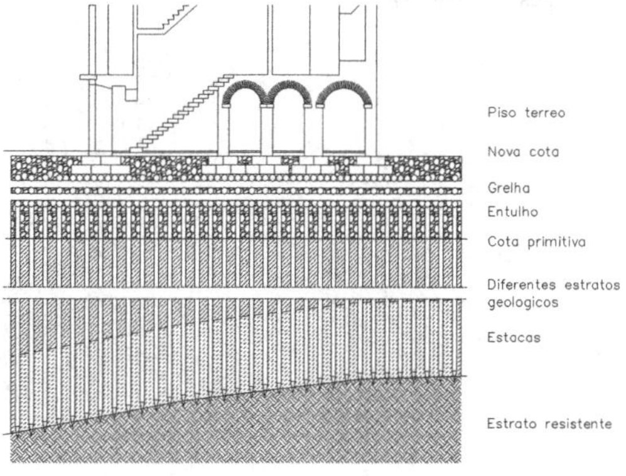

Figure 2. *Pombalino* system: Constructive detail (credits: V. Lopes dos Santos).

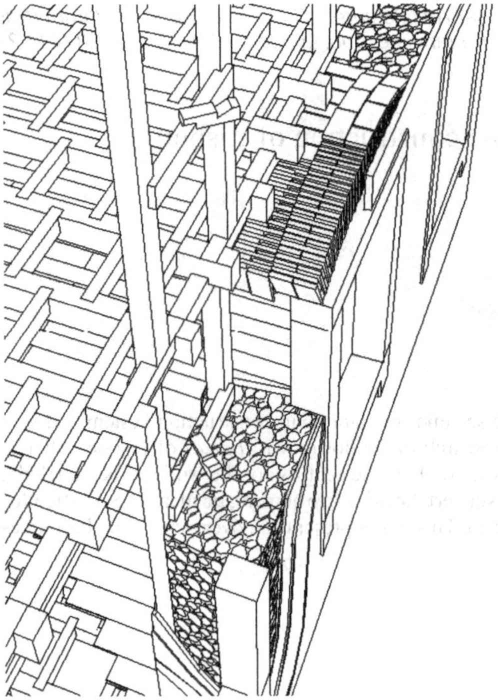

Figure 3. *Pombalino* system: Constructive detail (credits: V. Lopes dos Santos).

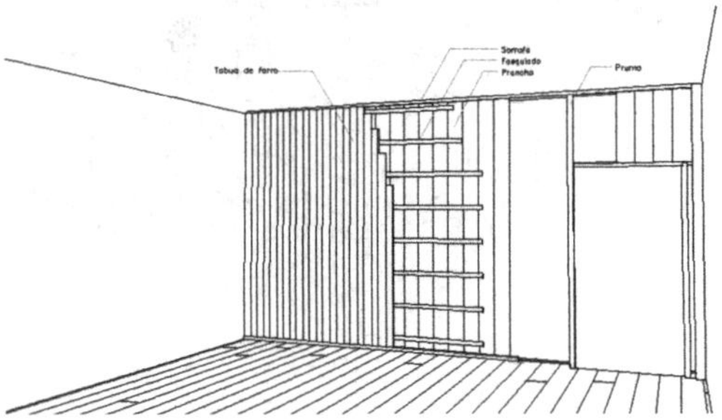

Figure 4. *Pombalino* system: Constructive detail (credits: V. Lopes dos Santos).

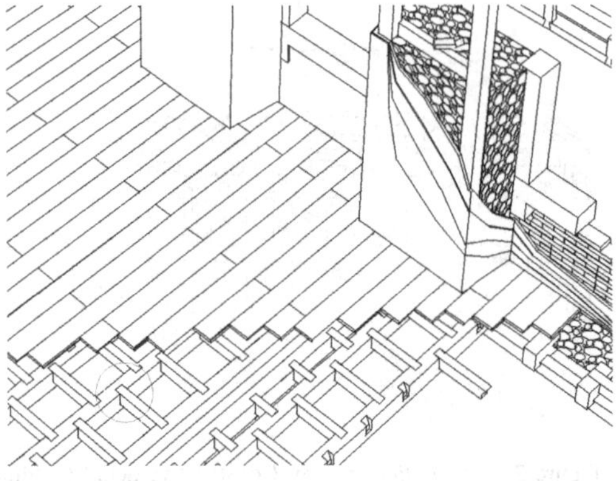

Figure 5. *Pombalino* system: Constructive detail (credits: V. Lopes dos Santos).

finishing. If they had a reduced thickness, the coverings of the interior and exterior resistant walls, would have been thickened, with a deeper covering.

The result was a structured masonry. The wood gives traction capabilities to the masonry walls, in order to resist much better to seismic actions.

Due to bigger absorption of seismic waves, and also the reduced propagation (due to the material discontinuity), the resistant walls have demonstrated a good behaviour during several earthquakes.

The used masonry does not resist very well to horizontal mechanical solicitations, for being only prepared to resist to compression efforts. Time demonstrated that resistance to horizontal and traction efforts was assured by the wood frame.

The elasticity of the wood, included in its application at non-structural interior walls (*tabique*), sets of beams and planks, allows the absorption of seismic waves, decreasing the propagation, at the same time that the stuck timbering in the main walls works like a structural element of the masonry, complementing it in its mechanical properties.

Modulated and repetitive building façade design located in the territory of implementation of the 1758's Plan (Baixa de Lisboa), and, in accordance with the diversity of space organisation, might be wrongly induced to think, on a general precast construction concept. It is verified that most of the times only jam stones, windows and tiles were elements subject to standardisation.

It becomes therefore evident, at the pombalino construction system, the identification of different solutions/ variants of the described elements. They should be based on the experience of the contractors and the workers.

2 DOGMA AND HARMONY

The dogma, if accepted, erases the desire to reach the understanding of the fact; and even more important, of the act that produces it.

Displayed in the simplest acts, causality is a part of the reality. Man exhibits himself by acts; especially when following the principles that rule all mankind. The compilation of concepts or principles, that in a fatalist form intends to organise what, at the beginning and by itself it is balanced, it is no more than a compilation, more or less extensive, but needed and with weak objectives.

The Harmony cannot be invented, because it always existed. The evidence of the 'new' might be seen only 'today', but it has always existed. Perhaps, because it had never been understood.

Every day, new elements, isolated or in association, are observed and understood. Harmony also is present in the mind of those who produce intellectually, and also in the work of who take in charge the materiality.

The main idea applied at the design conception of the *Baixa Pombalina* does not leave many doubts about the existence of a strong humanistic culture of the social group, who leads the projects.

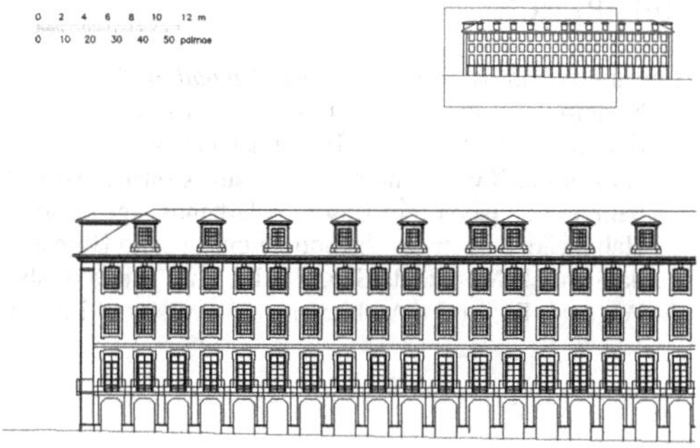

Figure 6. Pombalino typology block (credits: V. Lopes dos Santos).

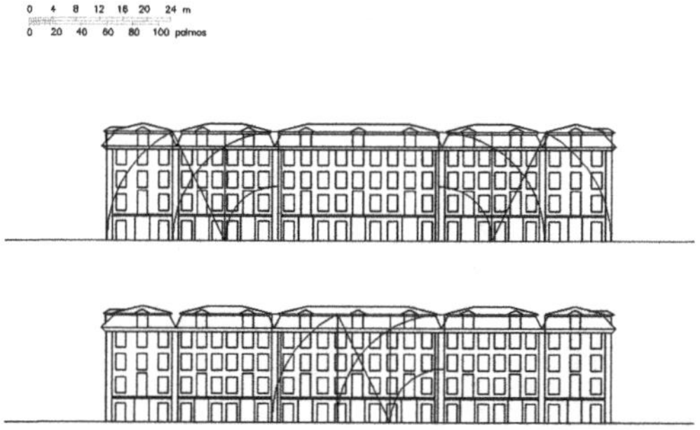

Figure 7. *Pombalino* typology block: Analysis (credits: V. Lopes dos Santos).

3 HUMANISTIC CONCEPT

This humanist main behaviour will punish the minds of those, who believe that the violent confrontation of ideas, almost always generating radical positions, is 'the' solution for the stability or the social evolution. However, the idea of the existence of an artificially imposed harmony, based on 'unquestioned truths', always denies the rationality of Man.

Although the strong cultural opposition between social groups affected the continuity of the established order; for doctrinal agents like the Church of Rome and others, it determined the reestablishment of a humanist order.

The confluence of efforts is verified. For selecting it will be the validity of the negation of the position, or the honour of who defends a collective ideal, against individual purposes with no altruism.

The effect is determined by its causes, with the same causes being able to determine a different effect, depending only on its chronological succession and the influence of the universe where they occur.

'Today' it can represent a new break to the sinuous line of the historical evolution.

The succession of the historical phenomenon and the cycle of its occurrence are comparable to a function identified for a wave and oscillatory movement of constant period, whose period decreases in accordance with an exponential rule.

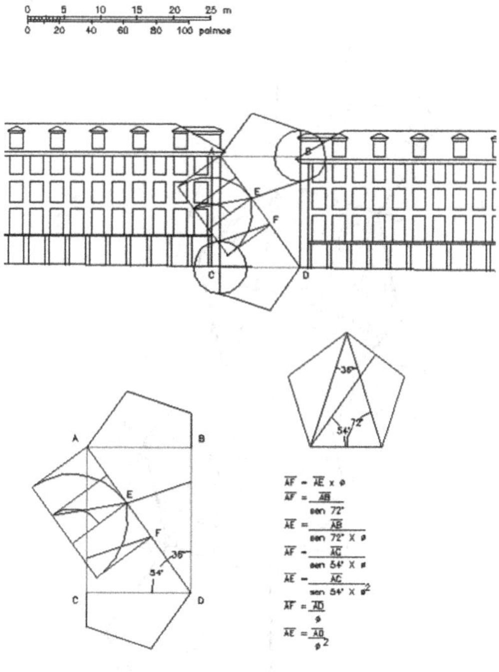

Figure 8. *Pombalino* typology block: Analysis (credits: V. Lopes dos Santos).

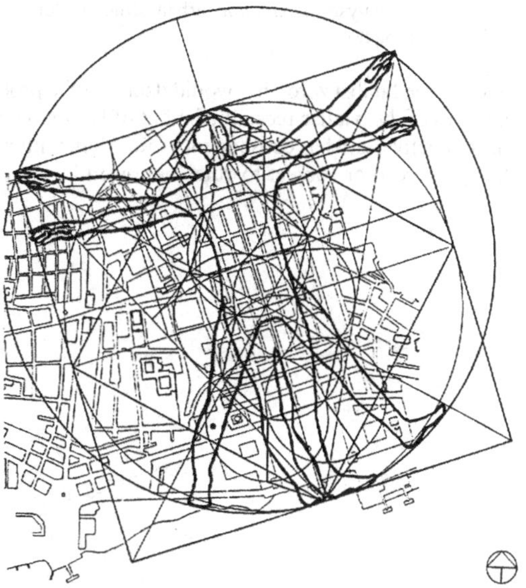

Figure 9. Analysis: *Pombalino* urban structure (credits: V. Lopes dos Santos).

However, the successive movements along History (time), even if periodically repeated, have an influence that could be established by a harmonic oscillatory movement, cushioned for each one of them.

Although the respective amplitude is cushioned each time if the period repeats, they do not hinder the appearance of new cycles. It will be preferable to consider for the limit of the asymptotes of the infinite,

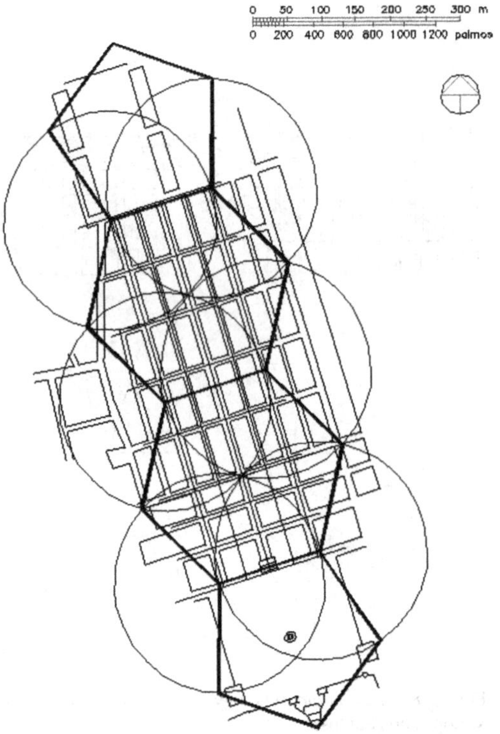

Figure 10. Analysis: *Pombalino* urban structure (credits: V. Lopes dos Santos).

and not the fatalist zero, that would decrease the probabilities of the pacific reestablishment of Harmony of the collective, and has consequences for Man, without being coerced or subject to rules engraved by social lobbies.

REFERENCE

Lopes dos Santos, V. (1994). The *Pombalino* Construction System in Lisbon: in Urban Grouped Buildings Collective Housing – Study of a Humanist legacy of the Second Half of the XVIII century [O Sistema Construtivo Pombalino em Lisboa: em Edifícios Urbanos Agrupados de Habitação Colectiva – Estudo de um Legado Humanista da Segunda Metade do Séc. XVIII]. PhD Thesis. Lisbon, Portugal: Faculty of Architecture, University of Lisbon.

Seismic Retrofitting: Learning from Vernacular Architecture – Correia, Lourenço & Varum (Eds)
© 2015 Taylor & Francis Group, London, ISBN 978-1-138-02892-0

Timber frames as an earthquake resisting system in Portugal

E. Poletti, G. Vasconcelos & P.B. Lourenço
ISISE, Faculty of Engineering, University of Minho, Guimarães, Portugal

ABSTRACT: Timber frames are commonly adopted as a structural element in many countries with specific characteristics varying locally, in terms of geometry and materials. Their diffusion in Southern European countries is linked to their good seismic-resistant capacity, but only in the last decade interest has grown for this structural typology, and studies have been performed to better understand their behaviour. In this contribution, a brief state of the art on existing timber frame building typologies is presented, focusing on their seismic-resistant characteristics. Additionally, an overview of possible strengthening solutions, adopted both in practice, and tested experimentally are presented. Their performance when applied to walls and connections is also discussed.

1 INTRODUCTION

Timber frame buildings are a common traditional construction present in many countries, particularly in local vernacular architecture, as they constitute an important cultural heritage worth preserving. They are characterised by a timber frame, which in some cases presents bracing members, filled with different infill materials such as earth mortar, canes, masonry. Portuguese *Pombalino* buildings represent a valuable construction typology, both for their cultural significance after the 1755's Lisbon earthquake and the creation of a new city, as well as for the introduction of new and innovative technological features. In fact, it is in this period that timber frame construction developed as an effective seismic-resistant system in regions of high seismicity, such as Southern European countries.

2 TIMBER FRAMES AS A LOCAL CONSTRUCTION SYSTEM

2.1 *Diffusion of the system*

Timber frames have been adopted worldwide for centuries. The first timber frames date back to more than 2000 years, as testified by Vitruvius, who in his *De architectura* describes the so-called Opus Craticium. It evolved from wattle and daub construction and it used for the first time squared structural timber elements and a more resistant infill (Ulrich, 2007). This kind of construction has evolved along the centuries as joinery got more sophisticated, craftsmanship improved, and the availability of wood changed in different countries.

In each country, a great variability exists in terms of geometry and materials, as the type of frame was influenced by local traditions, and tendency for this cheap construction was to build with local materials. The availability of wood and the length of the elements dictated the geometry of the frame. Infill was also based on the available materials and could vary from wooden planks, to skin, to wattle and daub, to earth mortar and to bricks.

Examples of timber framed construction can be found all over Europe, in North and South America, and in Asia. In England, timber frame construction evolved with the adoption of curved elements, and from the XV century the use of close studding and decorative panels, to show off the owners' wealth, (Fig. 1) became common. After the great fire of London, this type of construction was viewed with scepticism, but it was later adopted for upper storeys in revival houses from the XIX century. In Germany, *fachwerk* construction can be found all over the country, and later emigrants took this tradition with them when they left the country for the USA, or for Eastern European countries. Commonly to other central European countries, timber frame buildings (*colombages*) were introduced in France during the Middle Ages and were used until the XIX century. The buildings evolved from frames adopting long timbers, to frames adopting shorter timbers, with a lighter and stiffer structure, and the introduction of jetties. The French exported too their know-how abroad, and examples deriving from *colombages* construction can be found in Haiti and the USA.

Similarly, in Turkey, *hımış* construction is a very common vernacular construction, usually with a timber-laced masonry ground floor (*hatıl*). Different infill typologies were adopting, such as rubble masonry, lath and plaster (*bağdadi*) and wooden planks (*dizeme*). Other examples of timber frame construction can be found in Kashmir (*dhajji-dewari*), Peru (*quincha*), Nicaragua (*taquezal*), El Salvador (*bahareque*), just to name a few (Langenbach, 2007). For more information on timber frame construction worldwide, see Poletti (2013).

Figure 1. A XV century timber frame house in York, England (credits: E. Poletti).

Figure 2. A traditional timber frame house in Chalkida, Greece (credits: E. Poletti).

2.2 *Earthquake-resistant solution*

Past earthquakes have shown the good seismic properties of timber frame structures, when compared to other construction present in the territory. As a result, timber frames have been specifically adopted in the past as a seismic-resistant solution. This is particularly true for Portugal, Italy and Greece, where timber frames were adopted in regulations and codes (in Italy, the first ever seismic code was drawn up suggesting the use of the Borbone system).

In Italy, after the 1783 earthquake that destroyed Reggio Calabria, a new system, called *casa baraccata* (literally, 'baracca' means shack) was adopted by the authorities, by imposing standardized construction methods. The same construction technique, with slight changes, was also adopted after the Messina earthquake in 1908. The government (at the time, the Bourbon dynasty was ruling in the south of Italy) appointed engineers to develop rules for the reconstruction of the region. In 1784, the Royal Instructions *("Istruzioni Reali")* were emanated, and they consisted of rules to be applied to the new buildings. The rules included instructions on the exterior aspect of the buildings, on the height of buildings, on the width of streets, rules on construction of balconies, on domes and bell towers; and on the addition of an internal timber skeleton. These rules constituted the first official norms for seismic design.

Even though timber framing was already common in Calabria, a standardised type of half-timbered building was introduced by Giovanni Vivenzio, the court's physicist. This choice was born by the observation of the good seismic behaviour of existing half-timbered buildings, such as the palace of Nocera, built in 1638 (Bianco, 2010). Vivenzio proposed a 3-storey building with a timber skeleton, aiming at reinforcing the external masonry walls, avoiding their premature out-of-plane collapse (Figure 2.7b). The timber frames constituted the shear walls, presenting a bracing system of S. Andrew's cross, similar to what can be found in Lisbon (Tobriner, 1997). This system was also adopted for exterior walls in this type of buildings. Similarly to the Portuguese example, Vivenzio also proposed a construction by blocks, in this case of three buildings. The idea is that the central building has a higher height, and the lateral ones act as buttresses. This disposition allows for symmetry in the two directions, ensuring a similar stiffness for both directions. Different dispositions were adopted for the timber frame, from a double bonded frame with masonry in the middle (used for public buildings) to a single frame embedded into the wall at different depths (Ruggieri et al., 2015).

Other important earthquakes that affected the south of Italy occurred in Messina in 1905 and in 1908, during which the casa baraccata system showed a good seismic resistance (Rugieri et al., 2014). Nonetheless, after these events, the new standards of 1909 included rules for foundations. The standards suggested to prepare the foundation ground with a foundation slab and carry out diggings if necessary. Rock foundations or a firm soil are preferred. Moreover the posts had to be well fixed to the stone or to the foundation slab for at least 80cm. They should be burned at the extremities to prevent decay (Bianco, 2010).

Another country that uses timber frame buildings as a seismic-resistant solution is Greece. Half-timbered buildings were common all over Greece in different periods, as reported by many authors (Vintzileou et al., 2007; Tsakanika, 2008). Examples of this system are the monastic buildings in Meteora and Mount Athos, the post byzantine (Ottoman period) buildings in Central and Northern Greece; and the traditional buildings in the island of Lefkas.

Timber has been used together with masonry in Greece since the Minoan period (Tsakanika, 2008). This constructive solution evolved from the use of only horizontal timber elements to tie masonry, and to prevent the propagation of cracks to a heavier timber frame, adopting both vertical and horizontal timber members to reinforce masonry.

The half-timbered buildings in the island of Lefkas are different from those present in other regions of Greece. Here, a local structural system was developed before the 19th century. It demonstrated to be able to sustain seismic actions after buildings, built with this system, had showed a good seismic performance during the earthquake of 1821 (Vintzileou et al., 2007). During the same earthquake, the existing masonry buildings collapsed. Based on this, British Authorities (which ruled the Ionian Islands at the time) imposed new rules, developed from the aforementioned local system (Code of construction, issued in 1827) (Vintzileou et al., 2007). The rules provided guidance on the selection of building materials, thickness of stone masonry walls (at the ground floor), storey height and distance between adjacent buildings, similarly to what happened in Portugal and Italy. An innovation present in these buildings is the existence, at the ground floor, of timber columns stiffened by angles that constituted a secondary load bearing system in case of failure of the masonry walls, since they were connected to the timber-framed structure of the upper storeys (Vintzileou et al., 2007).

These buildings proved to be able to efficiently resist to earthquakes in 2003, when the island was hit by a strong quake. The main damages observed were not only due to this latest earthquake but also to previous ones, and the decay was due to poor maintenance. Additionally, modifications done in the existing structure, such as removing the ground masonry walls, led to large permanent horizontal displacements, since the secondary bearing system did not have sufficient stiffness (Vintzileou et al., 2007). In fact, the damages observed were sometimes due to poor or no maintenance, and to modifications done to the buildings, pointing out the importance of preservation. But timber frame buildings are present all over Greece. Usually consist of a timber-tied ground floor, and timber-framed upper storeys (see Fig. 2).

But even in countries where this system was not standardised, this structural typology has proven its very good seismic-resistant capacity. Of course, all observations should be taken with caution, since they depend on the level of maintenance of the traditional timber frame structures analysed. In Turkey, for example, after the big earthquakes that hit Kocaeli and Duzce in 1999, reports show that, in many regions, well-preserved traditional timber frame houses suffered less severe damage when compared to reinforced concrete and masonry structures (Gülhan & Güney, 2000). Similar reports also came in after the earthquakes in India and Peru. Because of their good performance, timber frame buildings have been used for reconstruction plans of vernacular buildings in rural areas hit by catastrophes, such as Haiti and Pakistan.

3 PORTUGUESE TIMBER FRAMES TRADITION

3.1 *Lisbon area*

After the devastating earthquake in 1755 that destroyed Lisbon Downtown, a reconstruction plan was put into action by the Prime Minister of the time, Marquis of *Pombal*, who appointed engineers and military architects to elaborate reconstruction plans of the city. The new regulations adopted provided rules for urban, architectonical and structural design, such as minimal distances between buildings, typology of façades, width of roads and sidewalks, height of the buildings, orientation of the buildings, structural system and creation of blocks (Mascarenhas, 2004). The new buildings designed took into account the seismic capacity of the structure, safety against fire, as well as a standardization of the structural elements, in order to achieve a cheaper and faster construction.

The buildings that derived from the proposed plan, called *Pombalino* buildings, were characterised by external masonry walls and an internal timber structure, named *gaiola* (cage), which is a three dimensional braced timber structure, similar to many half-timbered buildings that can be found in several European countries. The gaiola consists of horizontal and vertical elements, and diagonal bracing members, forming the typical X of St. Andrew's crosses (Fig. 3), which have a dissipative function (Cóias, 2007). The timber frame walls are usually filled, either with rubble or brick masonry, or even earth mortar and hay. Plaster was applied to frontal walls, creating small cuts in the timber elements so that mortar, generally lime-based, could adhere better. The adoption of a weak mortar and infill allowed flexibility to the wall joints, which could dissipate a higher amount of energy in case of an earthquake.

The ground floor consists of stone masonry columns supporting stone arches and vaults made of clay bricks, above which, in the first floor, the gaiola develops, reaching up to 5 storeys. This solution was adopted in order to prevent fire propagation to the upper floors. Early *Pombalino* buildings had a constant width of the stone walls of the façade, while in earlier or later buildings the width decreases along the height (Mascarenhas, 2004). The typical width of the external masonry walls was of 90 cm at the ground floor. A simplified internal timber structure was embedded into the external masonry walls on the inner side, to facilitate and to improve the connections with the floors, and the inner timber frame walls. The connections between the external masonry walls and the internal timber frame walls varied and it depended mainly on the number and

Figure 3. Internal timber frame walls in a *Pombalino* building (credits: MONUMENTA).

length of timber elements embedded in the external walls (Moreira et al., 2014).

Additionally, masonry walls perpendicular to the façades divided the buildings and avoided fire propagation in adjacent buildings.

This construction typology was not completely new to Lisbon, as in the oldest parts of the city, near the Castle; a similar simplified construction can be found. The innovation consists of the improvement of the system, as well as the standardization of the constructive practice in Lisbon (Mascarenhas, 2004).

The internal walls of the *gaiola* (*frontal* walls) may have different geometries in terms of cell dimensions and number of elements, as it depended greatly on the available space and the manufacturer's practices. The timber elements are notched together or connected by nails or metal ties. Traditional connections used for the timber elements varied significantly in the buildings: the most common ones were mortise and tenon, half-lap and dovetail connections. Variability exists in the sectional dimensions of the elements themselves: the diagonal members are usually smaller (10 × 10 cm or 10 × 8 cm), whilst the vertical posts and horizontal members are bigger (usually 12 × 10, 12 × 15 cm and 14 × 10 cm or 15 × 13, 10 × 13 and 10 × 10 cm respectively). The thickness of frontal walls can vary from 15 to 20 cm (Mascarenhas, 2004; Cóias, 2007).

The peculiarity of this type of buildings is that under a seismic event it is admissible that the heavy masonry of the façades falls down, as well as the tiles of the roof, and the plaster of the inner walls, but the timber skeleton should remain intact, assuring the resistance of the timber floors and keeping the building standing (Mascarenhas, 2004).

The *Pombalino* timber frame walls were built to provide an adequate resistance to seismic loading. However, it should be pointed out that their seismic efficiency has never been tested under onsite real conditions, as no other great earthquake has hit Lisbon since their construction. A number of experimental studies exist proving the good seismic resistance potential of *Pombalino* walls (Meireles, 2012; Poletti & Vasconcelos, 2015).

Figure 4. Timber frame walls in a traditional house in Guimarães (credits: E. Poletti).

Approximately a hundred years after the earthquake the building practice changed, getting worse from a seismic point of view. Pombalino buildings, where a complete *gaiola* structure is present and it is expected to be efficient against seismic actions, were progressively replaced by *Gaioleiro* buildings, where the timber structure does not exist and therefore structurally represent a worse construction quality (Mendes & Lourenço, 2010).

3.2 *Other Pombalino developments*

After the beginning of the reconstruction of Lisbon, other similar developments were created in locations of the country that were thought to have a strategic position and a potential for economic growth.

Vila Real de Santo Antonio in Algarve was redesigned in 1773 by military engineers, who follow plans similar to those adopted for the reconstruction of Lisbon.

Similarly, Porto Côvo in Alentejo was financed by a merchant, to transform a small fishing village into a trading centre (Mascarenhas, 2004). The interesting point is that the existing traditional houses were adapted to the organized scheme introduced by the *Pombalino* construction.

3.3 *Northern regions*

Pombalino-type constructions can also be found in medieval city centres in the north of the country, for example in Porto, Vila Real, Chaves, Braga and Guimarães, even if the seismic hazard is very low in these locations. Their construction dates back to the XV-XVII centuries adopting a traditional half-timbered wall typology (*taipa de rodizio)*. The ground floor consisted of stone masonry, typically granite, while in the upper storeys the external walls were half-timbered, adopting diagonal bracing members. The internal walls were either half-timbered or partition timber walls, typically lath and plaster walls (*taipa de fasquio*). Sometimes, apart from brick and rubble masonry, hay was used as infill. Considering the elements dimensions, in the buildings surveyed in

Guimarães, the posts and main beams had a width of 15 cm, while the diagonals and the secondary elements of 7 cm (Fig. 4). *Taipa de rodizio* was typically adopted for bourgeois houses, while noble houses had external walls built in granite and timber frames were only used for internal walls (Ferrão & Afonso, 2001).

4 APPROPRIATENESS OF RETROFITTING TECHNIQUES

Timber frame buildings constitute an important portion of many historical city centres in the world. Many of these buildings have known little or no care during their life, or they have been modified without taking into account the seismic response of the structure after the alterations had been made. Indeed, a common and extremely invasive practice has been the demolition of the inner part of the building, which is substituted by a reinforced concrete one, keeping only the original masonry façades (Cóias, 2007), therefore actually losing the original timber frame structure.

Following good practices, and as the result of the neglect of these constructive elements and their consequent progressive decay, namely at the connections, it has been often decided to retrofit these structures. Here, a brief summary of strengthening techniques adopted and their efficiency is presented.

4.1 *In-situ applications*

Many examples are available on restoration works done in traditional half-timbered buildings (Cóias, 2007; Appleton & Domingos, 2009). Numerous *Pombalino* buildings in Lisbon have been retrofitted using FRP sheets in the connections of the frontal walls (Cóias, 2007), or damping systems linking to frontal walls and to the outer masonry walls through injected anchors and providing additional bracing (Figure 5) (Cóias, 2007). Another practice is to project reinforced shortcrete onto the timber frame walls (Appleton & Domingos, 2009), but such a solution could effectively have an overly stiffening effect on the joints.

4.2 *Experimental solutions*

Various strengthening techniques have been tested experimentally specifically for *Pombalino* timber frame walls (Gonçalves et al., 2012; Poletti et al., 2014; Poletti et al., 2015). From previous experimental studies (Meireles, 2012; Poletti & Vasconcelos, 2015) it was seen that the cyclic behaviour of timber frame walls depends mainly on the response of the connections. Therefore, strengthening should be concentrated at the location of the connections.

Strengthening with steel plates (Figure 6) led to an increase in strength and stiffness of the wall, also greatly improving its dissipative capacity (Gonçalves, 2012; Poletti et al., 2014). When linking the diagonals to the main frame through the plates, some out-of-plane problems were encountered, as the connections

Figure 5. Timber frame to masonry wall connection by means of injected anchors and steel plate (credits: MONUMENTA).

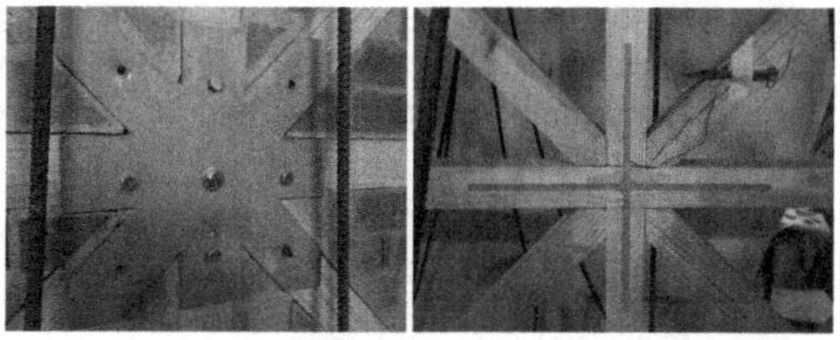

Figure 6. Steel plates strengthening (left) and NSM bars (right) (credits: E. Poletti).

were greatly stiffened when compared to the original configuration. This problem was overcome by not connecting the bracing elements to the plate, but leaving the original connection intact (Poletti et al., 2014). Steel plates led to an increase of the lateral capacity of the walls between 50% and 200%, also improving their post-peak behaviour. This technique is simple to implement and it is removable.

Near-Surface-Mounted (NSM) steel flat bars were also applied to the connections (Figure 6) and tests showed a good response in terms of stiffness and energy dissipation, particularly for weak infill (Poletti et al., 2015). Additionally, this technique is potentially invisible, even though not removable.

Tests performed on walls, on which reinforced render was applied on both sides (Gonçalves et al., 2012), showed that this solution greatly increases the stiffness of the walls, not taking advantage of the dissipative capacity of the connections.

Another possible strengthening solution consists of an elasto-plastic steel damper acting along the diagonal of the wall (Gonçalves et al., 2012). This solution showed a good dissipative capacity when the damper was in tension, while buckling was observed when in compression, therefore leading to a problematic application in practice.

Other examples of retrofitting techniques applied mainly at the connections with great potential are Glass Fibber Reinforced Polymer (GFRP) sheets, as well as a simple solution, such as the adoption of self-tapping screws (Poletti, 2013).

5 CONCLUSIONS

This chapter presents a brief state of the art of timber frame buildings, particularly when specifically built for their seismic-resistance capacity. In Portugal, this typology has been specifically adopted for the reconstruction of Lisbon, due to its large deformation capacity with limited damage.

Timber frame buildings have shown a good seismic response in recent events worldwide, and have been chosen for the reconstruction of rural areas.

An important aspect is that, for them to achieve their full potential, they have to be preserved and well maintained. Therefore, interventions have to be well thought out. They should improve the behaviour of the building without overly altering its original structure. Some of the interventions described here, such as steel plates and NSM bats, are believed to be appropriate techniques for timber frame walls, while others need further research.

REFERENCES

Appleton, J. & Domingos, I. (2009). *Biografia de um Pombalino. Um caso de reabilitação na Baixa de Lisboa.* Editora ORION, Lisbon

Bianco A. (2010). *La casa baraccata. Guida al progetto e al cantiere di restauro.* Rome, Italy: GBeditoria.

Cóias, V. (2007). *Reabilitação estrutural de edifícios antigos.* Lisbon: ARGUMENTUM, GECoPRA

Ferrão, B., Afonso, J.F. (2001). *A evolução da forma urbana de Guimarães e a criação do seu patrimonio edificado.* Guimarães: Câmara Municipal de Guimarães.

Gonçalves, A., Ferreira, J., Guerreiro, L. & Branco F. (2012, September). Seismic retrofitting of Pombalino "frontal" walls. *15th World Conference on Earthquake Engineering,* Lisbon.

Gülhan, D. & Güney I.Ö. (2000). The behaviour of traditional building systems against earthquake and its comparison to reinforced concrete frame systems: experiences of Marmara earthquake damage assessment studies in Kocaeli and Sakarya. *Proceedings of Earthquake-safe: Lessons to be Learned from Traditional Construction,* Istanbul, Turkey.

Langenbach, R. (2007). From "Opus Craticium" to the "Chicago Frame": Earthquake-Resistant Traditional Construction. *International Journal of Architectural Heritage,* 1(1), 29–59.

Mascarenhas, J. (2004). *Sistemas de Construção – V.* Lisbon, Portugal: Livros Horizonte.

Meireles, H.A. (2012). *Seismic vulnerability of pombalino buildings, (PhD Thesis, IST Instituto Superior Técnico, Lisbon, Portugal).*

Mendes, N. & Lourenço, P.B. (2010). Seismic assessment of masonry "Gaioleiro" buildings in Lisbon, Portugal. *Journal of Earthquake Engineering,* 14: 80–101.

Moreira, S., Ramos, L.F., Oliveira, D.V. & Lourenço, P.B. (2014). Experimental behavior of masonry wall-to-timber elements connections strengthened with injection anchors. Engineering Structures, 81, 98–109.

Poletti, E. (2013). *Characterization of the seismic behaviour of traditional timber frame walls (PhD Thesis, University of Minho, Guimarães, Portugal).*

Poletti, E. & Vasconcelos, G. (2015). Seismic behaviour of traditional timber frame walls: experimental results on unreinforced walls. *Bulleting of Earthquake Engineering,* 13, 885–916.

Poletti, E., Vasconcelos, G. & Jorge, M. (2014). Full-Scale Experimental Testing of Retrofitting Techniques in Portuguese "Pombalino" Traditional Timber Frame Walls. *Journal of Earthquake Engineering,* 18, 553–579.

Poletti, E., Vasconcelos, G. & Jorge, M. (2015). Application of near surface mounted (NSM) strengthening technique to traditional timber frame walls. *Construction and Building Materials,* 76, 34–50.

Ruggieri, N., Tampone, G. & Zinno, R. (2015). In-plane vs Out-of-plane "Behaviour" of an Italian Timber Framed System: the Borbone Constructive System. Historical Analysis and Experimental Evaluation. *International Journal of Architectural Heritage: Conservation, Analysis, and Restoration.* doi: 10.1080/15583058.2015.1041189

Tsakanika-Theohari E. (2008, May). The constructional analysis of timber load bearing systems as a tool for interpreting Aegean Bronze Age architecture. *Proceedings of the Symposium 'Bronze Age Architectural Traditions in the Eastern Mediterranean: Diffusion and Diversity', Munich.*

Ulrich, R.B. (2007). *Roman Woodworking.* Yale University Press.

Vintzileou, E., Zagkotsis, A., Repapis, C. & Zeris Ch. (2007). Seismic behaviour of the historical structural system of the island of Lefkada, Greece. *Construction and Building Materials,* 21, 225–236.

Part 4: Portuguese local seismic culture: Assessment by regions

Seismic Retrofitting: Learning from Vernacular Architecture – Correia, Lourenço & Varum (Eds)
© 2015 Taylor & Francis Group, London, ISBN 978-1-138-02892-0

Lisbon: Downtown's reconstruction after the 1755 earthquake

G.D. Carlos, M.R. Correia, G. Sousa, A. Lima & F. Gomes
CI-ESG, Escola Superior Gallaecia, Vila Nova de Cerveira, Portugal

V. Lopes dos Santos
Faculdade de Arquitetura da Universidade de Lisboa, Lisboa, Portugal

ABSTRACT: The 1755 earthquake is the most striking example of the national scene, especially by the violence of its effects in the Portuguese capital city. The lower part of Lisbon has been reduced to rubble, due to the action of quakes and the fire spread accordingly. The process of reconstruction of Lisbon's downtown, according to the original model of the "House of risk", will be the main national reference of earthquake-resistant construction. The elected solution is not referred to only in what the need to improve the structural performance of buildings is concerned, as it also promoted the integration of several pro-active strategies to the post-disaster scenario. Its influence will be seen in the implementation of a specific urban plan, and the adoption of real architectural typologies. Technological innovation of building systems, applied in a systematised way, using standardization and prefabrication, is one of its most remarkable features. The piloting system of foundations, the regular application of vaulted ceilings on the ground floor and the implementation of structural walls with combined locking feature, determined a national unparalleled paradigm.

1 INTRODUCTION

1.1 *1755 Earthquake*

The Lisbon and Tagus Valley regions are the surveyed ones of the country regarding the historical seismicity. Given the scale and impact it had in the capital, the occurrence of November 1, 1755 significantly dominates the research conducted in this area.

Lisbon was devastated by an earthquake of intensity X, followed by a tsunami and numerous fires that caused the destruction of much of the city, and caused tens of thousands of fatalities.

Pombalino downtown, located at a lower level, in the grounds of an ancient estuary, had increased seismic vulnerability, as evidenced by the reports of past events, such as those of 1356 and 1531 (Moreira, 1991). Its concentrated economic activity, its high occupational density, the general state of disrepair of buildings and the saturation of the soil, caused by poor sanitation, were relevant concerns even before the 1755's catastrophe.

1.2 *Downtown's reconstruction*

The selected process is basically the strategy to overcome the drawbacks of the devastated urban fabric. The previous city was heavily conditioned by an organic matrix, a cluttered grew and it was founded on unstable alluvial ground. The issues identified as priorities bound up with the necessary increase in quality of movement, health, safety and structural reinforcement of the buildings. The downtown rebuilding process was based on the application of a concept of Enlightenment city, fully replacing the existing urban fabric. The approach objectively valued the upward social class, the bourgeoisie, focusing on revenue building at the expense of the noble palace, the public or religious program equipment, seeking to expeditiously enhance the economy of the capital city. The process would not be conceivable without a set of duly regulated measures and additional procedures, as the prohibition of the construction outside the city limits, or control of the materials price to avoid their speculation. The operability of this process is naturally related to the authority and the reformer dynamism of the administration, supervised at that time by the inevitable Marquis of *Pombal*. The reconstruction proposal was coordinated by the Engineer-Major of the Kingdom: Manuel da Maia. This was derived from an exercise organised in progressive stages, involving a large team of technicians, who deliberated upon the more viable alternatives.

2 URBAN PLANNING

2.1 *The full redefinition of the existing fabric*

The elected solution assumed the implementation of a completely new town planning, with a fully orthogonal layout, based on the urban proposal of Captain Eugenio of Santos e Carvalho's team. The layout is

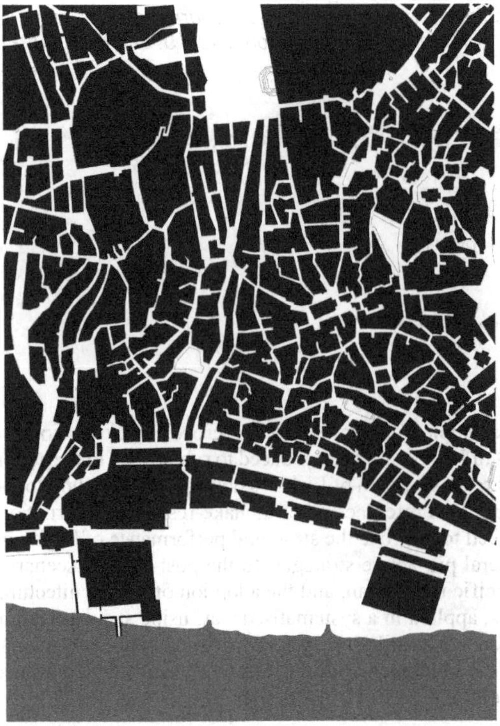

Figure 1. Plan prior to the 1755 earthquake (credits: CI-ESG, based on the plan of 1650 by João Nunes Tinoco).

Figure 2. Plan of the proposal (credits: CI-ESG, based on the plan by Eugénio dos Santos and Carlos Mardel).

defined by the regular repetition of rectangular blocks, mainly displaced in the north-south direction, placing the shorter facades to the usual directions of the quakes occurred. The creation of perpendicular streets exclusively, of significant width (main streets 60 spans and secondary 40 spans), determine the exact alignment of the facades. The maximum height of buildings is limited to the width of the main streets; reducing the effects of an eventual collapse of the facades. The squares are determined by the full removal of modules of blocks. Unlikely what was being done so far, greater importance was given to the private property, leaving the public buildings to the Commerce Square. The buildings, disjointed by transverse separation walls that rise above the roof, are grouped in closed blocks, ripped in the inside by a long and narrow central patio (the *alfugere*), introduced also as an hygienist innovation.

The most common aggregations comprise two apartments per floor, commonly referred to as left/right, and sharing a single intermediate stairway, usually displaced from the geometric centre to avoid intersections in common structural walls. The number of singular buildings with only one apartment per floor is also significant.

The top buildings present a larger set of alternatives, in order to solve the transition between facades, the angle of the block, and to adapt to the dimensions of side/back streets. In this case, the apartments did not link to the *alfugere*, although complying with the main rules of composition.

3 DEFINING A TYPOLOGICAL MODEL

3.1 *The Pombalino income building*

The reconstruction process makes use of a building "model", whose implementation is done systematically with small compositional variations. The building consists of a ground floor, for trade or industry, and three upper floors (later four), intended for housing. It was called *Pombalino* income house. The partitioning corresponds to an orthogonal grid, arranged according to the alignment of the facade. The facade had solid plans, with symmetrical, repetitive and regular composition, producing uniform construction volumes to the scale of the block. The covers, with interior exploitation, are inclined with longitudinal ridge, parallel to the main streets, integrating dormer windows.

The spans are vertically aligned; the doors of the shops corresponding in height to the height of the first floor bay windows. The remaining floors present sill windows, whose dimension gradually decreases in height. No decorative elements are allowed, thus avoiding structural dissonances in the facade plans.

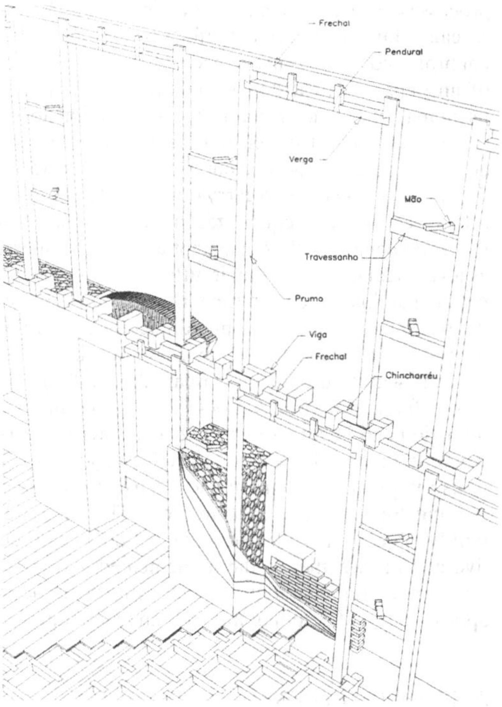

Figure 3. Pombaline cage (three-dimensional picture) (credits: Vitor Lopes dos Santos).

4 CONSTRUCTIVE CHARACTERISATION

4.1 *The constructive system and the integration of seismic-resistant elements*

The foundations, locked together by massive brick arches, are based on wooden vertical piling grids, with significant depth, superiorly topped with cross-platforms.

The ground floor is constructed on stony masonry and has a generally domed cover system or sustained by arches, functioning as a structural stability point. The use of master-walls in masonry, cross the facade, enables the regular separation between buildings and apartments. The wooden beams of the floors are tied to the master walls with metal clips. The *Pombalino* Downtown is characterised thereby, by the development and systematic use of the "cage" system, in all interior facings.

The cage consists of a structure comprised of vertical elements (uprights), horizontal (the windowsill) and diagonal, forming cross matrix modules (according to the repetition cross of Saint Andrew).

The three-dimensional timber structures are articulated with filler panels in cut stone masonry, allowing the implementation of lighter and flexible walls, generally referred to as Front walls. The resulting panels were locked by the floor beams, assisted by the other wooden elements of the flooring. The floor beams were always perpendicular to the facade, where they were

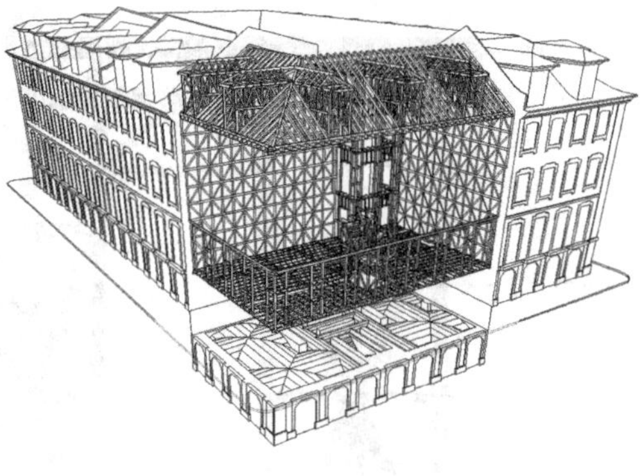

Figure 4. Pombaline block model (credits: Vitor Lopes dos Santos).

recessed, and thereby articulating the whole (Santos, 1995).

The exterior walls, leaning to the street and the *alfugere*, are constructed in stone masonry, originally nominated as wall of frontal, with wooden structures placed through the inner face. These were less developed (without diagonal elements), since they do not contribute actively to the structural support of the outer cloth. Their binding was performed by specific connection elements, the "hands", placed along the cage structure. The spans vertically and horizontally aligned, were located in the spaces between props, they were constructed in stone masonry and attached to the cage structure through metallic devices (cats).

The facades plan integrates multiple reinforcing elements in square massive brick masonry, such as proviso arches, compression arches and cornices.

The cornices of the roofs involve the entire block without breaks between different buildings, and thus serving as a tie strap (Mascarenhas, 2009).

The corners of blocks have a special treatment, working as compact elements, constructed in stone masonry and coated by fixed elements with metallic devices (cats).

5 SEISMIC BEHAVIOUR OF A POMBALINO BUILDING

Regardless of the specific contribution of a given system or building element, it should never be forgotten that the unit of the downtown reconstruction plan is the block. However, and in order to ensure a greater objectivity of results, most recent studies are based on the structural behaviour analysis, of conservation nature, based on computer modelling of unique building and its different components. These simulations contemplate the mechanical properties of the materials; the geometry of the structural elements and the associated behaviour; as well as the type of links between the various systems. In order to simulate the occurrence of 1755, usually type 2 earthquake's waves are chosen, with high magnitude and significant

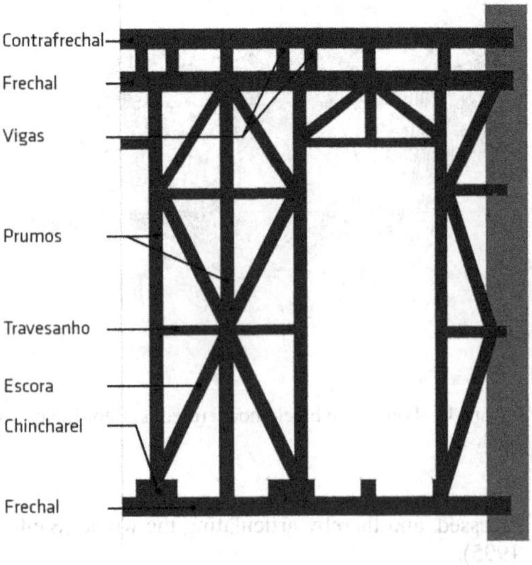

Figure 5. Elements of an internal wall (credits: CI-ESG).

focal length, characterised by low frequency and long duration (Pena, 2008, p. 80).

According to this type of analysis, the biggest benefit is conferred by the cage efficiency, which acts both as locking and dissipation system of the energy produced by the earthquake, providing the necessary toughness to the built unit (ibid, p. 102). The behaviour of pavements, determined greatly by the employed framework system, and its delivery to adjacent walls, is also referred to as positive, yet its performance is far from the frontal walls level. Of all the elements analysed are also the connections between the front and the masonry facades, the "hands", which have the most negative structural results, showing a pronounced level of breakage. This condition often appears to induce a failure mechanism, reportedly taking the overthrow of the facades of buildings, in favour of the interior security.

6 FINAL REMARKS

The rebuilding intervention of the *Pombalino* Downtown is a complex phenomenon that must be considered beyond the applied technology. The intervention shows great depth on reactive disaster response, articulating different approaches and processes in a consistent progression scale. It also reveals a huge management effort, combining different professional areas in a concerted collective action, of an unprecedented dimension. Also to refer to is its humanist character, with special attention to the social and cultural dimensions, expressed from the analysis process, whose accuracy was crucial to the development of action to carry out. Despite the scholarly aspect and consequent technological innovation that is associated with it, it cannot ever be considered as a vernacular example; their influence will cross the national architecture. At a time, in which even before the disaster, extreme economic difficulties were felt, the process of reconstruction will involve the former existent scarce resources (Mascarenhas, 2009). The labour force mobilised not only determined the participation and integration of the best master builders, where traditional know-how played a major role, as it also implied an extensive and intensive process of training for many recruited workers. This condition will be objectively reflected in the national constructive culture, whether in the implemented actions by the current administration, as well as in the regional appropriation of resulting methods and techniques.

ACKNOWLEDGMENTS

The authors gratefully acknowledge the support by the Portuguese Science and Technology Foundation (FCT) to the research project 'SEISMIC-V – Vernacular Seismic Culture in Portugal' (PTDC/ATP-AQI/3934/2012).

REFERENCES

Lopes, M. S., Bento, R. & Monteiro, M. (2004). *Análise Sísmica de um Quarteirão Pombalino*. Lisboa, Portugal: IST.

Mascarenhas, J. (2009). *Sistemas de Construção-V: O Edifício de rendimento da baixa pombalina de Lisboa. Técnicas de Construção.* Lisboa, Portugal: Livros Horizonte.

Moreira, V. J. S. (1991). *Sismicidade Histórica em Portugal Continental. Revista do Instituto Nacional de Meteorologia e Geofísica*, 29–63.

Pena, A. F. (2008). *Análise do comportamento sísmico de um edifício pombalino* (Master Thesis, IST/UTL, Portugal).

Santos, V. L. (1995). *O sistema Construtivo Pombalino em Lisboa em Edifícios Urbanos Agrupados de Habitação Colectiva – Estudo de um Legado Humanista da Segunda Metade do Séc. XVII (PhD Thesis, FA/UTL, Portugal).*

Seismic Retrofitting: Learning from Vernacular Architecture – Correia, Lourenço & Varum (Eds)
© 2015 Taylor & Francis Group, London, ISBN 978-1-138-02892-0

The 1909 earthquake impact in the Tagus *Lezíria* region

F. Gomes, A. Lima, G.D. Carlos & M.R. Correia
CI-ESG, Escola Superior Gallaecia, Vila Nova de Cerveira, Portugal

ABSTRACT: Tagus *Leziria* is characterised by an intense seismic activity. 1909's seism, which registered IX (MCS), is known for the damage caused in *Benavente*, where part of the built structure was destructed. Facing these evidences, and the impact on specific areas of the region, *Benavente* and *Coruche* were identified as case studies. These case studies were selected considering the following premises: reactive / proactive approach of the population before the earthquake due to the built structure of vernacular character; earthquake-resistant elements; typological characteristics; preservation of the strategies; identifying a local seismic culture.

1 INTRODUCTION

1.1 *Context: Tagus Leziria Region*

Tagus Leziria Region locates in the centre of the country, bounded by the districts of Leiria (nortth and northwest), Portalegre (East), Évora (Southeast and South), Setúbal (South) and Lisbon (West). As a structuring physical element, it stands out the Tagus River and part of its expansive watershed, both granting, to the majority of the territory, a landscape of great floodplains (*lezírias*). Considering its clime, Tagus *Leziria* is located in a transition area, presenting a Mediterranean climate, dampened by the tempered configuration of the coastline, where the average temperature is around 16°C.

Figure 1. Municipal Neighbourhood in *Benavente, Portugal* (credits: CI-ESG).

The land occupation, north of the Tagus *Leziria*, is based on concentrated populations marked by small settlements. South of the Tagus River, the cores are compact, distant and of defined structure. Amongst them, voids arise, only annihilated by farms and hills. In contrast, arises another form of occupation, the widespread settlement, which occasionally appears in the voids between the compact cores.

Traditional typologies are characterised by a constructive and morphological simplicity, marked by the direct connection to the economic activity, the agriculture. Local populations based their construction in the use of brick, tuff and adobe. The buildings present a simplified plan of a single floor, and through the horizontal aggregation appears the shed, the oven and the winery. In correspondence to the widespread settlement, another typology arises using rammed earth and adobe as construction techniques. Morphologically, the house is characterised by presenting a set of volumes, which has a single elongated floor, with few openings. To the central core of the house, compartmentalised by light wooden partitions, are added the agricultural support areas.

2 HISTORICAL SEISMICITY: TAGUS *LEZÍRIA* REGION

The Tagus *Leziria* Region situates within the area covered by the tectonic fault of the 'Lower Tagus Valley (VIT)', in a highly intense seismic activity area, characterised by the occurrence of great earthquakes in the last centuries. To be referred the 1909's seism in this region.

1909's seism, of IX intensity (MCS), is known as the *Benavente* seism. The quake originated in the failure of the 'Valley of the Lower Tagus' being the

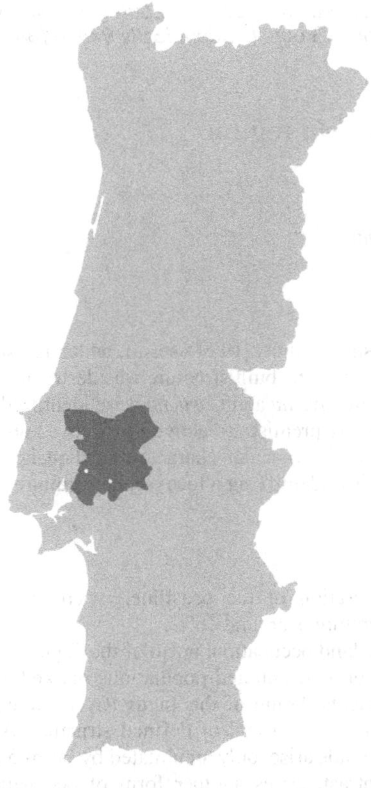

Figure 2. Portugal Map: Region 1 – Tagus *Lezíria* (credits: CI-ESG).

Table 1. Main earthquakes in the Tagus *Lezíria* Region (LNEC, 1986).

Year	Intensity | MCS
1531	IX
1755	VII
1909	IX
1914	VII
1969	VI

most important earthquake occurred in the Portuguese territory during the twentieth century.

Considering its degree of destruction, the town of *Benavente* was the most devastated, where forty percent of the houses were completely destroyed, many others were unable to be inhabited, and only twenty percent were recoverable.

The seism caused 42 deaths and 75 injured, being the register of deaths distributed as follows: 30 in *Benavente*'s parish, 7 in *Samora Correia*'s parish, 3 in *Santo Estêvão*'s parish and 2 in the parish of *Salvaterra de Magos*. In addition to the human losses, the seism caused extensive material damage in housing and other buildings of municipal heritage, such as the Mother church, Santiago's Church, S. Tomé's Church,

Figure 3. Effects of the 1909's seism (credits: Ilustração Portuguesa).

Nª Srª da Paz's chapel and the Municipal Building (Vieira, 2009).

Given the huge damage in the region, study commissions of the phenomenon emerged, which developed action plans at the urban and typological levels. The 'General Plan of Roads' was prepared for the creation of measures to be applied to the existing tracing ('Old Town'), and to the structuring of the new expansion zone (Vieira, 2009).

In the 'Old Town' the enlargement and the amendment of the roads were suggested. The expansion zone, the 'New Town', was structured according to a rational array, in which wide streets in an orthogonal logic prevailed.

The blocks designed by the 'Streets General Plan' were slowly occupied through an autonomous process, having just as a connection, the street plan of village expansion. The produced housing types encompassed concerns on the seismic prevention, which were clearly expressed by the introduction of some principles, such as the symmetrical plant, and the use of a wood skeleton (from the interpretation of the '*Pombalina* cage' – an internal wooden cage).

Before all the evidence and the impact caused, *Benavente* was selected as the first case study.

The area of the Tagus *Leziria*, was entirely affected by the 1909 earthquake damage, with higher incidence in *Benavente* village. The built structure in the area was seriously damaged. Throughout the field surveys accomplished, several vernacular character types with the application of earthquake-resistant elements were identified. It is a clear indication of the Seismic Culture of the region during the beginning of the twentieth centuries.

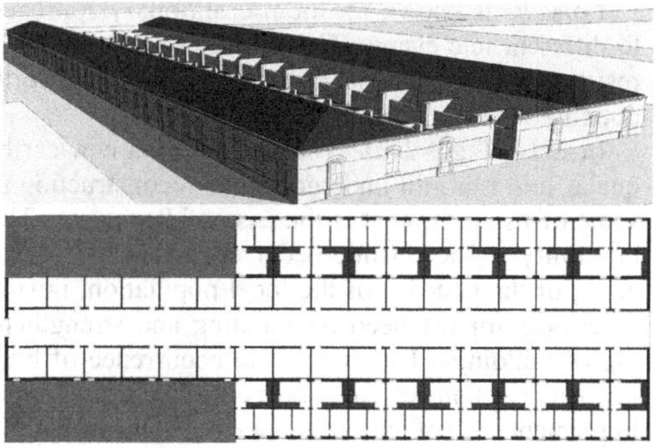

Figure 4. Scheme of the Municipal Residential area: *Benavente* (credits: CI-ESG).

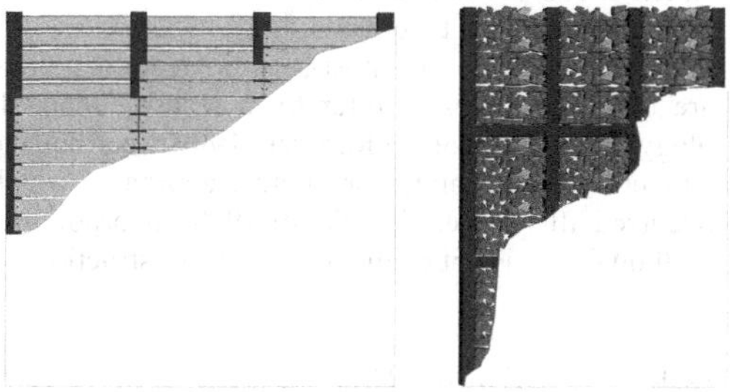

Figure 5. Different configurations of the *'gaiola pombalina'* applied in the typology post 1909's seism (credits: CI-ESG).

Therefore, the village of *Coruche*, was selected as the second case study by its significant sample of buildings of vernacular character with the systematic incorporation of earthquake-resistant principles in the constructions, such as the construction of terraced houses, the symmetric plant, the strengthening of the indoor walls and the usual placement of buttresses and tie rods.

3 CASE STUDY: BENAVENTE

3.1 *Constructive characterisation and integration of seism-resistant principles*

The village of *Benavente* is typologically marked by the buildings created after the earthquake of 1909, with the ground floor, a rectangular plant, gable roof and a uniform facade. The buildings arise conformed by uniform blocks with terrace houses aggregated in symmetrical bands, in two parallel alignments, only separated by an interior street. They present, mostly, a street front and a backyard (Fig. 3–4).

The integration of earthquake-resistant principles applied by the population after the earthquake of 1909 arises as a reactive and preventive approach.

The reactive approach was evident in the buildings that suffered superficial damage, where contingent solutions were applied, such as the use of tie rods.

Figure 6. General view of the built set in *Coruche* (credits: CI-ESG).

Figure 7. Scheme of the housing typology in *Coruche*. (credits: CI-ESG).

The preventive approach, applied in new buildings, exposed concerns at the level of the village planning and of the typologies. An aggregation system through the linear structure blocks and the widening of the road system were established.

Typologically, the introduction of earthquake-resistant elements was based on the use of a symmetric plant and on the articulation of the walls in stone masonry (structural) with the inner walls, these characterised by its wooden skeleton derived from the simplification of the "*Pombalino* cage" (Fig. 5).

4 CASE STUDY: CORUCHE

4.1 *Constructive characterisation and integration of seism-resistant principles*

The vernacular typology identified in *Coruche* is mostly integrated in agricultural holdings. The typology is characterised by single storey buildings, of small dimensions, with a regular rectangular plant and a gable roof. They are generally aggregated in terrace houses, added by the tops, in orthogonal alignments. They present two street fronts, both with direct access to the outside. Its partitioning results from the subdivision of a single space, dominated by the space of the main entrance, whose positions are fixed and set up the matrix of the whole (Fig. 7).

The exterior walls, in stone masonry, are structural, supporting the simple truss of the roof. The voids are scarce and small. The interior wattle and daub walls do not play any structural function, having the particularity of not reaching the ceiling, taking advantage of the resulting interstitial space for storage.

The inclusion of seismic-resistant principles by the local population emerges as a preventive approach. The studied typology shows the introduction of principles, such as the aggregation system (available in terrace houses), and elements, such as buttresses and tie rods (Fig. 7–8).

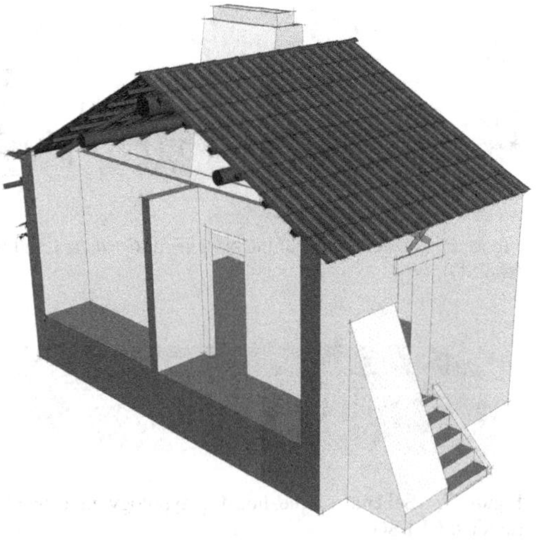

Figure 8. Constructive cut of the housing unit in *Coruche*: identification of seismic-resistant elements (credits: CI-ESG).

5 SYSTEMATISATION OF THE DATA COLLECTED

As aformentioned, the Tagus *Leziria* Region is an área of intense seismic activity, covered by the failure of the 'Lower Tagus Valley'. In the Tagus *Lezíria* were selected, as a case studies, Benavente and *Coruche*. The 1909's seism affected seriously the populations of *Benavente* and *Coruche*, where construction methods are concerned by the apparent weaknesses shown in resistance to ground oscillations, a weak cohesion capacity and the high degree of fragmentation.

It emerged the need to improve the existent construction systems (pre-earthquake) by ínserting earthquake-resistant principles.

In *Benavente*, in the years of reconstruction, intervention matrices based on Lisbon plan for reconstruction after the earthquake of 1755 were implemented, focusing mainly on the structure of symmetrical and linear blocks, as well as on the application of the '*Pombalina* Cage'.

Coruche is known for the precautionary approach to different icic events. The insertion of earthquake-resistant principles in new constructions is also worth noticing.

In 1914, Tagus *Lezíria* suffered from a new earthquake, and it is still undergoing the reconstruction in some of its areas, as it is the case of *Benavente*. The proximity of the seismic occurrences led to a worsening of the concern of the local population, raising awareness for the need of planning and strengthening of buildings. The systematic occurrence of high intensity earthquakes, in a short period of time, led the implementation of an Earthquake-resistant Culture in the Tagus *Lezíria* region in the early twentieth century.

Coruche and *Benavente* hold currently a local seismic culture, though not active, and marked by the evident forgetfulness by the populations of the ic risk, which led to the abandonment of the application of earthquake-resistant elements.

With respect to the post-seismic reaction, there are two approaches to refer to: a reactive methodology, as prevention before the 1909's earthquake was not equated; and a preventive approach, which occurred after the earthquake of 1909, incorporating earthquake-resistant elements in new construction.

ACKNOWLEDGMENTS

The authors gratefully acknowledge the support by the Portuguese Science and Technology Foundation (FCT) to the research project 'SEISMIC-V – Vernacular Seismic Culture in Portugal' (PTDC/ATP-AQI/3934/2012).

REFERENCES

Vieira, R. (2009). *Do Terramoto de 23 de Abril de 1909 à Reconstrução da vila de Benavente – um processo de reformulação e expansão urbana.* Bevanente: Câmara Municipal de Benavente.

LNEC (1986). *A sismicidade histórica e a revisaÞo do Cataìlogo Sísmico.* Lisboa: Laboratório Nacional Engenharia Civil

Seismic Retrofitting: Learning from Vernacular Architecture – Correia, Lourenço & Varum (Eds)
© 2015 Taylor & Francis Group, London, ISBN 978-1-138-02892-0

Costal *Alentejo* region: Identification of local seismic culture

F. Gomes, A. Lima, G.D. Carlos & M.R. Correia
CI-ESG, Escola Superior Gallaecia, Vila Nova de Cerveira, Portugal

ABSTRACT: The Costal Alentejo region presents a historical seismicity of high intensities. The earthquake of 1858 hit the region with an intensity of IX (MCS), causing severe damages in the region, especially the partial destruction of *Melides* and its surrounding areas. Alcácer do Sal also suffered relevant damage with the earthquakes of 1858 and 1903. Preliminary missions were taken in the region, and several earthquake-resistant elements were observed, in both villages. Given the facts, *Alcacer do Sal* and *Melides* were identified as case studies for the research project SEISMIC-V and a deeper analysis was then addressed for each village.

1 INTRODUCTION

1.1 *Context: Costal Alentejo region*

The Coastal Alentejo region is bordered to the north by the *Setúbal* Peninsula and the Alentejo North, to the East by the *Alentejo*, to the south by the Algarve, and to the west by the Atlantic Ocean. This region, given its geographical situation, has very particular climatic conditions, being extremely dry, and presenting average temperatures during the winter season, and high summer temperatures.

The Coastal Alentejo region comprises the *Sado* watershed, characterised by *Grândola* and *Cercal* mountain ranges, of a low population density, with only a populated trail along the coast. Populations arise as concentrates, a typical feature of settlements to the south of *Sado* River, which extends along the Coastal Alentejo region, being punctuated by the emergence of fragmented occupations, the Alentejo hills.

The typologies that mark the region are characterised by their constructive and morphological simplicity. Local populations used rammed-earth to build their homes, which comprises a simplified houseplant, reinforced by the presence of buttresses and stone benches (*poial*). The houses have a single room and rooms in alcove; also presenting a gable roof and few openings. These houses can arise together in groups of six units (row housing), or as isolated dwellings.

Another way to inhabit the territory was through the Alentejo *Monte*, characterised by settlements of one or more houses, and buildings to support the subsistence activity. Morphologically, they present a larger number of inner partitioning and complexity.

Figure 1. Map of Portugal: Region 2 – Coastal Alentejo (credits: CI-ESG).

Table 1. Main earthquakes that occurred in the Costal Alentejo Region. (LNEC, 1986).

Year	Intensity | Mercalli Scale (MCS)
1755	IX
1858	IX
1903	VIII
1903	V
1911	V
1926	V
1966	VI
1969	VII

2 HISTORICAL SEISMICITY: COSTAL ALENTEJO REGION

The Coastal Alentejo is an area of high seismic intensity. During the eighteenth and nineteenth centuries it was affected by two earthquakes – in 1755 and in 1858 – both with IX intensities (MCS) and causing significant damage. The 1858's earthquake hit the region with a significant intensity: IX in Mercalli scale (MCS) in the epicentre (probably in the sea). The area of Setubal was severely affected, being worth noting the partial destruction of *Melides* village and its surrounding areas. This occurrence led the population to reflect upon, and to be concerned about the improvement of their housing behaviour towards earthquakes.

50 years later, the region suffers once again from significant earthquakes: in 1903, with an intensity of VII MCS; 1909, 1911 and 1926, with an intensity of V MCS. At the end of the twentieth century, the Coastal Alentejo is again hit by two significant shocks: one in 1966, with an intensity of VI (MCS); and another one in 1969, with an intensity of VII (MCS).

Considering the evidences and the impact of these events in some specific locations, *Alcacer do Sal* and *Melides* are selected as the first case studies. Through the fieldwork accomplished, it was possible to identify some typologies of vernacular character, presenting earthquake-resistant elements, a clear hint of the seismic culture in the region during the early twentieth century.

In *Alcacer do Sal*, buildings of vernacular nature were identified, which presented earthquake-resistant elements, such as *Pombalino* walls reinforcement, buttresses and tie rods.

In the surroundings of *Melides*, it was identified local seismic culture based on the improvement of constructive systems, and the use of earthquake-resistant reinforcement elements. In response to the effects of earthquakes, the population used reactive solutions, by applying structural reinforcement elements, such as buttresses and tie-rods.

3 CASE STUDY: *ALCÁCER DO SAL*

3.1 *Constructive characterisation and introduction of seismic-resistant elements*

The vernacular typology identified in *Alcacer do Sal* presents a rectangular houseplant, with a ground floor

Figure 2. Casa dos Romeiros in *Alcácer do Sal* (credits: CI-ESG).

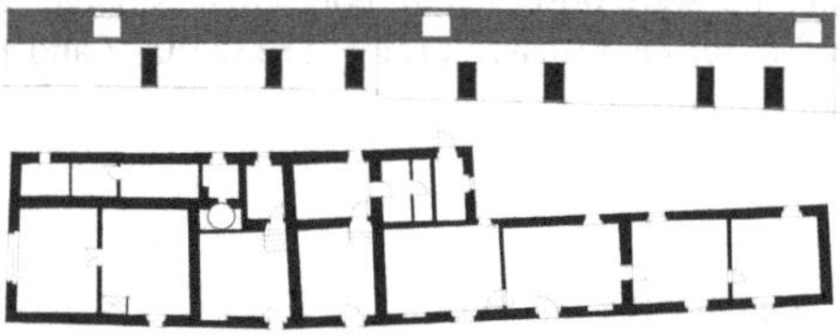

Figure 3. House-plant and main elevation of the Casa dos romeiros in *Alcácer do Sal* (credits: CI-ESG based in Correia & Merten, 2000).

of small dimensions, and a gable roof (Correia & Merten, 2000). These constructions are in a continuous row, generally with a significant extension. The houses are connected by the tops, in orthogonal configurations (Fig. 2). Both longitudinal façades have openings, with access to the outside. The façades are constructed by a mixed technique, using or mortared joint stone masonry or rammed earth walls, usually reinforced with solid brick rows (Fig. 3).

The use of earthquake-resistant elements, applied by the local population, emerges through the application of two approaches: a) a preventive reaction, observed after the 1755's earthquake, with the integration of earthquake-resistant elements, such as a simplifying '*Pombalino* cage' in new constructions (Fig. 4); b); and a reactive approach, characterised by incorporating reinforcing elements (tie rods) in the stabilisation of several buildings of vernacular character. These actions are also, possibly, associated with the occurrence of subsequent seismic events (earthquakes 1858, 1903, 1969) (Correia et al., 2014).

The following earthquake-resistant elements were identified: Reinforcement of the openings, using thick columns of massive brick and arched lintel in the same material; the stone benches between the housing units; walls reinforcement, by the interpretation of the "*gaioleiro*" system, and thus combining wood skeletons and cloths (or fills) of massive brick masonry, generally tied with metallic tie-rods placed in the upper edge, and therefore reinforcing the interlocking of the longitudinal façades (Fig. 4).

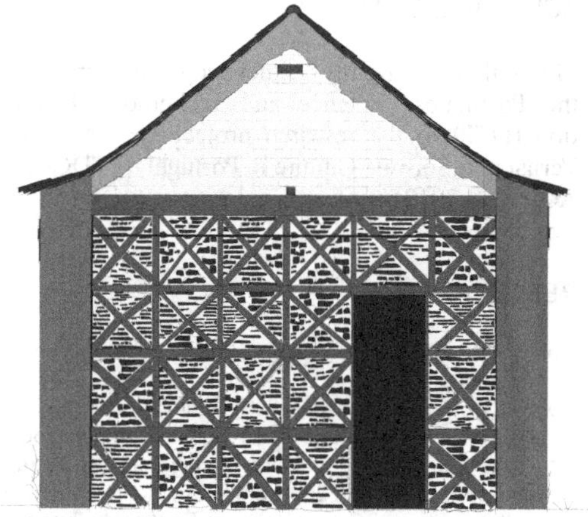

Figure 4. Constructive section of Casa dos Romeiros in *Alcácer do Sal* (credits: CI-ESG, based in Correia, 2007).

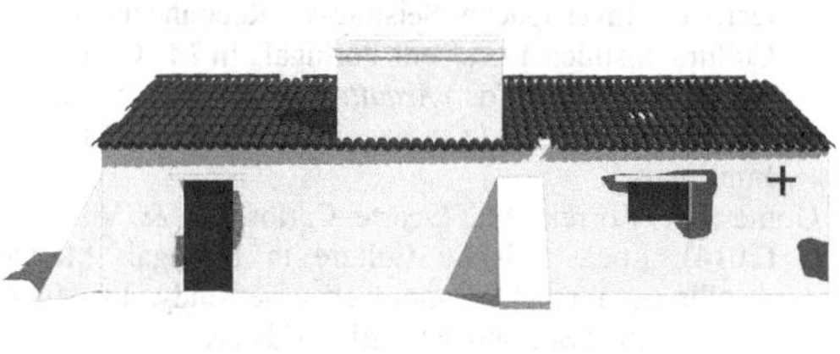

Figure 5. Representative typology of *Melides* (credits: CI-ESG).

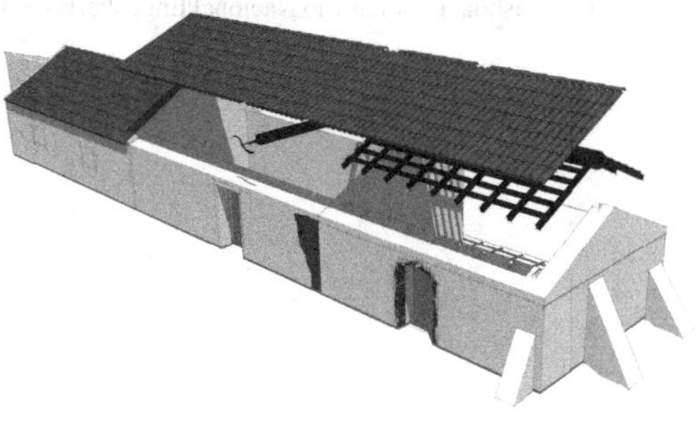

Figure 6. Axonometric of a typical typology of *Melides* (credits: CI-ESG).

4 CASE STUDY: *MELIDES*

4.1 *Constructive characterisation and insertion of seism-resistant elements*

In the region of *Melides*, 12 vernacular buildings of the late nineteenth century and early twentieth century were identified, all contemplating earthquake-resistant elements (Fig. 5 & Fig. 6).

The identified buildings arise, in the territory, isolated, or aggregated, lengthwise and in continuous row arrangements. The buildings present a ground floor of small dimensions, with a rectangular houseplant, and a gable roof. The chimney locates on the main façade. It is also in the main façade that the scarce and reduced building openings are found.

Figure 7. General perspective of a building in *Melides* surroundings (credits: CI-ESG).

These buildings do not define any particular or complementary outer space. The inner subdivision, if any, is defined by the subsequent placement of wattle-and-daub walls. The exterior walls are built with rammed earth, forming a monolithic structural perimeter (Fig. 5 & Fig. 6).

The earthquake-resistant characteristics in-use in *Melides* region are based on the application of 2 to 3 buttresses on the same façade (Fig. 8); the application of tie rods in the interior walls; and in the application of four tie rods around the four façades of the building. Also to consider is the existence of the reinforced plinth course in some of the façades. It is also to highlight the *poial*, a stone bench in the main façade, with a boost function. This was identified in a significant number of dwellings (Gomes et al., 2014).

A deeper analysis to the constructive systems makes noticeable that all counters were added to the original wall after its original construction. The observation is due to the identification of whitewashed plaster beneath the damaged buttresses. The same happens in the reinforced plinth. As for the tie-rods, these are reinforcing elements obviously introduced at a later stage. All elements addressed are clearly elements of a reactive approach by the population of *Melides* (Gomes et al., 2014).

5 FINAL REMARKS

Around *Melides*, the identified vernacular houses were characterised by the instability of the building systems, when facing the effects of earthquakes. This fact led the population to develop techniques, and to consider earthquake-resistant elements, so as to improve the behaviour of their housing in the occurrence of new earthquakes.

The population developed a preventive approach, applying earthquake-resistant elements in new constructions. However, due to the region scarce resources, the reactive approach was the most notorious, leading the population to repair the damaged buildings, by insertion of specific reinforcement elements in the existing structures. These elements are, mainly, two to three buttresses in the façades and a reinforced plinth course.

Years later, the region suffered, significant earthquakes: 1903, 1911, 1926, 1966 and 1969 earthquakes caused minor damage to the buildings, but generated fear among the population. The seismic frequency reminded the population of the damage occurred in the past, and of the possibility of future occurrences. This was probably the period, when the earthquake-resistant reinforcing elements were applied to the constructions, and the local seismic culture became active.

Currently, this local seismic culture is no longer active in *Melides*, as the buildings identified with earthquake-resistant elements are no longer in-use as well. This also indicates forgetfulness by the population of the seismic risk.

Alcácer do Sal was also seriously damaged by the earthquakes of 1755 and 1858. In response to the damage found in the built structure after the various earthquakes, two approaches were applied by the locals: a preventive reaction and a reactive reaction. The preventive approach arises after the earthquake of 1755, through the integration of earthquake-resistant elements, such as a simplification of the '*Pombalino* cage' in the new buildings (Fig. 4). The reactive approach appears associated to the stabilisation of several buildings of vernacular character, through the integration of earthquake-resistant elements (tie rods and buttresses). The two events mark the development of a local seismic culture during the post-event period, i.e. after the earthquake of 1755 (XVIII century), and after the earthquake of 1858.

Currently, *Alcacer do Sal*, holds a non-active local seismic culture, where the abandonment of the earthquake-resistant elements, or their lack of maintenance, is notorious.

ACKNOWLEDGMENTS

The authors gratefully acknowledge the support by the Portuguese Science and Technology Foundation (FCT) to the research project 'SEISMIC-V – Vernacular Seismic Culture in Portugal' (PTDC/ATP-AQI/3934/2012).

REFERENCES

Correia, M. (2007). *Rammed earth in Alentejo*. Lisboa: Argumentum.

Correia, M. & Merten, J. D. (2000). Restoration of the Casas dos Romeiros using traditional materials and methods—A case study in the southern Alentejo area of Portugal [preprint]. In TERRA 2000: 8th International Conference on the Study and Conservation of Earthen Architecture. Torquay, UK, May 11–13, 2000. Torquay, UK: James & James, 226–230.

Correia M., Gomes, F. & Duarte Carlos, G. (2014). Projecto de Investigação Seismic-V: Reconhecimento da Cultura Sísmica Local em Portugal. In M. Correia, C. Neves, R. Núñez (Eds.) *Arquitectura de Tierra: Patrimonio y Sustentabilidade en regiones sísmicas*. El Salvador: Imprimais.

Gomes, F., Correia M., Duarte Carlos, G. & Viana, D. (2014). Local Seismic Culture in Portugal: Melides dwellings, a reactive approach case study. In Mileto, Vegas, García Soriano & Cristini (Eds), *Vernacular Architecture: Towards a Sustainable Future*. London, UK: CRC/Taylor & Francis.

LNEC 1986. *A sismicidade histórica e a revisão do Catálogo Sísmico*. Lisboa: Laboratoírio Nacional Engenharia Civil.

Seismic Retrofitting: Learning from Vernacular Architecture – Correia, Lourenço & Varum (Eds)
© 2015 Taylor & Francis Group, London, ISBN 978-1-138-02892-0

Seismic-resistant elements in the Historical Centre of *Évora*

G.D. Carlos, M.R. Correia, G. Sousa, A. Lima & F. Gomes
CI-ESG, Escola Superior Gallaecia, Vila Nova de Cerveira, Portugal

ABSTRACT: This article discusses an analysis of the Historical Centre of *Évora* regarding the identification of a local seismic culture. A comparative synthesis of vernacular techniques of the region and the degree of exceptionality of this urban complex, conditioned by their geo-strategic importance, a factor that contributed to its historical notability and economic development will be established. The conducted study focuses on the characterisation of its urban structure, identifying the architectural typologies and construction techniques behind this problematic. The main purpose is to frame and to understand the relationship established between traditional construction and the main structural reinforcement elements. The study also aims to contribute to the consolidation of a hypothesis to justify the multitude and diversity of the elements referred to.

1 INTRODUCTION

1.1 *The region and its seismic activity*

Évora is located almost in the geographical centre of the Central Alentejo region. This region is demarked by its hot, dry climate with scant, though intense rainfall in the winter. The region consists of vast plains, articulated by a mild relief terrain, and occasionally interspersed with mountains and plateaus lines of low altitude.

The hydrographic system has little expression, as it bestows a rarity of sizeable waterways. The traditional building is primarily related to the agro-livestock, which structured the territory through large properties of extensive crops and "assembled" (cork oak) forest, causing concentrated population agglomerations, dispersed and spaced on a disseminated background.

The region has the lowest population rate in the country, being the city of *Évora* is one of the exceptions, given its high concentration. Central Alentejo seismicity is characterised by the high number of seismic events, of medium intensity, more likely to produce little damage to buildings, although more consistently.

Constructively, the dominant traditional technique in the region is the rammed-earth, especially applied in isolated typologies of ground floor, although the existence of stony soils focus the use this feature in specific areas, with greatest expression within the territory. The use of masonry appears thus in combination with rammed-earth or in an exclusive application in specific larger settlements, in particular those which relate to the extraction of this raw material. A particular area of the city of *Évora* is located on a granite massif. This feature determines the abundance of techniques and constructive elements based on its application. The fitting of the spans and the coating of the wedges, of granite stonework, are examples of its use as the primary structural reinforcement feature of the constructions.

Table 1. Main earthquekes in Central Alentejo (LNEC, 1986).

Year	Intensity*
1755	VII
1859	VII
1917	VII
1926	VII
1969	VII

*Mercalli Scale (MCS)

Understanding the implication of seismicity in constructive culture in this area is hampered by the scarce existing documentation. One of the exceptions is linked to the national survey process carried out by local parishes. Ordered by the Marquis of *Pombal, prime-minister at the time*, this process was implemented due to the survey damage caused by the 1755 earthquake, in order to ascertain the actual state of the rest of the country. The accounts recorded in the parish memories reveal the collective consciousness of the occurring impact.

However, the reading of the damage is very asymmetrical. Most of the evaluation material falls naturally on religious buildings or property of the kingdom, in this case the buildings are identified and the pathologies verified are described in detail. The damage steeples and cracking of the vaults of the naves of churches constitute the bulk of registered consequences. As for the residential buildings, especially of rural character, the reading of the damage is superficially or indirectly made. The damage recorded, even if occasional; is referred to the vulnerability of construction techniques used by its inhabitants. In certain settlements are only references to the non-existence of casualties, despite the level of destruction observed in the housing. This irrelevance given to the impact on residential construction also indicates a less catastrophic result than the one recorded in other parts of Portugal. Thus, it is possible to state that a significant part of the local building culture would be able to regenerate or replace the affected structures.

"26) Se padeçeo alguma ruina no terromoto de 1755 e em que e se está reparada? Respondo que alguma ruina padeçeraõ alguns edefiçios mas foy muito leve e algumas se achaõ reparadas e outras naõ." (Pinto, 1758, p. 97)

In the majority of the situations, the existing reference has to do solely with the need to address the financing of repair works, given the financial difficulties of religious institutions, and consequently all the technological nature of information is often missing.

2 GENERAL CHARACATERISATION OF THE CONSTRUCTIVE CULTURE

2.1 *The urban fabric of the city of Évora*

The central area of the city of *Évora* is structured through a medieval genesis matrix, providing a consolidated urban fabric, with a public domain of a particular uniqueness, namely the definition of its main squares. The city is noticeable by the quantity and relevance of the existing monuments, whose implementation is related to the logic of the plan. As in most cases with similar circumstances, the influence of the implementation process of these works is expressed in a greater relationship between classical architecture, traditional architecture and vernacular architecture.

Both in terms of spatial design and in terms of construction techniques it is common to see an interdependent contamination, in what appears to be a specific regional ownership (SNA 1961). Residential buildings, particularly those located in areas of the central core, next to the most important axes, refer to this feature. They have an elongated plant configuration, arranged transversely to the perimeter of the block, forming at least one front street, where the facades are to be refined (Simplicio, 2009).

In the main axis of communication, crossing the Giraldo Square, some fronts have the particularity to integrate continuous circulation galleries, linking the ground floors of different buildings, and usually taking a commercial function. Despite the arcades recourse to architectural locking solutions, such as pointed arches and vaults edges, the significant difference of the applied settings, and its non-compliance with the internal solution of the ground floors, leads to a rather heterogeneous structural performance.

At this characteristic incurs the frequent application of reinforcement, especially in border elements, and in the most dissonant situations of the arcades. The columns, especially those of a resolution of greater

Figure 1. Giraldo Square in 1848 (credits: Luís Coelho in Rosa, 1926*).

divergence settlement quotas, and those of edging, are directly reinforced with buttresses. The development occurred in *Évora* Centre determined that most of these elements should present a greater approach to the setting of their own arcades, with less emphasis on the assembly reading, and a greater departure from its original appearance (Rosa, 1926).

Although many buildings display the application of continuous cornices, their length is limited. In most constructions the heights of the surrounding buildings is divergent and of independent coverage, with different dimensions and a misaligned volumetric. In these areas it is common to use structural reinforcement elements, transverse to different units, such as internal arcs or prominent foundations, contributing to the structural solidarity of the aggregation unit. The internal arc arises as the outer locking element, applied in the constructions of greater height that are spaced apart by streets of reduced section.

2.2 *The analysed typology: The residential building of regional character*

According to the municipal inventory of typological classification, the traditional dwellings of the Historic Centre of *Évora* can be divided into four traditional categories: bourgeois residential architecture; noble residential architecture; royal residential architecture, and architecture of regional nature. The Bourgeois residential building is divided into three sub-categories: housing, wealthy home and palace. The noble residential architecture can take the form of Palace or Solar. The royal residential architecture takes the form of the Royal Palace. The forms mentioned previously have a significant architectonic formalism, which is translated into a more erudite technology component, possible by the high resources of the owners. The architecture designated as regional type is, of course, the closer to the local vernacular building culture, whether in formal terms or in terms of technology. The subtypes identified were: a courtyard home; the house with a chimney at the front; and the arcade building. Of the latter, the house with a chimney at the front; and the

Figure 2. Internal arcs multitude (credits: CI-ESG).

arcade building, correspond to the majority, presumably by further optimisation area that characterizes the urban context.

2.3 *The arcade building*

The arcade buildings are of urban character, between 3 and 4 floors, corresponding to the type with further developments in height in the Historical Centre. They present a narrow and elongated rectangular shape, arranged transversely to the lot, with gable roof. They make a clear distinction between the ground floor and the others. The main facades have a regular vertical alignment in the placement of the spans, ranging from balcony or sill openings, and significantly ripping the facade plan. On the other facades the spans are smaller and of scarce and irregular application. The inner partitioning is varied and irregular, and the larger compartments are generally adjacent to the main facade. The main feature of these buildings is the integration of circulation galleries on the basis of their main facades. Although it does not match the same technology solution and their configuration is not identical, the arcades arise usually articulated between several buildings. The arcades are executed in granite stone masonry with roofs supported by pointed arches or vaults edges.

Figure 3. Integrated reinforcement in the wedge of an arcade building, *Giraldo* Square (credits: CI-ESG).

The strengthening of the outer support columns section is frequent, with higher expression in the more dissonant elements of the volumetric composition. These outer ribs have a section variation, with larger basis, and with the same appearance as buttresses. The upper floors are executed in stone masonry, of simple cloth, observing sometimes the inclusion of internal walls with wooden structures, cross to the building design. The application of turnbuckles can arise occasionally, especially located at the level of the upper pavements and in the corners of buildings. Much of the observed internal arches are associated with side locking of all these buildings. Applied in the cross streets to the location axes of the arcades, the internal arches articulate the closer gables. Their largest application generally occurs in situations of higher geometric irregularity, both in the alignment of facades and in the compositional logic of its openings. Performed in granite stonework, they usually lie in the alignment of the floor structures, at least at one end, thus enabling the internal locking supplement the built structures. However, it is presumed that its main function is to avoid the impact between independent structures, contrary to the usual effect of "pounding".

2.4 *The house with a chimney at the front*

The house with a chimney at the front comes as a variant, developed in height, of a typology slightly present throughout the Alentejo region, with higher expression in smaller compact settlements. In terms of construction these end up expressing the same technological characteristics of the region, with the exception made to the use of stony materials, which are gradually imposed as they approach the highest concentration areas, where this raw material is more available (SNA, 1961). Although rammed-earth constitutes the constructive vernacular system of greater multitude in the wider Alentejo territory, it is important to note that shale, marble, and particularly in *Évora*, granite, are also of great expression. In the particular case of *Évora*, and despite its large amount, this typology comes in a dispersed manner in the city, farther from the core, usually taking smaller portions. Some examples of the use of interior roofing systems supported by arches, domes and vaults, executed in brick *'lambaz'* were identified (Ribeiro, 2013). These elements are usually applied in single-storey houses, directly on the garret of the attic, usually a gable roof. The development in height, occurred in more compact areas, this solution comes only associated with the ground floor, and not always taking its entire area.

The structural framework of the floors and the roof is mainly parallel to the facade, following the direction of the shorter span, discharging most of the weight on the side walls of the building. The beams, of square or circular section, are usually recessed in the masonry of these walls. The sidewalls assume therefore their structural nature, justifying its higher section in relation to the others. The units are assembled from these walls, to which are abutted the corresponding walls of adjacent buildings, a feature that clearly contributes to the overall solidarity of the urban set. Both in its configuration, narrow and elongated plant, and in the integration of reasonable structural strength elements, such as arches and covering domes and the fitting of the spans, appear to contribute to an interesting structural performance in general to this typology. The characterising element, the chimney, is almost entirely in solid brick, in order to better withstand higher temperatures; using also lintels and granite quoins in the connection points to the perimeter masonry. This element assumes an autonomous configuration with respect to the building façade, by intersecting the main outer surface, and forming a prominent protrusion on their volumes. The size of the chimney in relation to the reduced width of the facade emphasizes its role.

In some examples the section of the chimney can occupy almost half of the building's street front to which it belongs. The chimney can be developed from the base of the building or just from the first floor. In this second option, the chimney's lower intersection features a small overhang, often bore by a discharging arch. When developed from the ground floor, these chimneys generally have a greater dimension at the base, although without the expression of other urban centres, such as *Arraiolos* or *Monsaraz*, in which the slope of this element is more pronounced, resembling the configuration of an integrated reinforcement system (poial). However, this similarity is only apparent, since structurally, their geometric and equipment dissonance eventually show a significant propensity for the development of structural pathologies.

Figure 4. Vernacular house with a chimney at the front and counter arch in *Monsaraz* (credits: CI-ESG).

Figure 5. Two-storey house with a chimney at the front. (credits: CI-ESG).

3 FINAL REMARKS

The city of *Évora* is located in the Central Alentejo region, characterised in seismic terms for a regular activity of medium intensity. Its geological uniqueness and its administrative relevance contributed to a unique urban development, contrasting with the evolution of the remaining territory. The constructive culture, though occasionally referring to the use of traditional and regional techniques, often incorporates knowledge and more complex solutions.

This technological appropriation is directly related to the profusion of buildings of monumental character, and to settlement in the city of the wealthiest classes. Its constructive reference system is based on the use of granitic stone in masonry walls, in the stonework fitting and in the reinforcing of the corners.

The relevance of seismic activity is reflected in the collective memory of the city, also manifesting in particular registration of certain events, which affected the monumental architecture of *Évora*, mainly buildings of religious character.

The earthquake-resistant elements identified in *Évora* come mostly associated to the scale of the aggregate, reinforcing the structural solidarity of the block. The counterfort and the internal arc are the bulk of the found elements, yet some prominent foundations were also recognised. Despite its abundance, it is not evident an objective systematization in its application.

Thus, it is assumed that they are the result of a reactive action to the damage caused.

ACKNOWLEDGMENT

The authors gratefully acknowledge the support by the Portuguese Science and Technology Foundation (FCT) to the research project 'SEISMIC-V – Vernacular Seismic Culture in Portugal' (PTDC/ATP-AQI/3934/2012).

REFERENCES

Aguiar, J. (2001). Évora, exemplar e pioneira. In *Centros Históricos*. Santarém: Associação Portuguesa de Municípios com Centros Históricos.

CME – CÂMARA MUNICIPAL DE ÉVORA (1980). Plano Director de Évora, relatório nº 28. Évora: CME

LNEC (1982). *Construção Anti-Sísmica: Edifícios de Pequeno Porte.* Lisbon: Laboratório Nacional de Engenharia Civil.

LNEC (1986). *A Sismicidade Histórica e a Revisão do Catálogo Sísmico.* Lisbon: Laboratório nacional de Engenharia Civil.

Moniz, M. (1984). *A Praça do Giraldo.* Évora: Gráfica Eborense

Pinto, J. N. (1758). Memória Paroquial da freguesia de Nossa Senhora da Consolação da Igrejinha, comarca de Évora. *Memórias Paroquiais*, 18(14), 95–98.

Ribeiro, O. (2013). *Geografia e Civilização. Temas Portugueses.* Lisboa, Portugal: Letra Livre.

Ribeiro, O. (1986). Évora. Sitio, Origem, Evolução e Funções de uma Cidade. *Estudos em Homenagem a Mariano Feio, Soeiro de Brito*, 371–390.

Simplício, M. D. (2009). *Evolução da Estrutura Urbana de Évora: o século XX e a transição para o século XXI.* Évora: CME

Simplício, M. D. (2003). Évora: Origem e Evolução de uma Cidade Medieval. *Revista da Faculdade de Letra – Geografia*, XIX, 365–372.

SNA- Sindicato Nacional dos Arquitectos (1961). *Arquitectura Popular em Portugal.* Lisboa: SNA

* Rosa, J. (1926). *Iconografia Artística Eborense* – subsídios para a história da arte no Distrito de Évora, Imprensa Nacional, Lisboa, p. 19.

Seismic Retrofitting: Learning from Vernacular Architecture – Correia, Lourenço & Varum (Eds)
© 2015 Taylor & Francis Group, London, ISBN 978-1-138-02892-0

Seismic-resistant features in Lower Alentejo's vernacular architecture

A. Lima, F. Gomes, G.D. Carlos, D. Viana & M.R. Correia
CI-ESG, Escola Superior Gallaecia, Vila Nova de Cerveira, Portugal

ABSTRACT: Lower Alentejo region historical seismicity is characterised as frequent, though of medium intensity (VI-VII MCS). *Santo Brissos*, *Trindade*, *Baleizão*, and *Serpa*, are the case studies selected from the preventive approach of the population before the earthquake damage in the built structure of vernacular character; earthquake-resistant elements; typological characteristics; preservation of the strategies; identifying a local seismic culture.

1 INTRODUCTION

1.1 *Context: Lower Alentejo region*

The Lower Alentejo sub-region is limited to the north by the Upper Alentejo region, to the east by Spain, and to the south by the Algarve. It is a Mediterranean, warm and dry climate area. Geographically the territory is very uniform, the landscape is characterised by the dominance of the plain and the Guadiana valley to the southeast; and to the centre by the Sado valleys. The low population density of the Lower Alentejo is categorized according to two forms of distribution in the territory: concentrated and disseminated (interim dispersion). Occupation is based on two types of settlements: compact or large estates (*herdades*) with isolated *Montes* that cover the housing needs of the villages.

The isolated *Montes* are defined by the spread of clusters of one or more housings, very compartmentalized (barns, stables, shelters, furnaces etc.), presenting an orthogonal and compact plant (Correia, 2002). A gable roof and few openings are also featured. Of small dimensions and without any stone furnishing, these constructions usually lack glass, only featuring wooden shutters.

Rammed-earth is the predominant constructive technique, as it is also the most common construction technique around the villages and hills, overlaid by brick. The walls are plastered and carefully whitewashed, highlighted in different color (blue and ocher). The wide outer walls are the support to the gable roof of tiles arranged on a simple wooden framework. The interior rooms are usually in brick, and the pavement presents tiles or dirt. The stalks appear with some regularity in certain southern Alentejo regions, juxtaposed, and replacing the lath and the roof lining, thus forming an air cavity, known as reed. It is usual the placement of counterforts of loose rock as a structural reinforcement of the building (Fig. 1).

Figure 1. Boleja *Monte*, Baleizão (credits: CI-ESG).

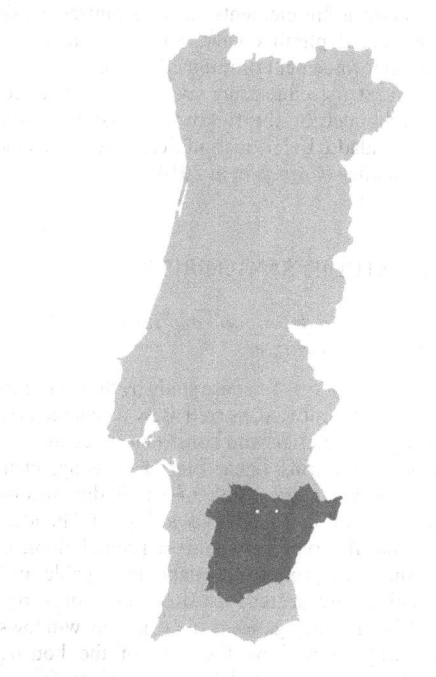

Figure 2. Portugal Map: Lower Alentejo region (credits: CI-ESG).

Table 1. Mainearthquakes in Lower Alentejo region. (LNEC, 1986).

Year	Intensity | Mercalli Scale (MCS)
1755	VII
1858	VII
1917	VII
1926	VII
1969	VII

Figure 3. *Santo Brissos*, linear cluster in Lower Alentejo (credits: CI-ESG).

Figure 4. *Santo Brissos, Beja*. Systemic placement of buttresses (credits: CI-ESG).

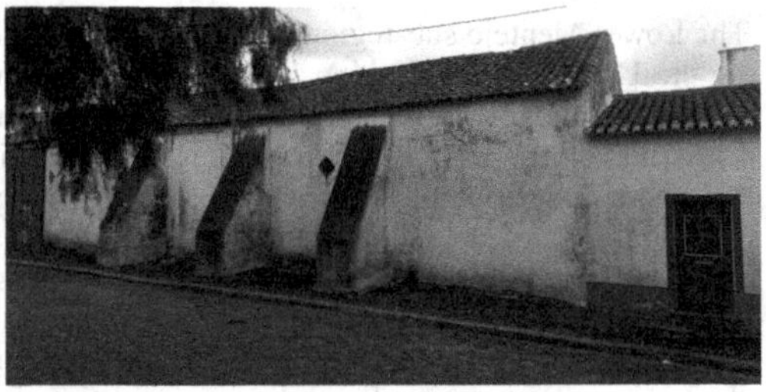

Figure 5. *Santo Brissos, Beja*. Systemic placement of buttresses (credits: CI-ESG).

2 HISTORICAL SEISMICITY: LOWER ALENTEJO REGION

The historical seismicity of the Alentejo though frequent presents a medium intensity (VI-VII MCS). Along the Lower Alentejo, and throughout the many fieldwork missions, was possible to verify the application of structural reinforcement elements in several buildings, which used rammed earth as a construction technique.

The early twentieth century is marked by the near occurrence of two earthquakes of medium intensity (seismic frequency), which reported minor damage to the built structure in the region. The fear generated by these events is reflected in the method of the population in the construction/rehabilitation of their houses, being noted a common concern in improving the constructive systems/techniques, and the introduction of reinforcing elements (earthquake-resistant).

At that time is possible to identify a local seismic culture, materialised in the typologies that host earthquake-resistant elements, such as buttresses, tie rods, reinforced plinth course, stone bench, and a structural reinforcement of rammed-earth.

After an extensive literature review and data collection, corroborated by the fieldwork missions, Santo Brissos, Trindade, Baleizão, and Serpa were selected as a case studies (Correia et al 2014).

3 CASE STUDIE: SANTO BRISSOS

3.1 *Constructive description and seismic-resistant elements integration*

Santo Brissos emerges as a case study by the reinforcement solutions identified, as well as by its characteristic typologies, materials and construction techniques. The village conforms linear cluster housing, characterised by the aggregation of small dimensions, longitudinal semi-detached houses (Fig. 3). The identified vernacular typology holds a ground floor, of regular and small proportions plant, and a gable roof. The openings are scarce, in most cases only represented by housing gateway that has no windows. The linearity marked by the sum of the housing facades, usually connected by the tops, conform a homogeneous front street, defined by interchanging window-door-window, or simply by the transition of doors.

Constructively it is characterised by the exterior walls of stone masonry or mixed, massive brick or rammed-earth, which are structural and support the roof. The inner walls are typically of brick, and the gable roof consists of tiles arranged on a single framework.

The earthquake-resistant reinforcing solutions identified in *Santo Brissos* indicate the existence of a local seismic vernacular culture over a given period of time. The settlement aggregation structure, of parallel and longitudinal alignments, reflects a smooth and consistent geometry, nonetheless spontaneous, in closed structures (blocks) usually reinforced structurally at the extremities with a masonry vaulted structure placed at the base of the building and stone benches. Another reinforcing element identified was the buttress of considerable dimensions, which in some cases occupied almost the entire facade of the housing, interrupted by the hollow door. (Figs. 4 to 7).

Figure 6. *Trindade, Beja*. Vernacular typology with seismic-resistant elements (credits: CI-ESG).

Figure 7. *Boleja Monte, Baleizão* (credits: CI-ESG).

4 CASE STUDIES: BALEIZÃO, SERPA AND TRINDADE

4.1 *Constructive description and seismic-resistant elements integration*

Trindade, Baleizão and *Serpa* appear as case studies by the identified enhancing solutions, by the typology presented, the materials and the construction techniques. The identified case studies are related to the typology of the Alentejo *Montes*, as they present a dispersed occupation of the territory: the isolated rural housing (Fig. 7).

They emerge spontaneously in the territory, isolated or in small semi-detached rows. The ground floor and the rectangular plant, with bearing exterior walls of whitewashed mud, and the gable roof of wood structure and straw tile are the main features of these constructions. The interiors are of solid brick, half-timbered or adobe, and the floor is of screed. The inner partitioning is simple and regular, distinguishing the agricultural and residential spaces. The most common construction technique is the rammed-earth, mostly marked by the reinforcement in-between the different rows.

The exterior walls are usually reinforced with several giants/buttresses of considerable size, built of stone masonry or massive brick. Their placement along the facades varies, but at least two are placed to reinforce opposed walls.

Another common element is the tether or tensioning rod, a connecting element and structural reinforcement that prevents the collapse and deformation of the outer

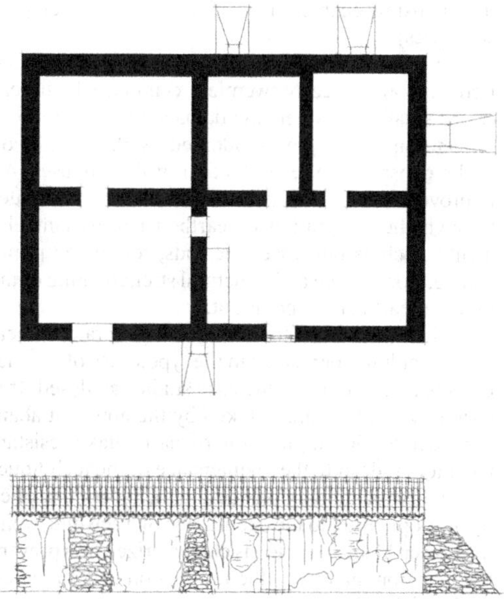

Figure 8. Vernacular typology in *Serpa, Beja* (credits: CI-ESG, based in Correia, 2007).

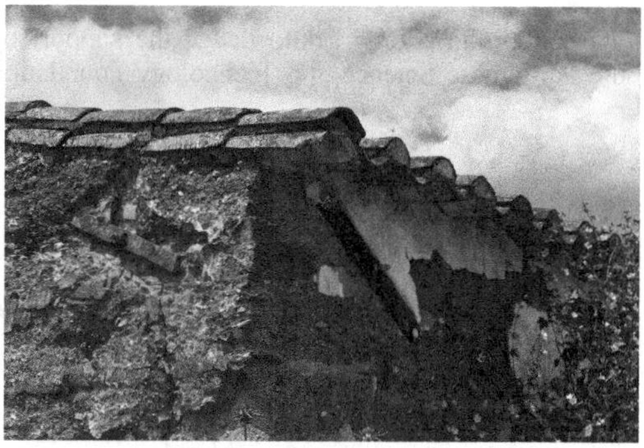

Figure 9. Vernacular typology in *Trindade, Beja*. Systematic introduction of tie-rods (credits: CI-ESG).

walls. The linear elements are responsible for the traction, by bars or steel wires fixed to two parallel walls by anchorage elements, interconnecting the structural elements. In the Alentejo region it is usual that the same housing presents both longitudinal and cross rods, and two anchors in the corners (Fig. 9).

5 FINAL REMARKS

The Lower Alentejo is characterised by the frequency of medium intensity earthquakes. Their uniformity trace similar characteristics to most of the buildings of vernacular character in the region. Thus, it is possible to typologically group thedifferent case studies (*Santo Brissos*, *Trindade* and *Baleizão*; and *Serpa*). Two distinct types emerge: the Alentejo *Montes*, arising from the dispersed occupation of the territory;

rural housing, characteristic of urban centers (compact structures).

The case studies identified are marked by seismic frequency in the early twentieth century, which generated small and systematic damage to the buildings, reflected in the approach adopted by the population in the construction/rehabilitation of their houses. An improvement of the construction systems/techniques through the integration of earthquake-resistant elements, such as buttresses, tie rods, reinforced plinth course, stone bench and structural strengthening of the rammed-earth are to point out.

In the early twentieth century was generated a local seismic culture, perceived in the typologies of vernacular character found in the case studies analysed. It is currently inactive, and marked by the apparent abandonment of the application of earthquake-resistant techniques, despite the maintenance of the techniques in specific locations. Considering the reaction applied by the population to the occurrence of a seismic event, the Lower Alentejo region is characterized by a preventive reaction, patent by the production of typologies, distinguished by the seismic prevention character.

ACKNOWLEDGMENTS

The authors gratefully acknowledge the support by the Portuguese Science and Technology Foundation (FCT) to the research project 'SEISMIC-V – Vernacular Seismic Culture in Portugal' (PTDC/ATP-AQI/3934/2012).

REFERENCES

Correia, M. (2002). 'A Habitação Vernácula Rural no Alentejo, Portugal'. In Memórias del IV Seminário Iberoamericano sobre Vivienda Rural y Calidad de Vida en los Asentamientos Rurales (Cap.4 – Vivenda Rural, Etnia, Cultura y Género). Chile: CYTED-Habyted y Universidad del Chile.

Correia, M. (2007). Rammed earth in Alentejo. Lisbon: Argumentum.

Correia M., Gomes, F. & Duarte Carlos, G. (2014). Projecto de Investigação Seismic-V: Reconhecimento da Cultura Sísmica Local em Portugal. In M. Correia, C. Neves, R. Núñez (Eds.) *Arquitectura de Tierra: Patrimonio y Sustentabilidade en regiones sísmicas*. El Salvador: Imprimais.

Gomes, F., Correia M., Duarte Carlos, G. & Viana, D. (2014). Local Seismic Culture in Portugal: Melides dwellings, a reactive approach case study. In C. Mileto, F. Vegas, L. García Soriano & V. Cristini (Eds.) *Vernacular Architecture: Towards a Sustainable Future*. London, UK: CRC/Taylor & Francis.

LNEC 1986. *A sismicidade histórica e a revisão do Catálogo Sísmico*. Lisboa: Laboratório Nacional Engenharia Civil.

Seismic Retrofitting: Learning from Vernacular Architecture – Correia, Lourenço & Varum (Eds)
© 2015 Taylor & Francis Group, London, ISBN 978-1-138-02892-0

Seismic vulnerability of the Algarve coastal region

G.D. Carlos, M.R. Correia, G. Sousa, A. Lima & F. Gomes
CI-ESG, Escola Superior Gallaecia, Vila Nova de Cerveira, Portugal

L. Félix & A. Feio
Universidade Lusíada, Vila Nova de Famalicão, Portugal

ABSTRACT: The Algarve is the national area, the most exposed to the action of the Bank of Gorringe, place to which are allotted the epicentres of higher magnitude earthquakes that ravaged Portugal in particular 60-63b.f., 1033, 1356 and 1755. However, despite being characterised by a reasonable geographical uniformity, the territory of the Algarve presents considerable variation at the level of impact occurred, both of a general historical analysis, and with respect to specific occurrences. The case studies chosen, regardless of the geological similarity, territorial proximity, and the high degree of devastation that they underwent, are primarily characterised by the divergence of the answers operated in the post-disaster context. From the complete abandonment of the settlements to the immediate reconstruction, the use of pre-existence to the full redefinition of the clusters structure, the Algarve offers a wide variety of approaches when considering how this region adapted to the seismic phenomenon.

1 INTRODUCTION

1.1 *The region and its historical seismicity*

The Algarve locates in the extreme south of the country. It is characterised by a predominantly Mediterranean climate influence. It presents a long and smooth coastline with solar exposure to the south. Its terrain is gradually developed from its mountain line, which protects it to the north from the Atlantic winds, forming the national area with lower rainfall index. Its coastal area has the densest occupation of the region, populated by clusters with intermediate dispersion. Its main urban centres are often developed from sea or estuary ports, displaying a clear presence of the Islamic occupation. The Algarve is also illustrated by a strong historical seismicity, presenting earthquakes that caused extensive damage in the built structure of the region. The earthquakes of 1722 and 1755 seriously affected the Algarve coast, with intensities of IX and X (MCS). Existing reports corroborate the significant seismic activity of the Algarve region, both in terms of quantity and in terms of intensity. The impact recorded affected its entire coastal strip, with no differentiation between Western or Eastern area, regardless of the Atlantic or Mediterranean influence, which also reinforces the theory of the prevailing spread from the Gorringe Bank (LNEC, 1986).

Its orography, exception made for the Atlantic angle and the mountainous line break, also enhances the vulnerability to tsunamis that in cases such as *Faro*, *Lagos* and *Santo António de Arenilha*, proved to be the main cause of devastation (Victor Mendes, 2006).

Table 1. Main earthquakes in the Algarve region (LNEC, 1986).

Year	Intensity*
1719	X
1722	IX–X
1755	IX–X
1856	VIII
1858	VI–VII
1969	VIII

*Mercalli Scale (MCS)

1.2 *The architectural diversity of the Algarve coast*

Although experiencing an improved variety throughout the region, the geology of the Algarve also turns out to condition the structural weakness of traditional nature buildings. The settlements implanted in sandy soils, of great erosive vulnerability and low density, generally have a high degree of destruction, as documented, more than once, in *Sagres*, *Vila do Bispo* or *Bensafrim*. In these cases it was frequent the use of rock bases, in sandstone, of dark reddish colour, usually called 'Silves stoneware'. These foundations were built in massive *socos*, existent in smaller buildings and of a single storey, generally clustered in row.

Constructively, the western end and the base of the mountain area are characterised by the use of non-specified stone walls, punctuated by the use of reddish sandstone and limestone masonry (with significant expression also in the floors and in the trim of the openings), with greater presence in urban areas. The composition of limestone and clay soils also

Figure 1. Rocky integration as structural *soco, Vila do Bispo* (credits: CI-ESG, 2013).

Figure 2. Trim reinforcement with *lambaz* brick (Credits: CI-ESG, 2013).

Figure 3. Rammed-earth building with reinforced plinth course, *Bensafrim* (Credits: CI-ESG, 2013).

determined broad application of lime as the main binder, and the profusion of ceramic materials such as tile and solid brick (SNA, 1961). These elements thus compensate the structural weakness of the crude masonry cloths, as well as the development of specific components in *lambaz* (solid) brick for use in openings reinforcement or discharging arches.

In structural terms, it should be further emphasised their combined application in the development of dome and vaulted roofing systems, which spread across the territory, and although considered generally, as it is the roof terrace, it is also often associated with specific locations given the observed systematic (for example in the case of *Fuseta* in *Olhão*). As in the neighbouring Alentejo regions using rammed-earth constituted, for many years, the traditional benchmark system for most of the rural areas, particularly in isolated buildings. The analysed architectural typologies are also analogous to that region, dominating the buttresses and the masonry vaulted structures placed at the base of the buildings as the main earthquake-resistant elements. As in the above-mentioned vaulted systems, the use of rammed-earth corresponds currently to a less observable construction technique, due to its pronounced disappearance or the significant altering of the existing examples (Caldas, 2010).

From the referenced seismic occurrences, this text will address the direct consequences of the earthquake of 1755, the depth of the physical and cultural impact that it left the region. However, it should be noted the occurrence of 1969 as part of a recent memory in the locations of the western area, although without explicit implications for their constructive cultures (Vasques, 2002).

Tavira encompasses the cases of high impact with moderate destruction, whose recovery of the building takes place almost immediately after to the occurrence. Within this condition can be classified, for example, the cases of *Loulé* and *Castro Marim*. At the other extreme, are found the cases of cities like *Silves*, *Faro* and *Lagos*, whose cities were severely destroyed, and the reaction of the population more time consuming and complex. Moreover, *Lagos* represents a situation of total abandonment, whose reconstruction occurs after a significant gap, and subject to distinct chronological phases (Victor Mendes, 2006). Finally, *Vila Real de Santo António* will be characterised as a unique case, representing a strategic paradigm at the national level, aimed at the implementation of the *Pombal* reconstruction system, both in urban and technological terms (Mascarenhas, 2009).

2 LAGOS

2.1 *General framework: The destruction of the city and its resurgence*

The development of Lagos, as most of the Algarve coastal cities is related to maritime activities, with its economic heyday at the time of Portuguese discovery. The urban core is divided into two parishes: the Parish of Santa Maria, where traces of Roman *Lacobriga* persevere, with orthogonal grid and blocks of approximate dimensions; and the Parish of San Sebastian, which has a more organic route, and an irregular urban fabric.

Figure 4. "Gaioleiro" system in Lagos (credits: Laura Felix, 2015).

Lagos took on great importance in the XV century by virtue of the Portuguese discoveries, becoming the most significant maritime base of the ships that sailed the Mediterranean and the Atlantic.

Over the centuries, several earthquakes affected this city, and the 1755's earthquake was the most devastating. With the earthquake and the tsunami that followed it, the city was practically devastated. As a direct consequence, this city is abandoned by the Governor and the army that was staying in Tavira, being no longer the capital of the Kingdom of the Algarve.

The delay in its reconstruction; the economic decline; the instability caused by the French invasion and the fighting between liberals and absolutists contributed to its economic and urban stagnation. Only amidst of the XIX century was resumed the process of urban reconstruction, aided by the economic prosperity of the expansion of the canning industry, and the growth of local businesses.

In the 60s, the tourism development will contribute to the beginning of the most intense urban expansion phase, which features significant part of the current form of the city.

2.2 *Characterization of Lagos historical centre*

The features of built area of the historical centre of *Lagos* were decisively affected by the 1755's earthquake, remaining only a few buildings. The most identified traditional buildings date back to the late eighteenth, nineteenth and twentieth centuries.

Currently, the historical centre of *Lagos* has a multitude of buildings formally and technologically dissonant. In terms of construction process, about 47% of the buildings are of reinforced concrete, being the remaining 53% of traditional construction. Of these, 40% are in stone masonry, while the remaining 13% are in brick or adobe mud.

The extramural northern area of the centre, bordering the old mesh, corresponds to a newer construction sector, with a totally different scale from the original. The blocks present variable forms and areas, strongly influenced by the steep topography and the access roads.

The north intramural area, adjacent to the walls, has elongated blocks with different characteristics from the rest of the city, adapted to the setting of the wall itself, and the adjacent access roads.

The Centre downtown area, where are confined the most relevant public spaces, has an organisation in compact blocks of smaller dimensions. The original nucleus of this area is the first walled perimeter, presenting very specific features and emphasising the regularity of the mesh that consists of small blocks. These blocks have sizeable patios, arranged laterally to buildings with exterior walls of great expressiveness. The central area has a more uniform mesh. The layout is structured by city permeation routes and their perpendicular branches, defining blocks with different dimensions. The area adjacent to the walls, but outside the walls, presents a mostly free building space. It is possible to observe the City Park, spaces with private gardens, still in a much more degraded area covered with ruins and poorly constructed. Finally, the extramural area, south of the centre, consists of isolated equipment and public gardens, and a significant portion is occupied by Military Camping Park.

With regard to volumetric, the historical centre of *Lagos*, and despite the transformation of the urban fabric that occurred between the late 60s and early 70s, there is a high prevalence of buildings of 1 and 2 floors, totalling approximately 75% of the total edified set. Despite greater number of floors buildings are a small percentage, it is noteworthy the visual impact they cause in relation to the remaining urban fabric. They become totally unrelated to the primitive type. These new buildings often result from the grouping of several plots, and the withdrawal of the patio, an element that was part of any type of housing in the city. The early alignment disappears, as well as building-street relationship. Construction materials are most often of low quality, having no correspondence with the vernacular building of the city. The vernacular buildings that characterise the core feature sloping roofs, varying from roof of only one water, a gable roof, hipped roof or several water and mixed coverage (terrace or roof terrace and roof of a water). The buildings present for the most part, rectangular plans in L or U, and visible wedge in clenching of several blocks. The spans of the main façades feature stone or mortar linings, and when the furnish is mortar, it is painted, as the *soco*. The spans are of small size, and sometimes misaligned. Unique elements are observed in these buildings, such as the verge (in buildings of 2 or more floors) or double eaves (common in 1 floor buildings). Sometimes it is also possible to observe structural reinforcement elements, either independent or transversal to the building systems.

2.3 *Seismic-resistance in Lagos*

Despite the massive destruction of the buildings in the historical centre of *Lagos* and its resulting abandonment, the research carried out allowed the identification of buildings prior to 1755. Some of these

Figure 5. 'Telhado de Tesouro' in Tavira (credits: CI-ESG, 2015).

buildings, slightly raised in relation to the streets and generally aggregated in row, show significant slope facades produced by gradually reducing the wall section height to its top trim.

As for the buildings constructed post-earthquake, reinforced plinth course, buttresses, bearing arches and 'gaioleira' (Fig. 4) walls were identified. However, despite its high number identified, the location of these elements is scattered and punctual, revealing no direct linkage with any constructive technique or architectural typology.

The analysis to some ruins also enabled the testing of a correlation between the construction method and its time of construction, allowing a better identification and understanding of the buildings of the same era. Some ruins presented the use of 'gaiola' systems on the walls of moiety. These systems were extremely simplified, generally without the presence of diagonal elements, and composed only by the vertical wooden elements (plumbs) and the horizontal elements (windowsill), sometimes reinforced with the inclusion of some metal elements.

The application tethers, transversely to the main façades, are also a frequent presence.

Inside of some buildings it was also possible to identify additional locking elements, such as transverse walls with wooden skeleton and brick filler, cross half-timbered walls and crossed floors.

3 TAVIRA

3.1 *General Context: The progressive rationalization of the medieval matrix*

Tavira is located on the Mediterranean coast line, shifted to its eastern side. It stems from its port vocation, given the natural protection of the salt marshes of the *Ria Formosa*. It is structured from two distinct cores, deployed in areas of rugged hillside, which develop along both banks of the River Gilão.

Hit violently by the 1755's earthquake, it quickly recovers thanks to the internal action of its population, which performed a process of spontaneous reconstruction process, resorting to the use and reuse of ruined buildings. In the geographical centre of Tavira, it is clear the sedimentation of the chronological strata in the buildings, with obvious intersections of medieval structures and post-earthquake construction. The centre is characterised particularly by the threading of the set of streets crossing the river axis. These areas, of a more dense occupation than the upper area, exhibit linear aggregates or elongate block with two street fronts, moving away progressively from medieval genesis matrix, which confers to it a more regular layout.

3.2 *The built typology: The height of the building cross to the plot*

The core of *Tavira* consists mainly buildings developed in height from two to three floors, with a narrow and elongated rectangular plan, which occupies the entire plot. They have dimensions slightly greater than the built area implanted in the highest parts of the slope. They have the particularity of presenting multiple roofs, of four waters, built on the same volume. It is called "treasure". According to Ribeiro (2013) these examples date back to the perpetuation of a constructive tradition of the golden age of its owners, expressing an external influence of the oriental, a common character of most dynamic port cities. The inner partitioning typically corresponds to a single cover unit, adopting an internal organization of cross fragmentation, which will lose the social role as the compartments move away from the main facade. The floors, always different, represent a great regular composition, by employing a significant number of arched solutions in the execution of their spans.

3.3 *From the regional building system to the specific coverage system*

From the buildings of the central core of Tavira, despite its proximity to the River *Gilão* waters, there are no known specific characteristics of implementation of its foundations. Yet, there is a clear tendency to enjoy the rocky integrations, of Jurassic limestone, for the start of buildings, and thus avoiding the vicinity of the alluvia marginal of the river. The exterior walls use a constructive system of stone masonry, of rough apparatus, determining a section of significant thickness. They present also, however, a more elaborate treatment in stone elements at the wedges of the buildings. They have structural interior walls, using the same building system, arranged transversely to the building, and lined with the inner beams of the covering. The peculiar system covers of "treasure", clearly a response to optimise the dimensioning of the spans woodwork, allows a more efficient distribution of its own load, as well as the increase of general fastening of the building. The spans of the main facades have stonework fittings, sometimes with slightly arched lintel. Independent structural reinforcement elements or transverse to the construction systems applied are not observed.

Figure 6. "Gaioleiro" system in Vila Real de Santo Antonio (credits: CI-ESG, 2015).

Figure 7. Covering system sustained by arches in Vila Real de Santo António (credits: CI-ESG, 2015).

4 VILA REAL DE SANTO ANTÓNIO

4.1 *General context: The implementation of the 'Pombalino' model*

The development of *Vila Real de Santo António* is part of the strengthening strategy of national occupation on the border with Spain. As the vast majority of the Algarve coastal region, it was based on fishing subsistence. However, it will be their customs status, which will give it a significant predisposition to trade, and accordingly its individual character. This area assumes the location of the former Village of *Santo António de Arenilha*, near the Guadiana bank, which has been devastated by the 1755's tsunami. The construction of the new village starts from the implementation of a core set by the *Casa do Risco das Obras Públicas*. It presents the direct application of the urban model of 'Pombalino' downtown, conveniently suited to the scale and resources of the region, in the mesh configuration, in the aggregation systems, and in the architectural typology. The urban mesh is implemented parallel to the river, providing the elongated blocks in a north-south direction. The blocks are compact, dominated by longitudinal fronts of great regularity, with roofing and façade of a single expression.

4.2 *The process of adaptation of the income building.*

Without the size and the resources of the capital city, the income building of 'Pombalino' downtown was too ambitious for the context of the Algarve village. Not only would it be needed to adjust the investment from the point of view of the real estate value, as it would also be appropriated to correct the functional program, as the occupational density was not justified, given the amount of available land. The buildings that acquiesce the urban front of the most important streets, near the dominant longitudinal axis or Central Square, follow more closely the application of the 'Pombalino' modular systematic, albeit with substantial reduction in the maximum number of floors.

They are usually of two levels, with a significant improvement of the coverage, which can vary between 4 and 6 water solutions. They present orthogonal rectangular plans of great symmetry and geometric systematization, which is expressed especially in the configuration of the facades. The plans, significantly narrower than the original model, vary between 1 and 2 subdivision levels, according to the parallel alignment to the main facade. The vertical difference between levels is evident, reflecting the typology used for the solution of the openings. Unlikely, in the horizontal level there is uniformity of treatment, both in the internal organisation and between different housing units. The side streets, especially the most remote of the central area, admit a typological variety, and may contain smaller residential buildings, or they may also integrate functional additions to the main building, which would be located on the opposite street. These streets lose therefore their block façade uniform character, assuming a large formal dynamism.

4.3 *The simplification of the 'gaiola' system*

The buildings, built according to the 'Pombalino' model, turn naturally to the cage system, with a significant reduction of the wooden skeletons sections and the corresponding stone masonry. The wedges and the spans are fully implemented with limestone fittings. The ground floor, only built in stone masonry, takes back the structural *soco*, using a covering system sustained by arches. The upper floor refers to the combined 'Pombalino' system (with the caveats already mentioned), integrating in several cases, the systematic application of metal tethers on the facade, especially in the anchorage of the upper floors pavements. The remaining buildings or their wall-backing supplements are of great constructive simplicity. These resort to ordinary masonry stone or brick systems. In these buildings, the application tethers across the main facade, is still often, though not carried out systematically.

5 FINAL REMARKS

Considered generally as a regional unit in terms of seismic vulnerability, the Algarve presents a great variety of processes and approaches from the impact caused by the major earthquakes.By its geographical and geological nature, the coastal strip represents the area of the most significant records.

Tavira, Lagos and Vila Real de Santo António represent three paradigmatic approaches in what the reaction to the 1755's earthquake is concerned. Tavira is a chaos of spontaneous reconstruction, regenerating and maintaining the structure built typologies and the local constructive technique. Lagos conveyed a traumatic process, conditioned by a long abandonment and a time consuming and asymmetrical reconstruction process, with traces of traditional elements and techniques, yet without the consolidation of a specific building culture that incorporates assumedly this problem. Vila Real de Santo António constitutes a rupture with the constructive culture, established by the full replacement of the typological model and the techniques employed in the region. However, the spread and permeability of the 'Pombalino' systems applied, along with the awareness of the hand labour, enabled a process of adaptation and adjustment to local conditions and resources. This contributed to the integration and effective dissemination of some elements in the processes of traditional construction of the Algarve region.

ACKNOWLEDGEMENTS

The authors gratefully acknowledge the support by the Portuguese Science and Technology Foundation (FCT) to the research project 'SEISMIC-V – Vernacular Seismic Culture in Portugal' (PTDC/ATP-AQI/3934/2012).

REFERENCES

Caldas, J. V. (2010). *Cidade e mundos rurais. Tavira e as sociedades agrárias*. Tavira: Câmara Municipal de Tavira.

GECoRPA (2000). Sismos e Património Arquitectónico – Quando a terra voltar a tremer. *Revista Pedra & Cal*, 8.

Gonçalves, A., 2009. Vila Real de Santo António: planeamento de pormenor e salvaguarda em desenvolvimento. In Monumentos, 30, 40-53.

Gonçalves, A. (2007). Caracterização do Núcleo Pombalino. *ECDJ*, 9, 18-35.

LNEC (1982). *Construção Anti-Siìsmica: Edifiìcios de Pequeno Porte*. Lisbon: Laboratoìrio Nacional de Engenharia Civil.

LNEC (1986). *A Sismicidade Històrica e a RevisaÞo do Cataìlogo Siìsmico*. Lisbon: Laboratoìrio Nacional de Engenharia Civil.

Mascarenhas, J. (2009). *Sistemas de Construção – V: O Edifício de rendimento da baixa pombalina de Lisboa*. Lisboa, Portugal: Livros Horizonte

Ribeiro, O. (2013). *Geografia e Civilização. Temas Portugueses*. Lisboa, Portugal: Letra Livre

SNA- Sindicato Nacional dos Arquitectos (1961). *Arquitectura Popular em Portugal*. Lisboa: SNA

Vasques, J. C. (2002). *Lagos e a Instabilidade Sísmica. Relatório do Centro de Estudos Marítimos e Arqueológicos de Lagos*. Lagos: CEMAL

Seismic Retrofitting: Learning from Vernacular Architecture – Correia, Lourenço & Varum (Eds)
© 2015 Taylor & Francis Group, London, ISBN 978-1-138-02892-0

The high and intense seismic activity in the Azores

F. Gomes, M.R. Correia, G.D. Carlos & A. Lima
CI-ESG, Escola Superior Gallaecia, Vila Nova de Cerveira, Portugal

ABSTRACT: The archipelago of the Azores is situated in a seismic area of high intensity characterised by the frequency of seismic events. The 1980 and 1998 earthquakes, according to the collected data, caused elevated damages and made possible the identification of seismic-resistant elements, as the prominent foundation and the reinforcement of the internal structure (wood box). *Terceira* and *Faial* Islands reveal a local seismic culture presented throughout their history, though inactive. There is an obvious oblivion of the seismic risk, which led to a progressive abandonment of the application of the seismic-resistant elements in the dwelling.

1 INTRODUCTION

1.1 *Context: The archipelago of the Azores*

The Azores region comprises nine islands of volcanic origin, with several dimensions and configurations. Its geographical location, stipulates the climate of the archipelago, presenting different climatic characteristics in the different islands, and in each island microclimates are also identified.

Manly linear, the Azorean settlements are, favourably, parallel to the coast. They can also appear with a perpendicular track, when accompanying water lines.

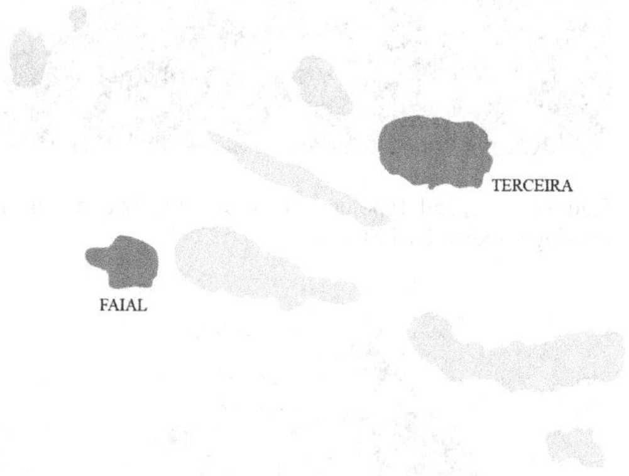

Figure 1. Map of the Azores archipelago. Identification of the case studies (credits: CI-ESG).

2 HISTORIC SEISMICITY: THE AZORES REGION

The archipelago of the Azores presents an important historical seismic activity characterised by the systematic seismic crises, the significant and elongated volcanic activity and its geographic location, near the triple point associated to the joining of the Eurasian, African and Northern American plates. In the course of the XX century, the seismic crises of elevated intensity affected greatly the *Faial* Island in 1973 and the Terceira Island in 1980. Later, in 1998, the most recent seism in the archipelago of the Azores, affected seriously the *Faial* Island. Thus, through the revision of the literature focused on the historical seismic occurrences and damages caused, appear as study objects the *Terceira* and the *Faial* islands.

Table 1. Main earthquakes in the Azores. (Nunes et al., 2004).

Year	Place	Intensity | Esc. Mercalli (MCS)
1522	S. Miguel	X
1614	Terceira	IX
1757	Pico/S. Jorge	X
1801	Terceira	VIII
1841	Terceira	IX
1852	S. Miguel	VIII
1973	Pico/Faial	VII
1980	Terceira	IX
1998	Faial	IX

3 CASE STUDIES: FAIAL AND TERCEIRA

Over the last centuries, the Terceira Island suffered from several earthquakes, the latest of which occurred on the 1st of January 1980.

During this earthquake, more than 3000 buildings were totally or partially destroyed, much of it on *Terceira* Island, particularly in the city of *Angra do Heroísmo*. 80% of the buildings in this city were destroyed and extensive damage occurred also in the

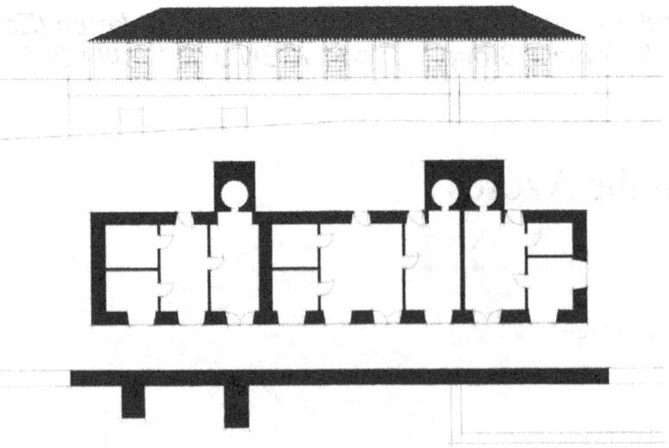

Figure 2. Typical Typology of the *Terceira* Island. Plant and facade of the linear Housing (credits: CI-ESG).

Figure 3. Typical Typology of the *Terceira* Island. Linear housing (credits: CI-ESG).

Figure 4. House with seismic-resistant element (reinforced plinth course) in *Faial* Island (credits: CI-ESG).

Figure 5. House with seismic-resistant element (reinforced plinth course) in *Faial* Island (credits: CI-ESG).

Figure 6. Traditional house with seismic-resistant element (reinforced plinth course) in *Terceira* Island (credits: CI-ESG).

village of San Sebastian, and in West and North West parishes of the island, particularly in the Parish of *Doze Ribeiras*.

During the field survey carried out in the *Terceira* Island it was possible to observe several buildings with at least one earthquake-resistant element. Many of these identified buildings were found on the outskirts of *Angra do Heroísmo*. In the northern part of Angra do Heroísmo case studies identified presented buttresses and reinforced plinth. An ancient brick oven with sharply sloping exterior walls was also there identified (Figure 4).

Between *Angra* and *Doze Ribeiras*, over 50 buildings presented property enclosure walls delimiters owned and exterior walls, which were wider at the base and narrower at the top. Some ground floors presented slanted walls, but the wall on the 1st floor had become closer (reinforced plinth) (Figure 6).

Faial Island was, throughout the twentieth century, repeatedly hit by destructive earthquakes. The 5th of April and 26th of August 1926's earthquakes, the seismic crisis of 1958, the earthquake of the 23rd of November 1973, and of the 9th of July 1998, all cause considerable material damage to the built structure of the Island, evidencing a high seismic vulnerability of the traditional construction.

Along the field survey, it was possible to observe several buildings containing a systematic integration of earthquake-resistant elements. The island holds several non-rehabilitated buildings (presenting post-earthquake damage), in particular in *Ribeirinha* and *Horta*. In Ribeirinha, half of *S. Matias* Church and its adjacent building were partially destroyed.

Some buildings in the historical centre of Horta were not restored as well, being, nowadays, in ruins.

In the island, it was also identified a building called 'Poial de Pólvora', whose origin is probably set in the early twentieth century, presenting a reinforced plinth course as well (Fig. 4).

The several queries made to the population also confirm that the ancient walls of property separation were

Figure 7. Traditional house with seismic-resistant element (reinforced plinth course) in *Terceira* Island (credits: CI-ESG).

built with a slight slope (wider at the base and narrower at the top), as it was identified in the Faial Island.

In the *Torre do Relógio*, in the historical centre of the city of *Horta*, there was the placement of various vertical tethers.

Buildings with earthquake-resistant elements, evidencing a clear preventive or reactive approach, were identified all across the island of *Faial*.

Some of the local interviewees indicated, informally, to be aware of the use of a seismic-resistant net, in current construction, which is placed between the plaster and the outer wall.

3.1 *Constructive characterisation and incorporation of seismic-resistant principles*

The traditional architecture of the *Terceira* and *Faial* islands is mostly defined by two different dwelling typologies: the linear house and the integrated house. The linear house presents a rectangular or in L house-plant, arranged perpendicularly to the street, and combining the dwelling and the agricultural adjoining. The integrated house appears like a massive block, developed in height, generally with two floors. The buildings are generally of volcanic stone masonry, and of elementary configuration. The roofing, either gable or a hip-roof, is supported by wooden trusses.

Considering the selected and analysed buildings, seism-resistant evidences were found, like the 'wooden box', that in spite of not constituting a regular and generalised use, it was noticed after the seism of 1980 and 1998 in rural dwellings. It is supported by a system of revetment covering of the dwellings, which detained behaviour towards the seismic happenings. In some cases, internal structures were perceived, executed in wooden panels, which are totally detached of their support, and thus surviving to partial downfall of the walls and covering. These structures form autonomous structural boxes, of great flexibility level, to minimize the risks of collapsing of structures of larger mass.

The prominent foundation constitutes one of the most expressive solutions, sometimes enfolding sometimes the whole perimeter of the constructions, though it is only promptly observed in the analysed villages.

4 FINAL REMARKS

Terceira and *Faial* islands are situated in a seismic area of high intensity, also noticeable for the frequency of seismic activity. The seisms of 1980 and of 1998, according to the collected data, caused elevated damages and made possible the identification of the incorporation of seism-resistant elements, such as the prominent foundation and the reinforcement of the internal structure (wood box), through a preventive action to the seismic activity taken by the local population. Both islands, *Terceira* and *Faial*, present a local seismic culture, but there is an obvious oblivion of the seismic risk, which led to a progressive neglect of the application of the seism-resistant elements in the dwelling.

ACKNOWLEDGEMENTS

The authors gratefully acknowledge the support by the Portuguese Science and Technology Foundation (FCT) to the research project 'SEISMIC-V – Vernacular Seismic Culture in Portugal' (PTDC/ATP-AQI/3934/2012).

REFERENCES

Correia, M., Gomes, F. & Carlos, G. D. (2014), Projecto de Investigação Seismic-V: Reconhecimento da Cultura Sísmica Local em Portugal in Arquitectura de Tierra: Patrimonio y Sustentabilidade en regiones sísmicas, eds. M. Correia, C. Neves, R. Núñez, El Salvador, Imprimais.

Neves, F., Costa, A., Vicente, R., Oliveira, C.S. & Varum, H. (2012). Seismic vulnerability assessment and characterisation of the buildings on Faial Island, Azores. *Bull Earthquake Eng* 10(1), 27–44.

Nunes, J. C., Forjaz, V. H., Oliveira, C. S. (2004). Catálogo sísmico da região dos Açores. Versão 1.0 (1850–1998). Ponta Delgada.

Part 5: Typology performance study

Seismic Retrofitting: Learning from Vernacular Architecture – Correia, Lourenço & Varum (Eds)
© 2015 Taylor & Francis Group, London, ISBN 978-1-138-02892-0

Seismic behaviour assessment of vernacular isolated buildings

J. Ortega, G. Vasconcelos & P.B. Lourenço
ISISE, Faculty of Engineering, University of Minho, Guimarães, Portugal

H. Rodrigues
School of Technology and Management, Polytechnic Institute of Leiria, Portugal

H. Varum
CONSTRUCT-LESE, Faculty of Engineering, University of Porto, Portugal

ABSTRACT: This paper presents the numerical seismic analysis of isolated vernacular buildings characteristic of the Alentejo region, which is considered a medium seismic hazard region in Portugal. A representative isolated building was selected from a database, and a geometric model was defined for the numerical pushover analysis. Subsequently, a parametric analysis was carried out to assess the influence of distinct parameters on the seismic behaviour of such buildings.

1 INTRODUCTION

The reasons behind the selection of the different vernacular building typologies, in order to assess their seismic performance are: (i) firstly, the buildings should represent typologies that can be encountered in regions that were previously identified as prone to have developed a Local Seismic Culture, i.e., where the seismicity is high, and earthquakes are frequent, even if of low intensity; (ii) secondly, traditional seismic strengthening solutions should have been already identified in some of the buildings belonging to the typology, revealing the possible existence of a Local Seismic Culture.

Additionally, in order to make the study more comprehensive, it was deemed necessary to include differentiated vernacular typologies in both, urban and rural contexts. Vernacular architecture in Portuguese rural environments, in contrast with urban vernacular constructions, usually consists of independent buildings of small dimensions, with no structural interaction between them. Therefore, the case study analysed herein involves isolated vernacular buildings.

2 SELECTION OF A CASE STUDY

The first vernacular building typology chosen as a case study consists of a representative vernacular rammed earth construction, commonly found in the South Portuguese region of Alentejo, where the seismicity is high in relation to other Portuguese regions, and is thus prone to have developed a Local Seismic Culture. Rammed earth construction, known as *taipa* in Portugal, essentially consists of the compacting of the earth using a timber formwork for the construction of free standing walls. This has traditionally been the most widespread technique in these regions and, even though its use decreased significantly in the last forty years, is still in use in some places.

Traditional dwellings in Alentejo have generally small dimensions, simple rectangular shape and only one floor, having predominant horizontal dimensions. They were also very simple, regarding their plan configuration, little compartmentalised, and using rammed earth walls also for the partition walls. They present massive shapes, with few or no openings, other than a single door, as a protection for the hot summers. Chimneys are the only relevant protruding non-structural element that can be systematically found in this type of buildings. Other materials are also used, such as stone or brick masonry, in order to reinforce the corners, and to build a base course or *soco*. This aims at protecting the rammed earth from the humidity and rain penetration, by preventing the action of rising damp, but it also helps reinforcing the rammed earth walls. Roofs are commonly mono-pitched roofs or gable roofs, usually presenting a low slope, and made with a simple framework of timber beams. Buildings were finally painted in white, in order to reflect the sunlight.

Satisfying the second important requirement to be chosen as a case study, traditional seismic strengthening solutions could be identified in several of these characteristic rammed earth constructions (Correia 2005, Correia 2007). Mainly, buttresses, known as *gigantes* in the region, could be usually observed attached to the exterior walls (Fig. 1). They perform an

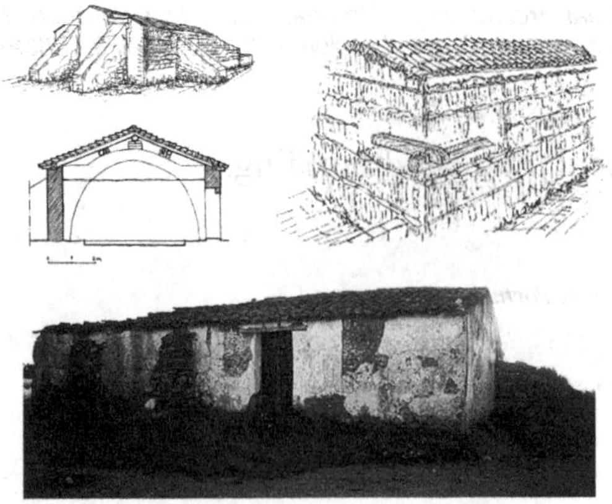

Figure 1. Traditional seismic strengthening solutions identified in rammed earth constructions of Alentejo (credits: Correia, 2007).

important task in the event of an earthquake, counteracting the horizontal forces exerted by the buildings. Their efficiency might be determined by their relative position within the building. Ties are also very commonly found in this type of buildings (Fig. 1), performing another important task by coupling the structural elements, such as parallel walls, and contributing to the achievement of a box-behaviour of the structure. Other traditional strengthening solutions observed consist in the introduction of timber elements to reinforce the connections of the walls at the corners (Fig. 1). Timber lintels and discharging arches over the openings are also among the reinforcing techniques observed.

3 STRUCTURAL ANALYSIS OF ISOLATED BUILDINGS

The study of the seismic behaviour of ancient constructions built with traditional materials is particularly challenging, given the multiple uncertainties regarding the material properties, connections between structural and non-structural elements, or even the uncertainties about the state of conservation of the construction. Rammed earth constructions are composed by generally thick load bearing walls, whose in-plane resistance is significantly higher, than its out-of-plane resistance. These thick walls can exhibit substantial structural ductility and deformations, larger than the expected for such a brittle material (Michiels 2014). This behaviour is associated with a highly nonlinearity, which complicate the structural analysis and safety assessment of these structures, particularly when submitted to seismic loading.

Aiming at having a better insight of the seismic behaviour of the traditional rammed earth construction found in Alentejo region, a numerical analysis was carried out based on representative building for the assessment of the seismic behaviour.

3.1 *Finite element model*

The Finite Element Modelling (FEM) is used for the global seismic analysis of a rammed earth vernacular building, following a common macro-model approach, which has already been extensively and successfully applied with the aim of analysing the seismic behaviour of complex masonry structures (Lourenço et al. 2011). However, in order to understand and accurately simulate the seismic behaviour of rammed earth constructions, it is important to describe accurately the nonlinear behaviour through advanced plastic constitutive models, since relevant deformation of the structural elements is expected. Few studies have focused on the finite element modelling of rammed earth buildings (Bui et al. 2008, Jaquin 2008, Braga & Estêvão 2010, Gomes et al. 2011, Angulo-Ibáñez et al. 2012, Gallego & Arto 2014, Miccoli et al. 2014), and most of them have adopted simple models, assuming simple constitutive laws, mainly linear elastic isotropic. Therefore, finite element modelling, based on nonlinear numerical analysis of rammed earth vernacular buildings represents a step forward in technical and scientific knowledge, as few results are available in literature.

This numerical simulation intends to understand, in a more detailed way, the resisting mechanisms of the different structural elements of this typology under seismic loading, based on nonlinear static (pushover) analyses. Pushover analysis has been already commonly used for the seismic assessment of existing masonry buildings (Lourenço et al. 2011), and mainly consists of simulating the seismic loading as static horizontal forces, which are applied incrementally on the structure. It allows determining the ability of the building to resist the characteristic horizontal loading caused by the seismic actions, taking into account the material nonlinear behaviour, while being simpler than other methods of analysis, like nonlinear dynamic analysis and, therefore, it was chosen for this study.

3.1.1 *Reference building geometry*

From the analysis of the buildings of the database found in the literature previously mentioned (Correia 2007), a reference model was built, which intends to be a simplified representative example of these constructions, gathering common characteristics in terms of dimensions and architectural layout. The reference building geometry was based on a specific example located in Vila Nova de São Bento, in the Beja district, in Alentejo (Fig. 2). However, some changes were adopted in the geometry and construction details, in order to typify more precisely the rest of the buildings belonging to this typology. The final plan and elevation views of the reference building used are shown in figure 3. The plan has a simple rectangular shape, symmetrical in both orthogonal directions, regarding also the distribution of the interior load bearing walls. The height of this type of buildings rarely surpasses 3 meters at the front and back walls. The gable walls are not very high either, keeping the roof slope low, between 15–20 degrees.

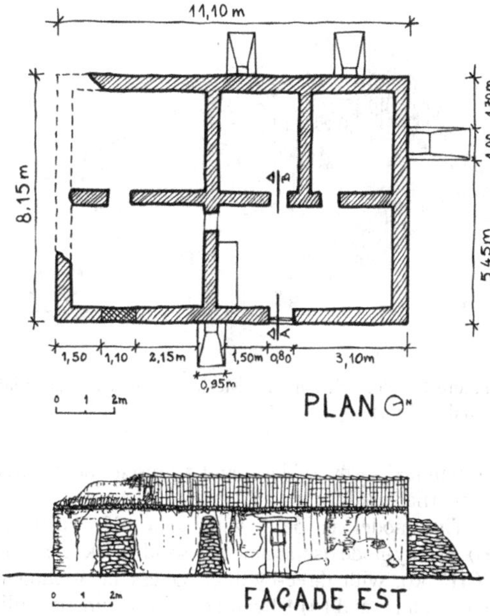

Figure 2. Original rammed earth building in Vila Nova de São Bento, in the Beja district, in Alentejo (credits: Correia, 2007).

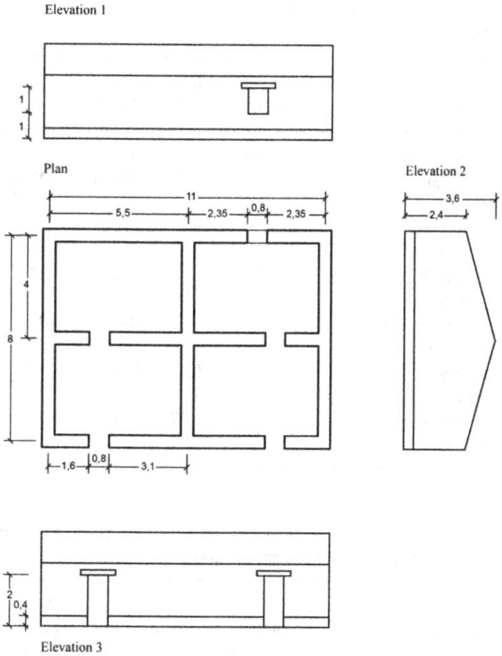

Figure 3. Plan and elevations of the reference building adopted for the construction of the numerical model (credits: J. Ortega).

The height of the stone masonry course at the base is very variable, but it was established as 0.4 meters. Regarding the openings, the position of the two doors and two windows have also a symmetrical configuration.

Timber lintels were considered over the openings, as this is also the common practice observed in almost every building of the database. Chimneys or other non-structural elements were not added to the reference building at this initial step.

3.1.2 *Reference numerical model*

The final reference numerical model was built based on the previously commented characteristics, using DIANA software (TNO 2009). Three different materials are considered. Stone masonry is used for the base course, which is usually built with irregular schist, or granite masonry and thus, poor material properties are assumed. Rammed earth is used for the structural walls, both interior and exterior. Timber is used for the lintels over all the openings. The roof is only considered as a distributed load on the top of the walls, and the displacements of the elements at the base are fully restrained.

The material model finally adopted to represent the nonlinear behaviour of the rammed earth and stone masonry, which are the two materials considered to present nonlinear behaviour; it is a standard isotropic Total Strain Rotating Crack Model (TSRCM), which describes the tensile and compressive behaviour of the material with one stress-strain relationship, and assumes that the crack direction rotates with the principal strain axes. An isotropic model was chosen, because despite its layered structure, experimental tests found in the literature have shown that the mechanical properties of rammed earth do not behave in an anisotropic way (Miccoli et al. 2014). This model is very well suited for analyses, which are predominantly governed by cracking or crushing of the material. The tension softening function selected is exponential, and the compressive function selected to model the crushing behaviour is parabolic. The detailed information required for the rammed earth material properties were obtained from data collected, from different authors, and it is shown in Table 1. It is noted that there is a big variability, bringing up even more uncertainties. For the timber lintels, only the elastic properties are considered, and an elasticity modulus of 10 GPa, and a Poisson's ratio of 0.2 were used (Gomes et al. 2011). Regarding the stone masonry elastic properties, a modulus of elasticity of 1500 MPa, and a Poisson's ratio of 0.2 were adopted. Its compressive strength and specific weight were obtained from reference values, given by the Italian code (NTC08 2009), assuming the lowest quality masonry class, an irregular rubble stone masonry composed of rubble and irregular stone units of different sizes and shapes. The remaining nonlinear properties of the masonry were computed directly from the compressive strength, based on recommendations given by Lourenço (2009). The compressive fracture energy was obtained, using a ductility factor of 1.6 mm, which is the ratio between the fracture energy and the ultimate compressive strength.

Table 1. Rammed earth material properties used for the finite element model found in the literature.

Author	W (kN/m^3)	f_c (MPa)	f_t (MPa)	E (MPa)	ν
Jaquin (2008)	–	0.60–0.70	–	60	–
Bui et al. (2008)	20	–	–	100–500	0.22–0.40
Braga & Estêvão (2010)	20	0.89	0.17	264	–
Gomes et al. (2011)	19	0.67	–	200	0.35
Angulo–Ibáñez et al. (2012)	20	–	–	500	0.2
Gallego & Arto (2014)	20	1.85	0.29	250	–
Miccoli et al. (2014)	–	3.7	0.37	4207	0.27

Table 2. Mechanical elastic properties adopted for the three materials used in the reference model.

Material	E (MPa)	ν	W (kN/m^3)
Stone masonry	1500	0.2	20
Rammed earth	300	0.3	20
Timber	10000	0.2	6

Table 3. Mechanical nonlinear properties adopted for the materials used in the reference model.

Material	f_c (MPa)	G_{fc} (MPa)	f_t (MPa)	G_{fI} (N/mm)	β
Stone masonry	1.5	2.4	0.15	0.012	0.05
Rammed earth	1	1.6	0.1	0.1	0.05

The tensile strength was estimated at 1/10 of the compressive strength. Finally, an average value of 0.012 N/mm is adopted for the mode I fracture energy. Concerning the rammed earth material elastic properties, an elasticity modulus of 300 MPa and a Poisson's ratio of 0.3 were used. A compressive strength of 1 MPa was adopted. The remaining nonlinear properties were again calculated directly from the compressive strength, following the same recommendations (Lourenço, 2009). The unique difference with respect to the stone masonry lies in the value used for the mode I fracture energy. According to Miccoli et al. (2014), the fracture energy of rammed earth should be obtained by increasing the compressive strength about ten times, as it is considered that rammed earth behaves in a different way, in comparison with stone masonry, which behaves as a brittle material. Due to its broad particle size distribution, which includes large particles that may have a significant contribution for the interlocking at the crack surface, by promoting its roughness, a value of 0.1 N/mm was adopted for the fracture energy

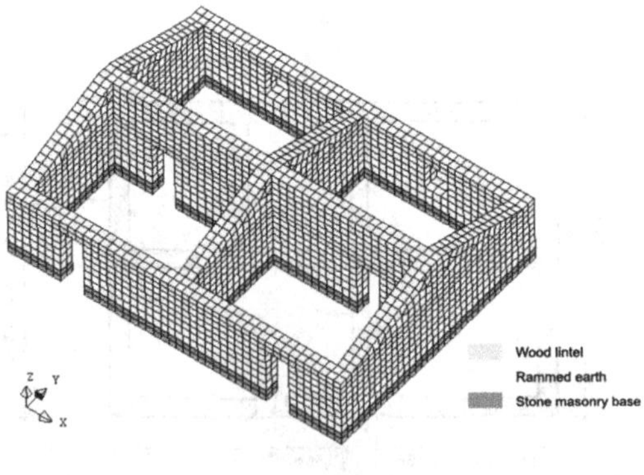

Figure 4. Numerical model: Mesh categorized by materials (credits: J. Ortega).

of rammed earth. Tables 2 and 3 present the material properties used for the analyses.

The model is built with solid 3D elements: (i) twenty-node isoparametric solid brick elements (CHX60), with three-by-three Gauss integration in the volume; and (ii) fifteen-node isoparametric solid wedge elements (CTP45), with a four-point integration scheme in the triangular domain, and a three-point scheme in the orthogonal direction, used to adjust the mesh to the geometry resulting from the triangular gable walls. The final reference model has two elements in the thickness direction of the wall and therefore, the resulting generated mesh has 31,264 nodes and 7,993 elements (Fig. 4). The total mass of the model is 150 tons.

4 SEISMIC PERFORMANCE OF THE CASE STUDY

As a first step, the main dynamic characteristics of the reference model are obtained and summarised in Table 4. Results show that the mass participation of the first modes is low, in comparison with common modern RC buildings with rigid floors, since the accumulated mass participation of the first 20 modes is around 45% and 65% in X and Y direction, respectively. Most of the modes are associated with local deformations, involving only specific structural elements at a time, and there is no global modes affecting the whole structure. This effect is derived from the fact that there is no rigid floor coupling the vertical structural elements, and the roof is not modelled. Thus, the walls get to vibrate independently. The first modes are associated with local out-of-plane deformations of the walls in the Y direction, particularly the taller inner walls, less resistant to local deformations. The sixth mode is the first one showing relevant displacements in the X direction. Figure 5 shows the shape of the first, third and sixth mode.

A nonlinear static (pushover) analysis was then carried out. First, only the dead weight and the distributed load on top of the walls, simulating the roof, are considered. After that, an incremental monotonic loading,

Table 4. Results from the dynamic analysis.

Mode	Period (s)	Frequency (Hz)	Mass participation* U_x	U_y	U_z
1	0.150	6.66	0.00 (0.00)	0.00 (0.00)	0.00 (0.00)
2	0.134	7.45	0.00 (0.00)	11.18 (11.18)	0.00 (0.00)
3	0.099	10.13	0.39 (0.39)	0.00 (11.18)	0.00 (0.00)
5	0.096	10.38	0.00 (2.56)	0.00 (11.18)	0.00 (0.00)
10	0.08	12.58	0.00 (40.76)	3.16 (31.73)	0.01 (0.02)
20	0.05	19.98	0.00 (43.98)	17.13 (63.59)	0.01 (0.06)

*Accumulated mass participation in brackets.

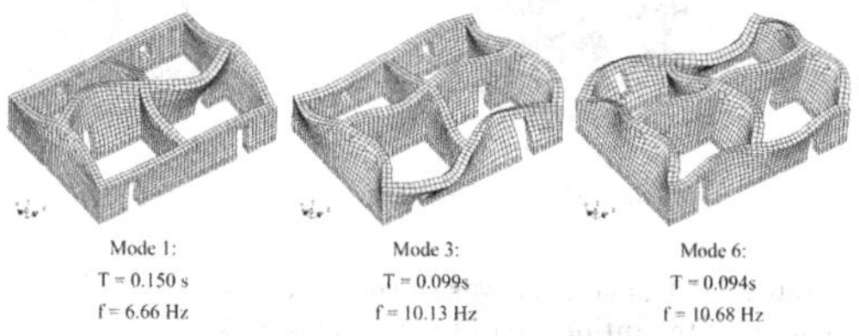

Figure 5. Shape of the first, third and sixth mode of the reference building (credits: J. Ortega).

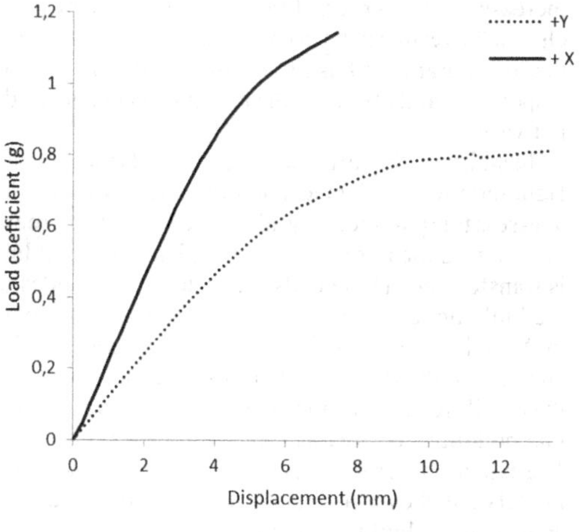

Figure 6. Capacity curve of the pushover analysis on the reference building in +X and +Y direction (credits: J. Ortega).

proportional to the mass, is applied on the structure in the main horizontal directions (X and Y), as recommended by Lourenço et al. (2011) for masonry structures. Given the local modes previously observed in the modal identification, a modal pushover analysis was disregarded. Only the positive directions are considered, since the behaviour of the building is practically symmetric.

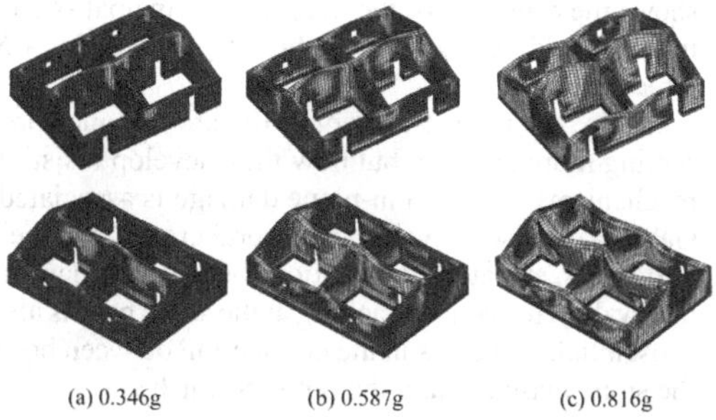

Figure 7. Evolution of maximum principal strains depicted on deformed mesh for the pushover analysis in +Y direction (credits: J. Ortega).

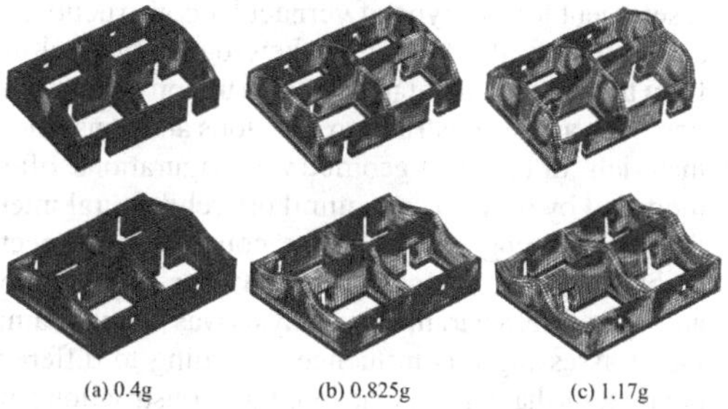

Figure 8. Evolution of maximum principal strains depicted on deformed mesh for the pushover analysis in +X direction (credits: J. Ortega).

Figure 6 shows the capacity curve for the reference building in both horizontal directions. The analysis shows that the structure capacity is higher than the expected for this kind of buildings, obtaining maximum load coefficients of around 0.8 g in +Y direction and over 1.1 g in +X direction. This might be due to the small dimensions of the reference building, particularly regarding the height of the walls, and the span between walls, but mainly because the structural elements of the buildings are considered to be perfectly connected among them, avoiding their premature local out-of-plane collapse.

Figure 7 presents the evolution of the maximum principal strains, which can be used as a cracking measure in the building for the pushover analysis in +Y direction. As it could be expected, the parts of the building presenting more damage are the middle walls, which should be attributed to their higher slenderness. These walls show flexural vertical cracks in the mid-span, and horizontal cracking at the base, and at the interface between the rammed earth and the stone masonry base course. The major damage takes place at the connections between perpendicular walls. The behaviour is improved by the cooperation of the orthogonal walls, which are bracing them. At the latest steps, the front and back walls also present some out-of-plane cracking and severe damage at the connections with the perpendicular walls.

The capacity in the +X direction is considerably higher than in the +Y direction, given the bigger amount of resisting walls in this direction. Figure 8 shows the evolution of the maximum principal strains in the building for the pushover analysis in +X direction.

Again, the middle interior walls are the ones presenting more damage, but now they develop resisting mechanisms, to which in-plane damage is associated. Gable walls also show heavy flexural cracking at their mid-spans, and at the connections between perpendicular walls. Horizontal cracking at the stone base is also substantial, as well as in the connection between both, the stone masonry, and the rammed earth.

5 NUMERICAL PARAMETRIC ANALYSIS

As previously mentioned, the seismic vulnerability assessment for this type of vernacular constructions is a difficult task, due to the great heterogeneity, resulting from the uncertainty of many construction characteristics, such as the construction solutions and constituent materials, or different geometry configurations, often modified by previous structural or architectural interventions, among others. These construction aspects highly influence the seismic behaviour of structures and, therefore, a parametric analysis was planned, aiming at assessing this influence according to different parameters that take into account the construction particularities of the representative vernacular rammed earth construction typology, chosen as a case study. A parametric analysis will also help understanding in a more detailed way the seismic behaviour of this typology. These parameters should be identified in order to be studied, and they were selected based on knowledge of the effects of past earthquakes (Blondet et al. 2011). The observation of earthquake induced damage has traditionally been a tool for the understanding of the structural behaviour of vernacular constructions in the sequence of earthquakes. With the inspection of constructions after earthquakes, it is possible to see the adequacy or inadequacy of construction practices, and to realize which are the main parameters affecting the seismic response of constructions.

Numerical nonlinear parametric analyses were defined, in order to assess the influence of the different parameters that were considered to have a decisive influence in the seismic behaviour of selected typology, and to try to quantify it. The initial configuration of the reference model was changed in terms of geometry and construction characteristics, and new models were built according to the parameters. The comparative analysis between the new models and the reference one is made in terms of capacity curves, and intends to identify the parameters that have a bigger influence in the building seismic response.

5.1 *Plan configuration*

Most of the buildings of this type show a very regular rectangular shape, but different length to width ratios may influence the seismic response of the building. Thus, the influence of the in-plan slenderness of the building was evaluated through two numerical models, namely by adding cells to the reference building to increase its slenderness. Figure 9 shows the new models and the results in terms of capacity curves, showing that this parameter has little influence on the seismic response, mainly because the resisting mechanisms do not vary.

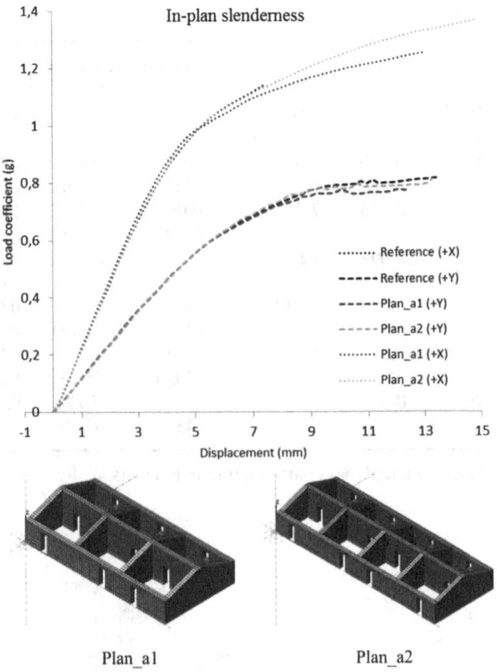

Figure 9. Capacity curves and new models built for the evaluation of the influence of the in-plan slenderness (credits: J. Ortega).

In addition, the influence of an irregular shape configuration was evaluated, and three models were built to assess this parameter by adding cells to the reference building, so the initial rectangular plan of the building is transformed. These cells are added at both sides of the building, changing the symmetry conditions both in X and Y directions. The size of the cells is also changed in the different analyses. The idea of adding these cells resulted from inspections of several vernacular buildings, where this configuration was observed. The results in terms of capacity curves and the new models are shown in Figure 10. The capacity curves show that the building response to horizontal loading is not very sensitive to the presence of projections, unless this part presents a significant size. The biggest difference can be found in the pushover analysis in +Y direction, in the analysis for the building 'Plan_b1'. In this case, the cell added is big enough to change the failure mode of the building (Fig. 11), which now takes place at the independent cell that is freer to deform and allow some torsion effects to take place. In terms of the X direction, buildings 'Plan_b1' and 'Plan_b2' show also a decrease in their capacity, which may be due to the bigger deformations taking place at the independent body. In every case, no significant contribution

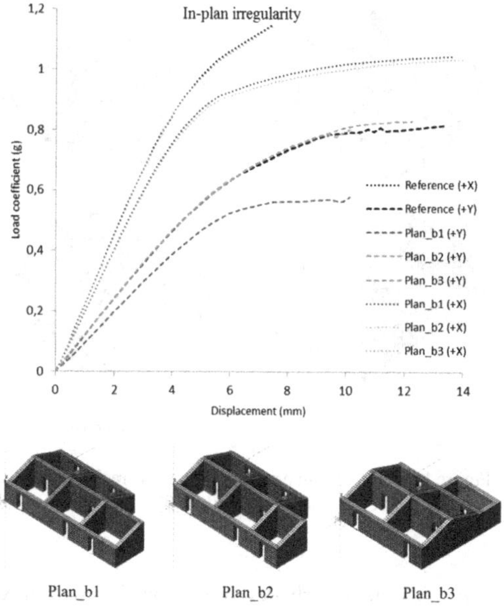

Figure 10. Capacity curves and new models built for the evaluation of the influence of the in-plan irregularity (credits: J. Ortega).

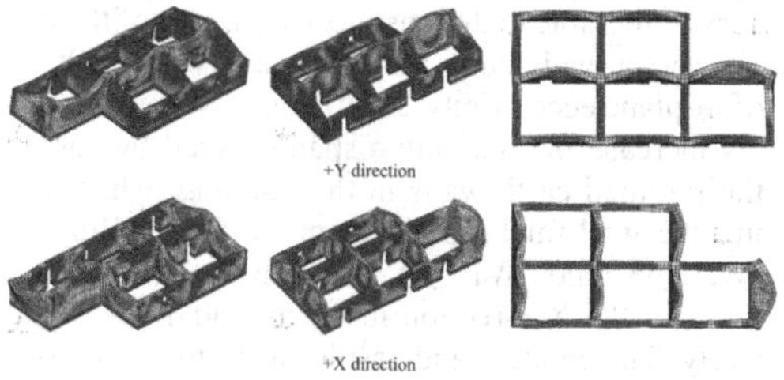

Figure 11. Maximum principal strains (a cracking measure) depicted on deformed mesh at failure for building 'Plan_b1' (credits: J. Ortega).

of a torsional response was found, resulting possibly from small dislocation of the rigidity center, in relation to the geometric center.

Another common source of plan irregularity showed by this type of buildings concerns the distribution of the resisting elements, and the lack of symmetry in one or both orthogonal directions. This may lead to an eccentricity of the stiffness centre with relation to the mass centre, enhancing the torsional effects in the event of an earthquake. Therefore, three more models were built, changing the internal distribution of the walls in relation to the reference building, in order to evaluate the influence of this stiffness eccentricity. Some interior walls were removed or added, aiming at avoiding the symmetry conditions used for the reference building, and at obtaining some eccentricity. Results in terms of capacity curves, as well as the new models constructed, are presented in Figure 12. It should be noted that for the building 'Plan_c1', only the +Y pushover analysis was carried out. because

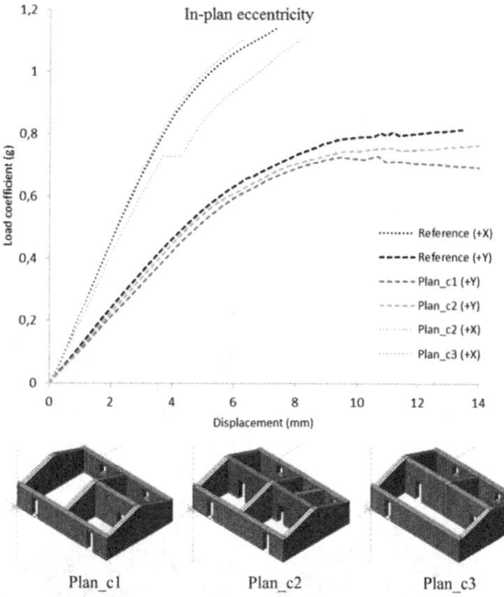

Figure 12. Capacity curves and new models built for the evaluation of the influence of the in-plan eccentricity (credits: J. Ortega).

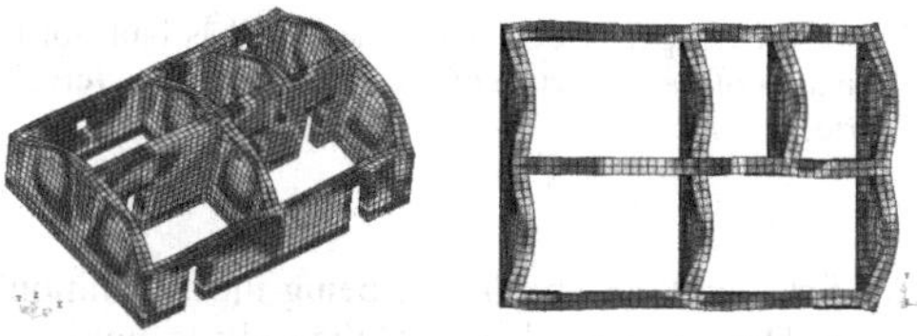

Figure 13. Maximum principal strains depicted on deformed mesh at failure for building Plan_c2 in +X direction (credits: J. Ortega).

conditions for the +X pushover analysis are much altered to be comparable, as one of the walls is now covering a much larger span. Similarly, regarding the building 'Plan_c3', only the pushover analysis in the +X direction is carried out. The analyses in the remaining directions will be performed as a next step, to assess the influence of the maximum distance between walls. Slight differences can be observed in terms of ultimate load or failure modes for both directions. A small decrease in the capacity of the buildings in the +Y direction is observed, which can be ascribed to this lack of irregularity and torsional effects. In terms of the capacity of the building in the +X direction, the biggest difference is observed for the building 'P2c_2', which may be due to the introduction of a new door in the wall, and the subsequent reduction of the size of the resisting piers (Fig. 13).

5.2 *Load bearing walls morphology*

The vertical resisting elements of this building typology consist of load bearing rammed earth walls. The wall thickness of this type of rammed earth buildings

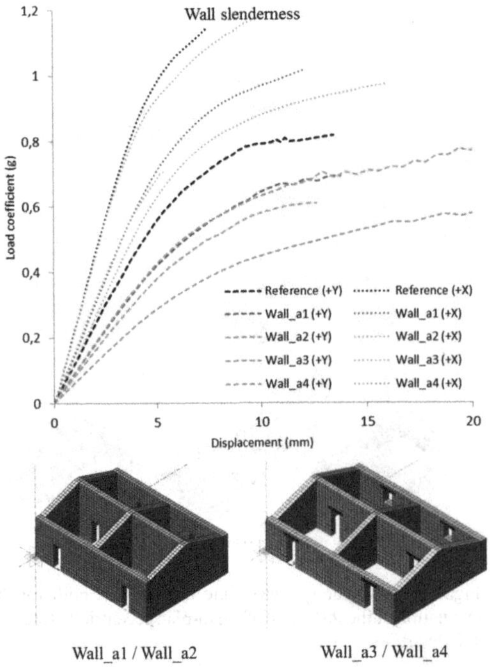

Figure 14. Capacity curves and new models built for the evaluation of the influence of the wall slenderness (credits: J. Ortega).

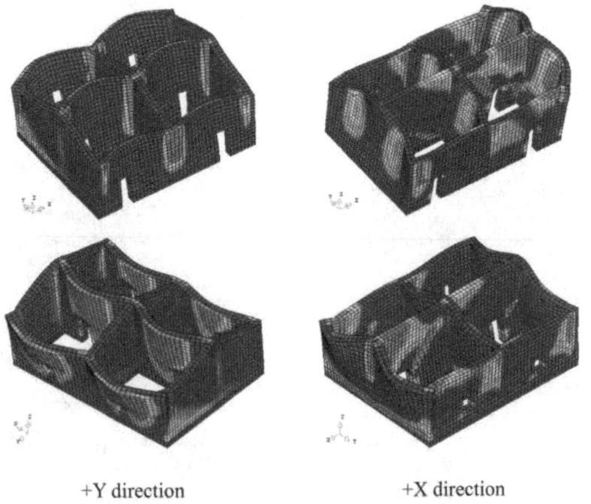

Figure 15. Maximum principal strains (a cracking measure) depicted on deformed mesh at failure for building 'Wall_a1'.

lies between 0.45 and 0.6 m, being most commonly 0.5 m. The maximum height of the walls is more variable, and the most slender elements can be much more vulnerable to the seismic action. Initially, two models were built, aiming at evaluating the influence of this parameter, by increasing the height of the walls in 0.5 m (building 'Wall_a1'), and in 1.0 m ('Wall_a2'), and keeping the wall thickness the same at 0.5 m. Secondly, another two models were built modifying the thickness of the inner walls, which are usually thinner than the exterior ones ('Wall_a3' and 'Wall_a4'). The four new models and results in terms of capacity curves are presented in figure 14.

As expected, the seismic capacity of the building in both directions decreases when increasing the height of the walls and, consequently, by increasing the flexural damage, the damage at the connections between perpendicular walls increase, both in the internal and the external walls. In terms of the failure mode in the +Y direction, failure does not take place only in the middle walls and at the base course, as in the reference building. Now, the front and back walls present also substantial flexural damage, mainly flexural cracks at the mid-span and at the connections with the perpendicular walls. There is barely any horizontal cracking at the base (Fig. 15). Regarding the failure mode in the X direction, damage is again mostly concentrated in the rammed earth walls, while the stone base now barely suffers from any damage. The common diagonal cracking, showed by in-plane shear failure of the walls, is more clearly developed, together with flexural damage at the connections between perpendicular walls (Fig. 15).

Four models were built aiming at evaluating the influence of the maximum free span of the walls, as the longest elements without intermediate supports can be very vulnerable to the seismic action. Two of the models were already built, in order to assess the influence of in-plane eccentricity, as previously stated. All models increase the maximum span covered by some of the rammed earth walls in the building, while keeping the wall thickness the same at 0.5 m. Buildings 'Wall_b3' and 'Wall_b4' increase the length of the walls in the X direction in 0.5 m, and 1.0 m respectively. The models and results in terms of capacity curves are presented in figure 16. The capacity curves show that, as it could be expected, when the span covered by the walls is doubled, as in models 'Wall_b1' and 'Wall_b2', the capacity of the building is highly reduced. The elements get to behave practically as free standing walls, highly reducing their horizontal resisting capacity. Failure takes place in these elements, consisting of an out-of-plane failure, a combination of flexural cracking at the mid-span of the walls and substantial horizontal cracking at the base. For the models 'Wall_b3' and 'Wall_b4', there is also a reduction of the capacity of the building in the +Y direction, when increasing the span covered by these walls.

The out-of-plane failure, also observed in the reference building, is more evident by increasing the span. However, regarding the pushover analysis in +X direction, the differences are smaller, and the response to horizontal loading is very similar, even though the area of resisting walls in that direction has been increased. The base course of the wall is built in stone masonry, usually schist or granite, and it is present almost in every building of this typology (Fig. 17). They have a variable height, which can usually vary from 0.45 up to 1 m. The influence of the stone masonry course and its height in the seismic behaviour

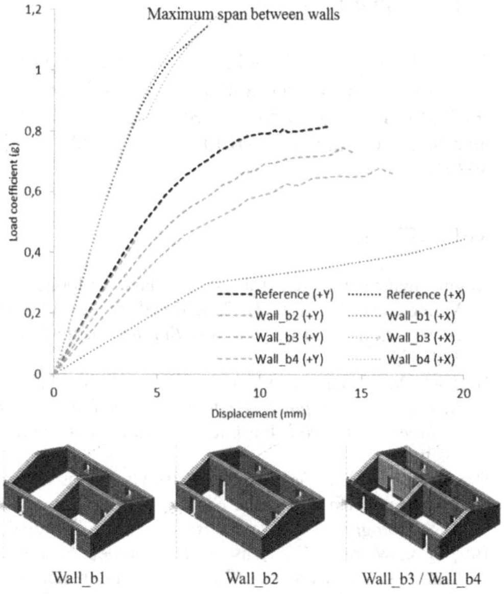

Figure 16. Capacity curves and new models built for the evaluation of the influence of the maximum span between walls (credits: J. Ortega).

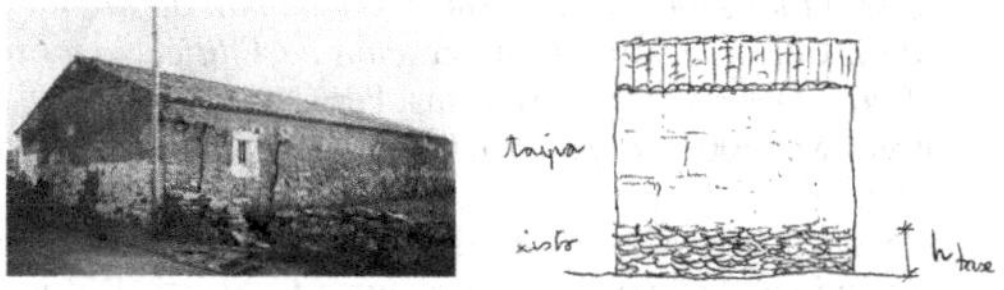

Figure 17. Stone masonry base course usually observed in this building typology (left) building in Outeiro (Correia 2007).

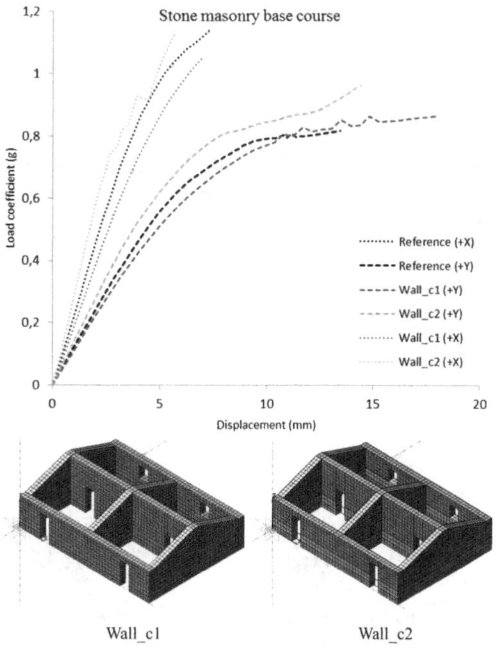

Figure 18. Capacity curves and new models built for the evaluation of the influence of the stone masonry base course (credits: J. Ortega).

Figure 19. Types of roofing systems usually observed in this building typology (credits: J. Ortega).

of the building was evaluated by constructing two different models.

First, the stone base was completely removed and the walls were considered to be built only with rammed earth. A second model was built with the stone base reaching a height of 1.0 m. The models and the results in terms of capacity curves are presented in figure 18.

The main difference in the results consists on the variation in the stiffness of the model, mainly resulting from the difference in stiffness between both materials. When there is no stone masonry base course, the building deforms more, but there is not a remarkable difference in terms of maximum load coefficient, or failure modes (both in +X and +Y directions). Similarly, when the height of the stone base is increased, a stiffer behaviour is observed, together with an increase in the capacity of the building in the Y direction. The failure modes are very similar, but the horizontal cracking, at the stone base, is enhanced.

5.3 *Type of roofing system*

Different types of roofs can be commonly observed in these buildings (Fig. 19). The type of roof has a decisive influence, since it modifies the geometry of the building.

The models with distinct types of roofs, and the results in terms of capacity curves, are presented in figure 20. For instance, if a truss roof is considered ('Roof_a1'), the height of the middle wall is significantly reduced to the same height as the exterior walls and, therefore, the capacity of the building may increase. On the other hand, these changes can lead to the formation of new vulnerable elements, such as the gable wall. The lack of a middle wall, bracing the gable end wall, increases the vulnerability of this element to out-of-plane loading, which becomes highly susceptible to collapse. Therefore, the seismic capacity of the building in +X direction decreases significantly. If the roof is composed of rafters, without a middle wall, so they can exert a thrust on the walls ('Roof_a2'), the capacity of the building in +Y direction is also highly compromised. If a middle wall is added, acting as a brace of the gable walls (third model – 'Roof_b1'), and simulating proper coupling between the parallel walls, a notable increase in the global stiffness of the model was observed in the +Y direction, since the walls are able to develop resisting mechanism to the horizontal action simultaneously. However, there is no

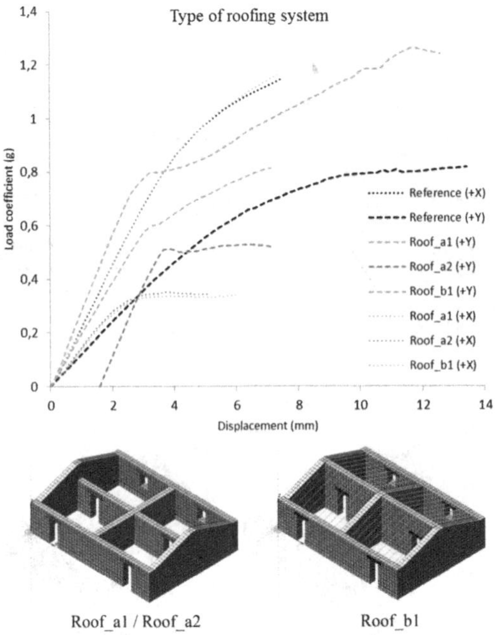

Figure 20. Capacity curves and new models built for the evaluation of the influence of the type of roofing system (credits: J. Ortega).

improvement in terms of ultimate load in relation to the reference model.

6 CONCLUSIONS

The effect of some geometrical and construction characteristics in the seismic response of the building has been evaluated by means of a numerical parametric study, using pushover analysis, where the loading was considered to be proportional to the mass. Different parameters that were initially assumed to have a relevant influence in the seismic behaviour were selected and adjusted for the specific building typology studied. The results obtained confirm that most of the parameters selected have a relevant influence in the seismic behaviour of the building, particularly the wall slenderness, and the maximum span between walls, which show the maximum differences in terms of peak loads, and even have an influence in the failure modes. The results of the analysis of the reference building show that the building is more sensitive to out-of-plane failure, which can be expected due to the height to thickness ratio of the rammed earth walls assumed. The connections between orthogonal walls are also very vulnerable, showing big concentration of stresses. This is particularly important, given the fact that a perfect connection between the walls was assumed in this first set of analyses. This is not usually true for this type of buildings, which are many times characterised by having poor wall-to-wall connections.

ACKOWLEDGEMENTS

The authors gratefully acknowledge the support by the Portuguese Science and Technology Foundation (FCT) to the research project 'SEISMIC-V – Vernacular Seismic Culture in Portugal' (PTDC/ATP-AQI/3934/2012).

REFERENCES

Angulo-Ibáñez, Q., Mas-Tomas, A., Galvañ-Llopis, V. & Montesinos, J.L. (2012). Traditional braces of earth constructions. *Construction and Building Materials* 30: 389–399.

Blondet, M., Villa García, M.G., Brzev, S. & Rubiños, A. (2011). *Earthquake-resistant construction of adobe buildings: A tutorial.* Earthquake Engineering Research Institute.

Braga, A.M. & Estêvão, J.M.C. (2010). Os sismos e a construção em taipa no Algarve. *Sísmica 2010 – 8º Congresso de Sismologia e Engenharia Sísmica*. Aveiro, Portugal.

Bui, Q.B., Morel, J.C., Hans, S. & Meunier, N. (2008). Compression behaviour of non-industrial materials in civil engineering by three scale experiments: the case of rammed earth. *Materials and Structures* 42(8), 1101–1116.

Correia, M. (2005). Metodología desarrollada para la identificación en Portugal de arquitectura local sismo resistente. *SismoAdobe2005, Seminario Internacional de Arquitectura, Construcción y Conservación de Edificaciones de Tierra en Áreas Sísmicas*, Lima, Peru.

Correia, M. (2007). *Taipa no Alentejo*. Argumentum, Lisboa, Portugal.

Gallego, R. & Arto, I. (2014). Evaluation of seismic behaviour of rammed earth structures. C. Mileto, F. Vegas, L. García-Soriano & V. Cristini (Eds.), *Vernacular Architecture: Towards a Sustainable Future.* London: CRC Group.

Gomes, M.I., Lopes, M. & Brito, J. (2011). Seismic resistance of earth construction in Portugal. *Engineering Structures* 33(3), 932–941.

Jaquin, P.A. (2008). *Analysis of historic rammed earth constructions (PhD Thesis, Durham University, UK).*

Lourenço, P.B. (2009). Recent advances in masonry structures: micromodelling and homogeneization. In U. Galvanetto & M.H. Ferri Aliabadi (eds), *Multiscale modeling in solid mechanics: computational approaches*: 251–294. Imperial College Press, London, UK.

Lourenço, P.B., Mendes, N., Ramos, L.F. & Oliveira, D.V. (2011). On the analysis of masonry structures without box behavior. *International Journal of Architectural Heritage: Conservation, Analysis, and Restoration* 5(4–5),369–382.

Miccoli, L., Oliveira, D.V., Silva, R., Müller, U. & Schueremans, L. (2014). Static behaviour of rammed earth: experimental testing and finite element modelling. *Materials and Structures*.

Michiels, T. (2014). Seismic retrofitting techniques for historic adobe buildings. *International Journal of Architectural Heritage*.

NTC08 (2009). *Circolare del Ministero delle Infrastrutture e dei Trasporti 2 febbraio 2009, n. 617. Istruzioni per l'applicazione delle nuove norme tecniche per le costruzioni di cui al Decreto Ministeriale 14 gennaio 2008*. Suppl. Ordinario n. 27 alla G.U. n. 47 del 26-02-2009. Italy.

TNO (2009). *DIsplacement methodANAlyser (DIANA) user's manual.* Release 9.4.4. Netherlands.

Seismic Retrofitting: Learning from Vernacular Architecture – Correia, Lourenço & Varum (Eds)
 ISBN 978-1-138-02892-0

Seismic behaviour analysis and retrofitting of a row building

R.S. Barros & A. Costa
RISCO – Department of Civil Engineering, University of Aveiro, Portugal

H. Varum
CONSTRUCT-LESE, Faculty of Engineering, University of Porto, Portugal

H. Rodrigues
RISCO – School of Technology and Management, Polytechnic Institute of Leiria, Portugal

P.B. Lourenço & G. Vasconcelos
ISISE, Faculty of Engineering, University of Minho, Guimarães, Portugal

ABSTRACT: Rammed earth is one of the oldest building materials in the world, and it is present in Portugal with a particular focus in the South of the country. The mechanical properties and the structural behaviour of rammed earth constructions have been the subject of study of many researchers in the recent years. This study is part of a broader research on vernacular seismic culture in Portugal. Numerical analyses were carried out to assess the influence of different retrofitting solutions in the behaviour and seismic performance of a rammed earth building, representative of the vernacular heritage of Alentejo region. Understating the structural fragilities of this type of constructions allowed determining the most appropriate retrofitting solutions.

1 INTRODUCTION

Vernacular constructions exist all around the world. Many of these constructions, besides their cultural and architectural heritage value, in many cases, present a pronounced level of degradation, urging for the need of conservation and strengthening actions.

Portugal may suffer from moderate seismic events, as evidenced in historical past events. The past seismic activity, particularly in the south of the country, have drove the implementation of some retrofitting measures in the built vernacular heritage, including the insertion of strengthening elements like ties, reinforcing rings, buttresses and other reinforcement elements. The insertion of these elements contributes to enhance the connection between structural elements, and to the improvement of the structural behaviour and performance of the buildings.

This chapter presents the main results of a numerical study of the influence and effectiveness of common seismic improvement measures, typically found in vernacular constructions.

In this study, it was selected a building, representative of the vernacular architecture in certain localities in the South of Portugal. The building model was calibrated with information on material properties and structural characteristics. With the calibrated model, parametric analyses are performed to assess the influence of different retrofitting solutions in their behaviour and seismic performance.

The analysis of the numerical results gives a first insight on the behaviour of these type of vernacular constructions, and points out which retrofitting solutions may be more efficient in their seismic performance enhancement. The numerical results may also contribute to a better understating of the structural fragilities of these constructions, namely in terms of seismic demands in the structural elements and structural damage distribution. (Vicente et al., 2011).

2 DESCRIPTION OF THE CASE STUDY

For the selection of the building studied were taken into account the objectives of the numerical study developed under the SEISMIC-V project. The case study is located in *Alcácer do Sal*, district of *Setúbal*, in Portugal. It is a building with eight spaces, and the vertical structural elements are principally rammed earth walls, with some elements in stone/brick masonry. The walls are 3.0 m high in average, and have a thickness of 0.50 m. The building has a regular plan (see Fig. 1–2). The roof is gabled, and possesses a wooden structure coated with ceramic tiles (Correia, 2007).

3 FINITE ELEMENT MODEL AND STRENGTHENING SOLUTIONS

To understand the structural behaviour of this traditional building typology, the selected building was

Figure 1. Building studied (Correia, 2007).

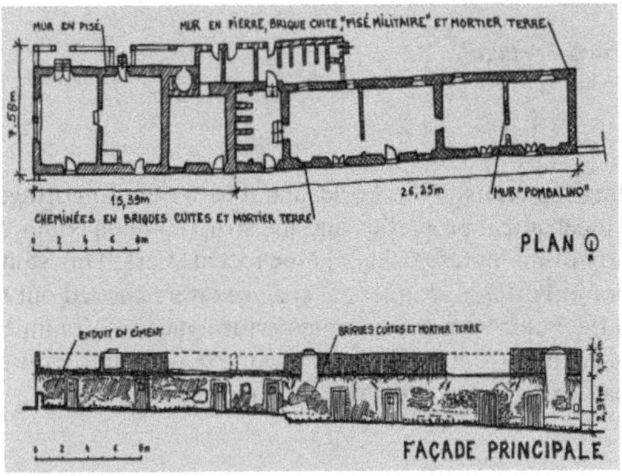

Figure 2. Plan and elevation (Correia, 2007).

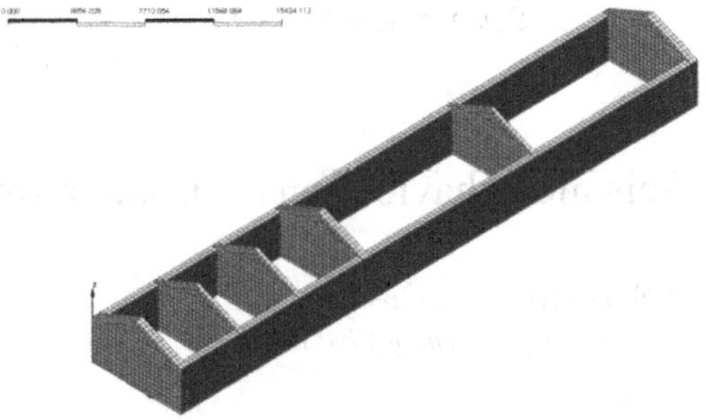

Figure 3. Model developed in MIDAS FEA software (credits: RISCO).

Table 1. Values adopted by other authors (credits: RISCO).

Reference	E (MPa)	f_t (MPa)	f_c (MPa)	ν
Limón et al. (1998)*	922	0.29	2.45	0.30
Gomes et al. (2011)*	200	0.13	0.67	0.35
Silva (2013)*	300	0.10	1.00	0.30
Gallego and Arto (2015)*	250	0.29	2.00	—
Miccoli et al. (2014)	4207	0.37	3.70	0.27

* Obtained experimental values

Table 2. Adopted values for the mechanical properties of the rammed earth in the numerical model (credits: RISCO).

E (MPa)	f_t (MPa)	f_c (MPa)	ν
300	0.05	1.0	0.1

modelled with a finite element tool. The results of the analyses with the numerical models will help in the vulnerability evaluation, on the identification of fragile parts of these buildings, and in the efficiency assessment of different retrofitting measures.

Numerical analyses were performed in the finite element program MIDAS FEA. The simplified model was developed, based on the available information on the selected building case study. The building model is symmetric in one direction, and has an irregular geometry in plan, defined by four interior partition walls, as represented in Figure 3. The walls have 3.0 m height, and 0.5 m of thickness, as indicated in the case study presented in the previous section.

In the model, all walls were considered to be of rammed earth. To define the material properties, it was made a literature survey, in order to obtain reference values (see Table 1). Based on the survey, the properties of the rammed earth adopted in the numerical models developed in this study are presented in Table 2.

For the finite element model, solid elements and a mesh with elements dimensions of 300 mm approximately were considered, having the walls two solid elements in their thickness (see Fig. 3). As boundary conditions, these were restricted to the horizontal translational displacements of all nodes at the base of the model, simulating the foundations. In the model it was not considered the roof structure, due to its weakness, but their weight was simulated with an equivalent distributed load of 0.001 N/mm^2 (benchmark for roofs with wooden structure and ceramic tiles as cover – Reis et al., 2005), applied on the top of the walls.

The demands are applied in two phases, first the vertical dead load, and then the lateral demand, simulating the seismic loading. The seismic loading is imposed as an equivalent horizontal distributed load, proportional to the element mass, for increasing seismic levels (0.2 g to 1.0 g), applied in two independent directions (X and Y).

To simulate the rammed earth walls, the total strain crack model was adopted in the non-linear analyses. For the material behaviour in tension was considered the Hordijk function (Fig. 4a), and the parabolic law for compression (Fig. 4b).

It was intended to analyse the influence of the most common strengthening measures applied in the typical vernacular buildings of Alentejo region. Therefore, three seismic reinforcement strategies were studied (see Fig. 5), two different configurations of buttresses distribution, and one solution based on the use of tie rods. For the buttresses, idealised with a composition

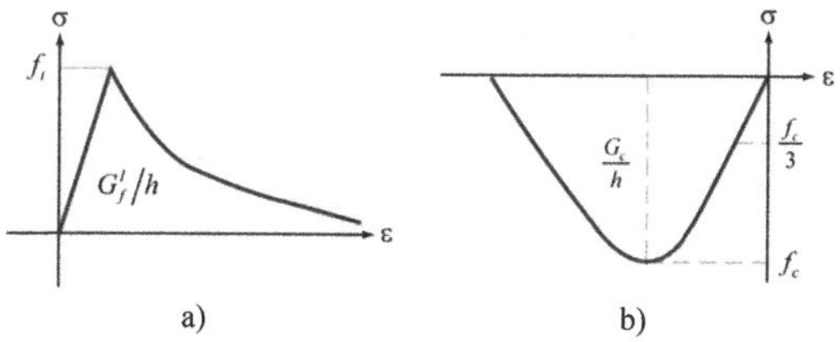

Figure 4. Behaviour laws for the rammed earth: a) Hordijk function in tension; b) Parabolic function in compression (credits: MIDAS FEA).

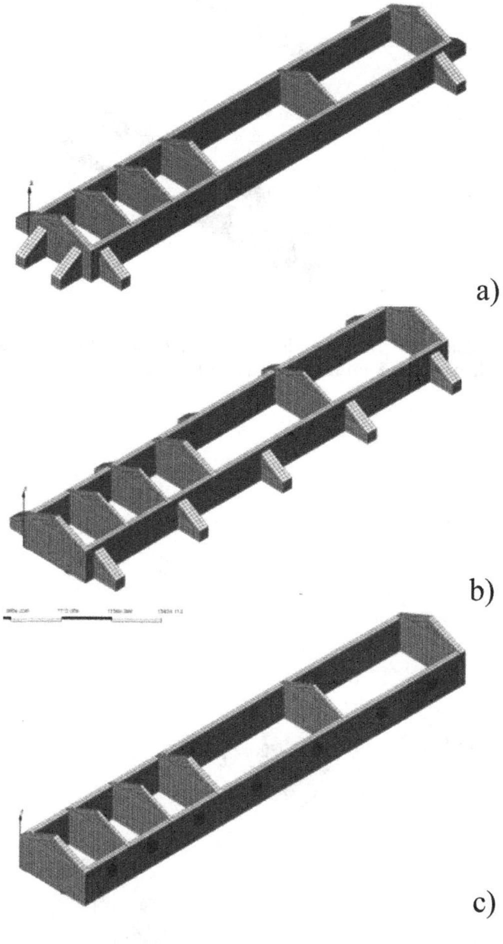

Figure 5. Reinforcement solutions studied: a) Buttresses only at building corners (B); b) Buttresses along the longer walls (BLW); c) Rods (R) (credits: RISCO).

similar to the rammed earth walls, the same material properties were considered in the numerical analyses.

The strengthening with buttresses, only at the building corners (B), shown in Figure 5a), is the most common solution in rammed earth constructions of the south of Portugal. It was also studied a distribution of buttresses along the long walls (BLW), and a retrofitting solution based on the use of tie rods (R),

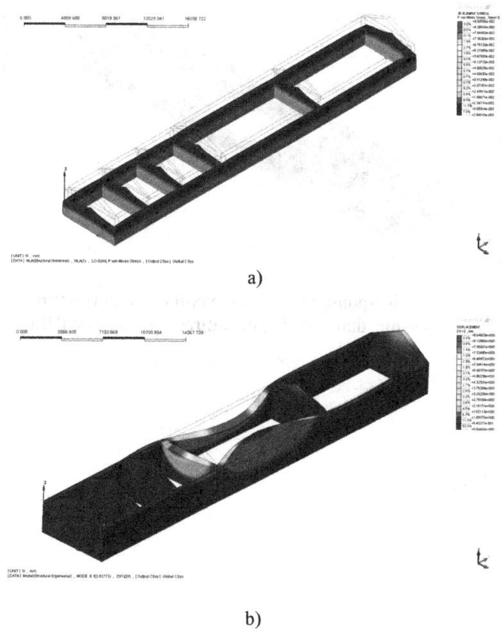

Figure 6. Results: a) response for the vertical loads; b) 1st mode of the structure (credits: RISCO).

perpendicular to the long walls, to improve the structural behaviour. With the latter retrofitting solutions, it is intended to reduce the potential development of out-of-plane mechanism of the longer walls.

4 RESULTS AND ANALYSIS

The results obtained with the numerical models are presented and discussed in three phases. Firstly, are analysed the results for the static vertical loads, and the first natural mode is discussed. Then, it is analysed the response of the structure for increasing seismic demands in each direction. Finally, are presented the results of the seismic vulnerability reduction, with the retrofitting solutions studied.

From the analysis of the results obtained with the non-linear model for the vertical static loads, it was observed, as expected, that the maximum vertical strain reaches reduced values (approximately 0.115 MPa, see Fig. 6a). From the modal analysis, it was observed that the first mode presents dominantly transversal displacements of the long external walls, highlighting the potential for the development of out-of-plane mechanisms of the façade walls (see Fig. 6b).

From the analysis of the results corresponding to an equivalent horizontal seismic demand of 1 g imposed in the X direction, it is observed a maximum displacement in this direction of 73.1 mm (Fig. 7). The maximum strain and damage develops at the base of the walls, perpendicular to the direction of the applied force, as well as in the connection between walls.

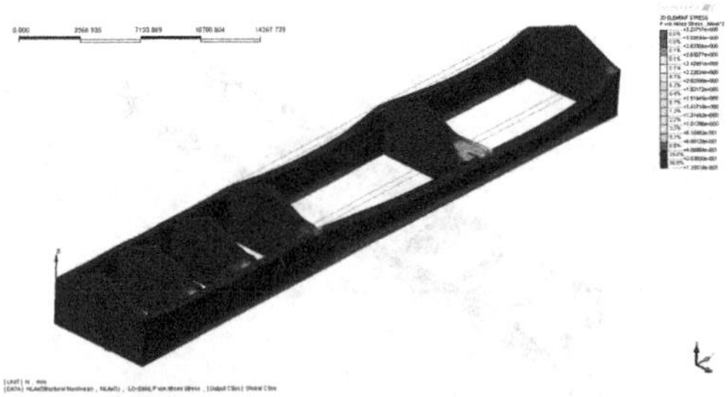

Figure 7. Response in terms of maximum strain distribution for the seismic demand (1g) in X direction (credits: RISCO).

0.2g

0.4g

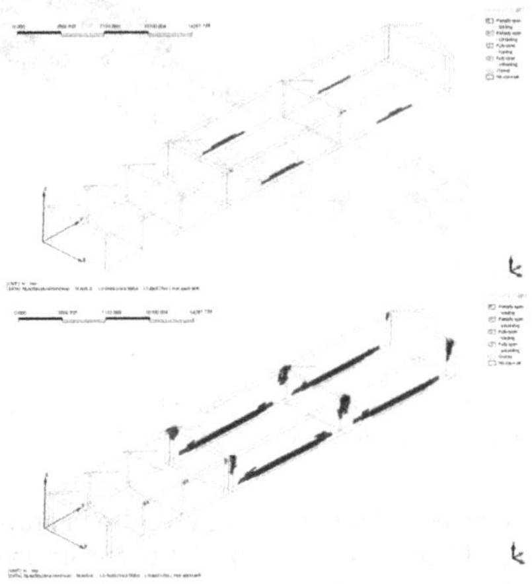

Figure 8. Damage distribution evolution for seismic demands in X direction (credits: RISCO).

When imposing the horizontal load (1 g) in the Y direction, a maximum displacement in this direction of 304.3 mm is reached. For the seismic demand in Y direction, the maximum strain and damages developed also in the base of the walls, perpendicular to the direction of the applied force, and also in corners and in the regions of connection between walls.

The damage evolution is presented in Figures 8 and 9, for increasing seismic demands in the X and Y directions, respectively. For the seismic demands in the X direction, the damage starts at the base and central areas of the longitudinal walls. Then, the damage evolves towards the perpendicular walls, until the collapse of the section connecting perpendicular walls (Fig. 8–9). For the demands in the Y direction, it was observed that damage start at the connection between the inner and outer walls.

With the increasing of the seismic demand, the damages develop towards the interior walls, until the collapse of the walls, perpendicular to the seismic loading imposed (Fig. 10–11).

Based on these prior analyses, it is clear that for the building studied, the out-of-plane collapse mechanism of the longer walls is one of the major vulnerabilities. Thus, the strengthening solutions studied, and presented next, correspond to the scenario of the seismic action in the Y direction.

0.6g

0.8g

1.0g

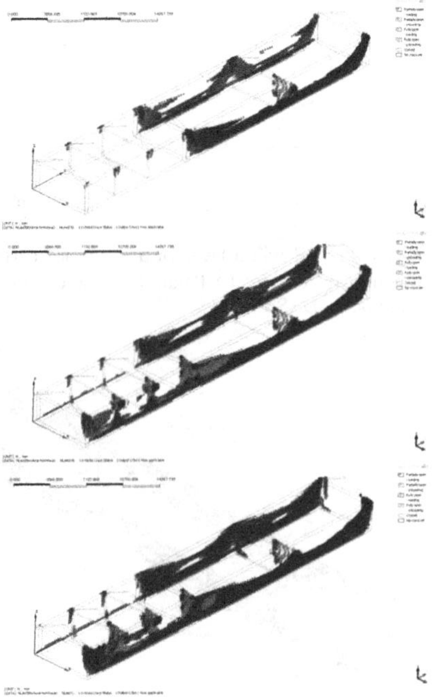

Figure 9. Damage distribution evolution for seismic demands in X direction (credits: RISCO).

0.2g

0.4g

0.6g

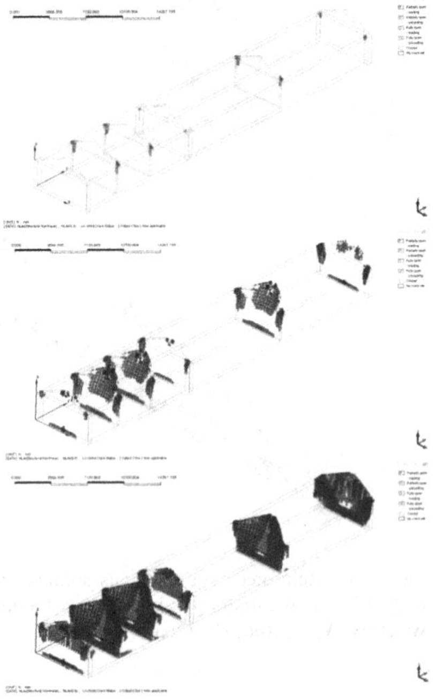

Figure 10. Damage distribution evolution for seismic demands in Y direction (credits: RISCO).

With the implementation of buttresses (B), for the horizontal seismic demand in Y direction, an increase in the maximum displacement in this direction is observed. In fact, the local stiffening of the

0.8g

1.0g

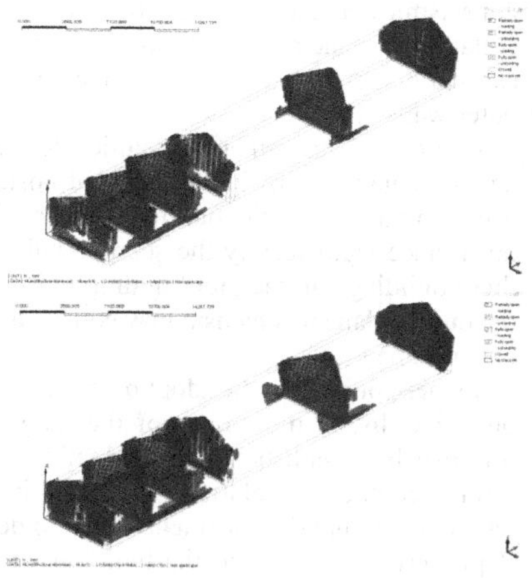

Figure 11. Damage distribution evolution for seismic demands in Y direction (credits: RISCO).

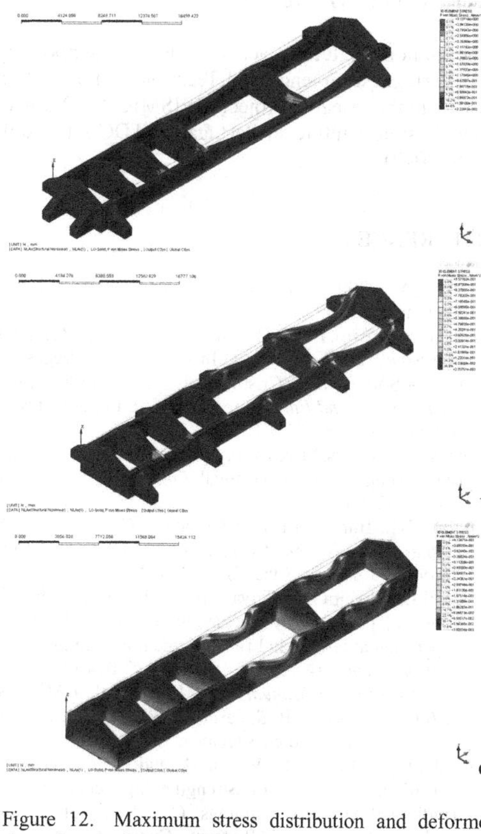

a)

b)

c)

Figure 12. Maximum stress distribution and deformed shape for accelerations of 1 g, for the models simulating the retrofitting solutions: a) B; b) BLW; c) R (credits: RISCO).

building corners with buttresses, result in a deformation demand that tends to concentrate in the walls, and in the central part of the building. This retrofitting solution leads to a maximum displacement of 90.6 mm,

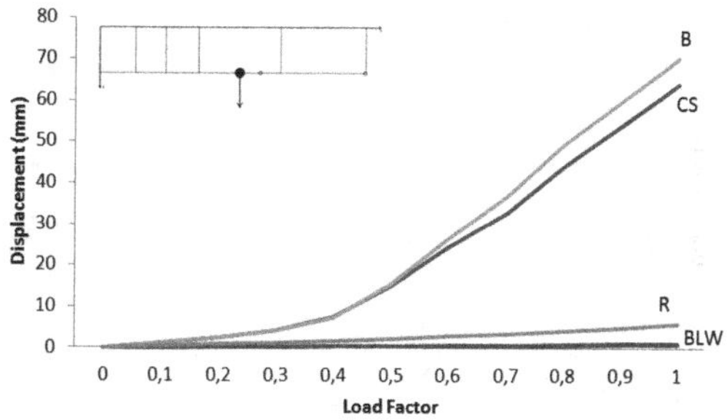

a)

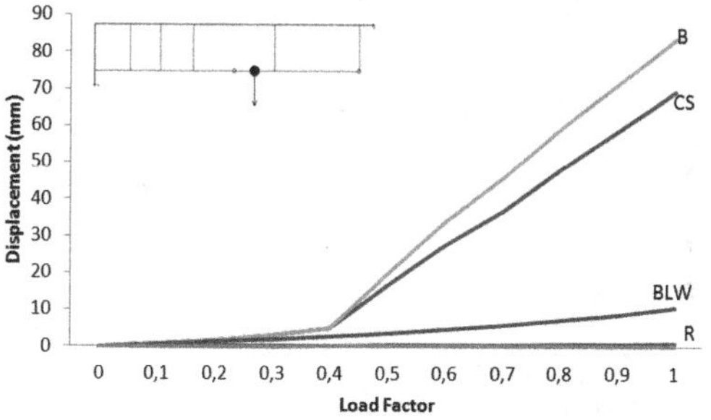

b)

Figure 13. Evolutions of the horizontal displacement for increasing seismic load; a) Central point; b) Middle point (credits: RISCO).

while for the non-retrofitted model a maximum displacement of 73.1 mm was observed, for 1 g seismic demand.

With the solution corresponding to the use of buttresses in the longest walls (BLW), the out-of-plane displacements decreases to 21.1 mm, and the displacement in these points is reduced to 7.3 mm in the model simulating the retrofitting with ties (R).

In terms of maximum stress, for the B model a maximum stress of 13.7 MPa is observed, while for the BLW model, the maximum stress decreases to 1.5 MPa, and for the R model to 0.4 MPa.

In Figure 13 and Figure 14 are presented the curves with the evolution of horizontal displacement with the seismic demand on three critical points, at the top of the walls, representative of the building response, which allows comparing the efficiency of the studied retrofitting solutions. Analysing the response of the walls' midpoint, it is observed that the BLW and R solutions improves largely the performance of these walls. At the building corners, all the reinforcement solutions improved the behaviour, either for the seismic demand in X and Y direction.

In Figure 15 are presented the damage distributions for two levels of seismic demand, for the three retrofitting solutions studied. Model R represents the solution that more reduces the damage, up to an equivalent seismic demand of 0.6 g. A retrofitting solution commonly adopted in rammed earth constructions in the south of Portugal was the installation of buttresses (model B).

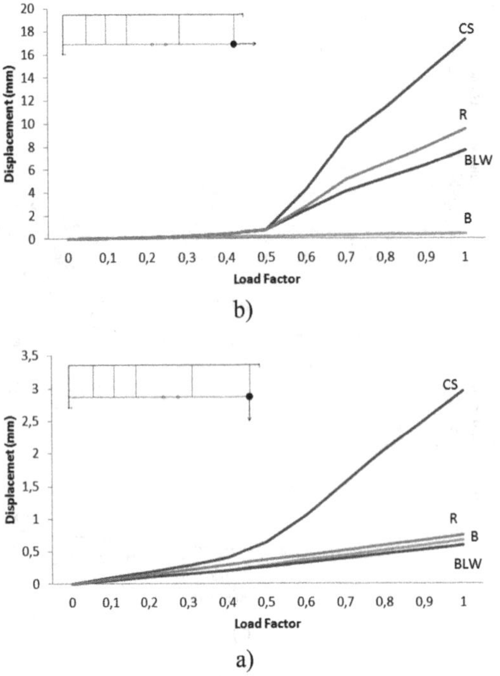

Figure 14. Evolutions of the horizontal displacement for increasing seismic load; a) Corner X direction; b) Corner Y direction (credits: RISCO).

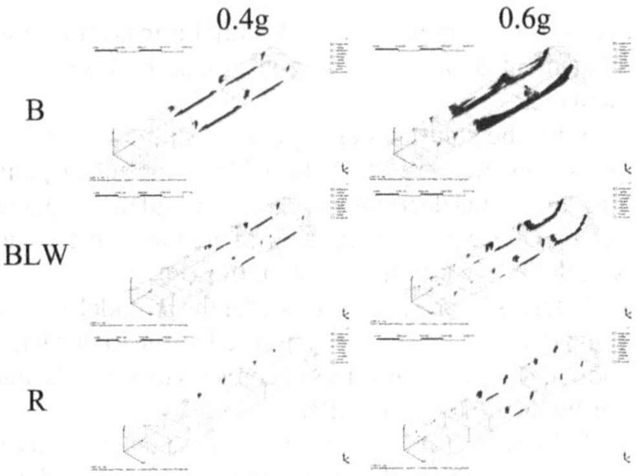

Figure 15. Damage distribution (credits: RISCO).

5 FINAL COMMENTS

The south of Portugal, and in particular the Alentejo region, is a region with potential moderate seismic activity. Through the results obtained in this study, it was possible to observe that rammed earth vernacular existing constructions, commonly found in the south of Portugal can suffer from substantial damages, when subjected to earthquakes with accelerations up to 0.4 g.

For seismic demand with larger peak accelerations, the collapse of principal structural elements may occur, especially associated to out-of-plane collapses of the outer walls.

A common strengthening solution founded in vernacular constructions in the South of Portugal, is the use of buttresses in the building corners. These may not reduce significantly the seismic vulnerability of these buildings. In fact, for buildings with long walls, the out-of-plane mechanism may control their seismic performance.

In these buildings, the adoption of buttresses or ties on the walls, in mid-points of the long walls, may improve the seismic behaviour and safety. These structural strengthening solutions, if located in the points, where maximum displacements tends to develop, and if properly connected to the building structure, can significantly reduce the displacement demand and damage.

ACKNOWLEDGEMENTS

The authors gratefully acknowledge the support by the Portuguese Science and Technology Foundation (FCT) to the research project 'SEISMIC-V – Vernacular Seismic Culture in Portugal' (PTDC/ATP-AQI/3934/2012).

REFERENCES

Correia, M. (2007). *Rammed earth in Alentejo*, Lisboa: Argumentum.

Gallego, R. & Arto, I. (2015). *Evaluation of seismic behavior of rammed earth structures*. In C. Mileto, F. Vegas, L. García Soriono & V. Cristini (Eds), *Earth Architecture: Past, Present and Future* (pp. 151–156). London: Taylor & Francis Group.

Gomes, M.I., Lopes, M. & Brito, J. (2011). Seismic resistance of earth construction in Portugal. *Engineering Structures*, 33(3), 932–941.

Limón, T.G., Burego, M.A. & Gómez, A.C. (1998). *Study of the materials from the factories in the tower of the Alhambra Comares*. Alhambra: CEDEX.

Miccoli L., Oliveira D.V., Silva R.A., Muller, U. & Schueremans, L. (2014). Static behaviour of rammed earth – experimental testing and finite element modeling. *Materials and Structures*. doi:10.1617/s11527-014-0411-7.

MIDAS FEA (2014). *Manual Midas FEA*. London: MIDAS.

Reis, A.C., Farinha, M.B. & Farinha, J.P.B. (2005). *Tabelas Técnicas*, Lisboa: Edições técnicas E.T.L., Lda.

Vicente, R., Rodrigues, H., Varum, H., and Mendes da Silva, J. (2011). Evaluation of Strengthening Techniques of Traditional Masonry Buildings: Case Study of a Four-Building Aggregate. J. Perform. *Constr. Facil.*, 25(3), 202–216.

Seismic Retrofitting: Learning from Vernacular Architecture – Correia, Lourenço & Varum (Eds)
© 2015 Taylor & Francis Group, London, ISBN 978-1-138-02892-0

Seismic vulnerability of vernacular buildings in urban centres—the case of Vila Real de Santo António

J. Ortega, G. Vasconcelos & P.B. Lourenço
ISISE, Faculty of Engineering, University of Minho, Guimarães, Portugal

H. Rodrigues
School of Technology and Management, Polytechnic Institute of Leiria, Portugal

H. Varum
CONSTRUCT-LESE, Faculty of Engineering, University of Porto, Portugal

ABSTRACT: Following detailed studies of Portuguese vernacular building typologies, this paper deals with buildings located in historical urban centres. An analysis of the history of the urban centre and, in particular, of some vernacular buildings is enhanced. Additionally, a discussion on the influence of changes of the geometry, and on added built volumes to original buildings in the seismic vulnerability of the buildings is also provided.

1 INTRODUCTION

Vernacular architecture has a distinctly open-ended and spontaneous nature, but it is also mainly characterised by its transformation process. Although there is a strong persistence of the vernacular form, resulting from a characteristic rigid tradition, vernacular constructions are subjected to continuous modifications, because of new needs of the users, or because of changed situations that force it to adapt. The effect of these alterations in the seismic response of the building can be of great importance, because it involves construction aspects such as: (i) the addition of new floors; (ii) the addition of new constructions, built in different materials and poorly attached to the original one; (iii) the enlargement of existing openings, and the opening of new ones; or (iv) the replacement of original structural elements with different ones, e.g. the common replacement of the old wooden floors with reinforced concrete ones.

At an urban level, the effect of these alterations does not affect only the building itself, but also the neighbouring ones, as it is common to have aggregate buildings. Therefore, buildings within an urban centre do not react independently to the seismic action but, on the contrary, they are much influenced by the adjacent buildings. The structural interaction among neighbouring buildings under seismic action depends on factors such as: (1) the position of the building within the block; (2) the difference in height between the buildings; (3) the presence of staggered floors, or (4) the difference in the stiffness of adjacent bodies, whether due to different constructive systems, or due to differences in the amount of openings of the façades.

Therefore, alterations over time on the urban fabric may increase the vulnerability of the whole block. For example, the replacement of buildings with different constructive systems should have a significant influence. Firstly, because different stiffness of the bodies leads to different responses to the seismic action of neighbouring buildings, causing damage at the connecting points. In this respect, for instance, reinforced concrete buildings, adjacent to masonry dwellings, introduce a severe risk of hammering actions to take place. Secondly, it can also modify the relative position of the buildings within the block, changing the uniform reaction of the block. Besides, the enlargement and opening of new windows and doors changes the stiffness of the buildings, and ultimately affects the whole block seismic response. The description of these alterations of vernacular buildings inside aggregates located in historical urban centres is addressed in this paper, selecting the city of *Vila Real de Santo António* as a case study.

The reconstruction of Lisbon after the 1755's earthquake was not an isolated phenomenon, and other contemporary developments took place following similar plans, and proposing similar urban, architectural and construction solutions. *Vila Real de Santo António* was a completely new planned town, erected at the end of the 18th Century, following a *Pombaline* plan and thus, sharing many characteristics with Lisbon, such as the inclusion of seismic-resistant measures at an urban and architectural level. The urban plan consists of an extremely regular rectangular grid, organised around a large central square, along the riverfront, and with strictly defined architectural building types.

It is a representative example of how an original rigorously designed plan, which took into consideration seismic-resistant provisions, has been continuously modified by anonymous and customary forces, without any principled changes, leading to the current unplanned spontaneous occupation and alteration of the urban layout. Most of the original buildings were substituted, or highly altered, and few buildings still preserve their original characteristics in terms of volume, construction characteristics, and opening distribution in elevation.

Because of this, *Vila Real de Santo António* was selected as a case study. Nevertheless, it fulfils another two main requirements: (i) it is located in the region of the Algarve, the southernmost area of Portugal mainland, which has a significant seismicity, and thus, it is prone to have developed a Local Seismic Culture; and (ii) traditional seismic strengthening solutions were already identified in the city, such as ties and, more specifically, *frontal* walls, which were one of the seismic-resistant provisions considered at the time of the erection of the city.

2 IDENTIFICATION OF POSSIBLE SEISMIC VULNERABILITY ASSESSMENT METHODOLOGIES – AN OVERVIEW

The seismic risk of a structure is determined by the combination of the seismic hazard, i.e. the probability of exceedance of a certain level of seismic activity of a certain intensity, and structural vulnerability. In some cases, other concepts such as the exposition of the elements at risk, involving factors like the density of the population, time of the day of the earthquake occurrence, and cost may be taken into account, when evaluating the risk. Since the seismic hazard is unavoidable, efforts should be placed on the assessment of seismic vulnerability, and on the subsequent reducing of the structural vulnerability, by the implementation of seismic strengthening strategies, in order to attain a reduced seismic risk.

The main objective of seismic vulnerability assessment methodologies is to measure and to predict the probability of reaching a given level of damage, in a given building, for a specific earthquake, and to estimate future losses (Calvi et al. 2006). However, they are able of identifying building fragilities, addressing an essential aspect, in which engineering research can intervene (Vicente et al. 2011), since the evaluation of the seismic vulnerability of existing constructions can be used to evaluate the need of retrofitting solutions, and to assess the efficiency in reducing the seismic vulnerability of proposed structural interventions.

Among the big variety of methodologies proposed by different authors, vulnerability index evaluation methods (Benedetti & Petrini 1984) are based on post-earthquake damage observation and survey data, focusing on the identification and evaluation of constructive aspects and materials that are more influent in the seismic response of the building. Qualitative and quantitative parameters are thus defined, and a vulnerability index is calculated, as the weighted sum of these parameters, classifying the building according to their vulnerability. This index can be used to estimate structural damage after correlation to a specified seismic intensity, supported by post-earthquake recordings and statistical studies. They constitute a reliable large-scale assessment and have been extensively applied in Italy (GNDT 1994), and were recently implemented for the seismic vulnerability assessment of Portuguese masonry structures in several historical city centres (Vicente et al. 2011; Neves et al. 2012; Ferreira et al. 2013), obtaining useful and reliable results, as a first level approach.

The parameters used in this type of methodologies that seem to be more decisive in terms of seismic vulnerability are similar, including construction aspects such as: (a) the presence and effectiveness of the connection between orthogonal walls; (b) the ultimate shear strength of the vertical elements, which manly takes into account the distribution of the structural walls and the area of resisting walls in each orthogonal direction; (c) type and quality of the masonry, which in the case of masonry buildings accounts for material properties, masonry arrangement, size of units or presence of mortar; (d) plan regularity and configuration; (e) efficiency of the connections of the walls to horizontal diaphragms and roofs; (f) roof typology, weight and thrust; (g) number of floors and floor height; (h) the type of foundations; (i) number and location of wall openings; (j) the previous damage, alterations and conservation state. Classes of vulnerability are associated to each parameter, as well as weights, which reflect the relative importance of each one of them on computing the vulnerability index that characterizes the seismic behaviour of a masonry building.

When characterising existing urban centres new challenges raise, as it involves buildings, which do not behave independently, but are structurally connected to the adjacent ones, and are the result of a progressive growth of the cities, eventually resulting in complex structural units continuously modified, with no specific constructive rules. They are commonly known as building aggregates, and their study implies taking into account new specific parameters (Vicente et al. 2011; Formisano et al. 2011; Cardani et al. 2015), which can include: (i) the position of the building within the aggregate; (ii) the relative height of the building within the aggregate; (iii) the presence of staggered floors; (iv) the typological heterogeneity among adjacent units; or (v) the percentage difference of opening areas among adjacent façades.

Nevertheless, identifying building fragilities and obtaining accurate knowledge of the building characteristics, and the specific local structural and constructive features that this type of methodologies requires, can only be obtained through detailed building-by-building inspection (Santos et al. 2013, Maio et al. 2015).

3 ANALYSIS OF VILA REAL DE SANTO ANTÓNIO URBAN CENTRE

Alterations on the urban fabric, such as the ones previously described, affect directly all the cited vulnerability parameters. Addressing the selected case of *Vila Real de Santo António*, the seismic vulnerability of the buildings has most likely been modified. In order to evaluate the effect of these alterations of the urban fabric on the seismic vulnerability of the constructions, a representative block was selected for a deeper analysis, including its constructive characterisation. Its structural history was also analysed, performing a comparison between original and current condition.

3.1 *The urban plan*

The distinctive feature of *Vila Real de Santo António* is the fact that it was born as a desire to design a city from scratch. It was an isolated sudden event, and not a transformation process with a continuation in time. The main reason for planning an entire new city was the attempt to boost the Algarve local economy through industrial development, and tighter tax. and customs controls (Mascarenhas 1996). This region was practically abandoned at the time, and much affected by the 1755's earthquake, but was viewed as having a considerable economic potential, and an official recovery program was enacted during the 1760s and 1770s by the Marquis of Pombal. The program included taxes on imports, the establishment of artisan-based industries, agricultural marketing reforms and, the setting-up of a new town: *Vila Real de Santo António* (Chester & Chester 2010). Because of its strategic position, at the extreme south of the Algarve, facing the Spanish border, it was created for the control of port transactions, and as a display of political power (Correia 1997).

The construction of *Vila Real* started in 1773, following the design of Reinaldo Manuel dos Santos, and it was finished in just a few months, using pre-fabricated and standardisation techniques and processes, applied also for the reconstruction of Lisbon (Chester & Chester 2010). The plan was also based on the ideas and criteria already applied in Lisbon, resulting in a similar urban design in terms of composition and rigorous geometric clarity, as well as in the social and industrial functionality. Moreover, it also included some of the seismic-resistant measures used in Lisbon, such as the use of the characteristic timber frame wall, known as *frontal walls*. Given these facts, the *Pombaline* design plan of *Vila Real de Santo António* is a unique exceptional case, worthy of preservation as cultural heritage.

The *Pombaline* city plan is formed by a rectangular area, with one of the long sides placed along the river, facing east, and consisting of a grid of seven by six urban blocks (Fig. 1). The grid is organised around the big central square, and oriented along the river, north to south direction. The buildings situated in the river front are composed as a sort of façade for the city, with the *Alfândega* or Customs House occupying the central

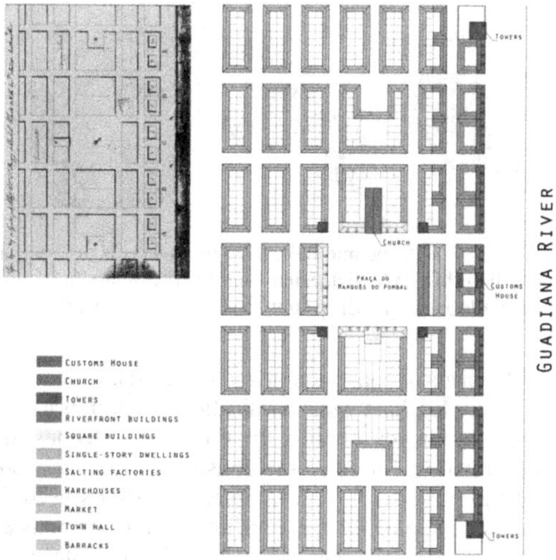

Figure 1. (left) first plan of the new city dating 1774 (Correia 1997); (right) Original plan of Vila Real de Santo António and main building types (credits: J. Ortega adapted from Rossa, 2009).

space, and two differentiated buildings at the edges, slightly higher than the rest, looking like two towers. The blocks are approximately 53 m long, and around 22 m wide, with the exception of the central one, which is slightly longer, approximately 55 m. The other significant urban element is the main square, each side of it being defined by one block of approximately 55 m long, and two streets of around 9 m wide. The façades of these four sides are exactly the same, with the exception of the North one, where the main church is located. Apart from the buildings situated along the river front and surrounding the main square, which have two stories, the rest of the town was occupied by single storey houses. The whole plan is characterised by its strong regularity, and also the streets have all the same width.

3.2 *The building typologies*

There are essentially four distinct architectural types that defined a clear hierarchy, at an urban level, and there are some unique buildings that break the rigid scheme in specific points, which are the Customs House, the church and the 'towers'. The four building types are: (a) buildings in the river front, which have two main stories, and a third attic floor with dormer windows; (b) buildings in the main square, which also have two main stories; (c) single story dwellings, which in turn present several different types, but all of them are characterised by their small scale and extreme simplicity; and (d) single storey salting factories and warehouses behind the buildings in the river front, which basically consist of a system of masonry arcades, perpendicular to the façade walls, organised around a patio. Figure 1 shows the different architectural types devised within the original urban plan.

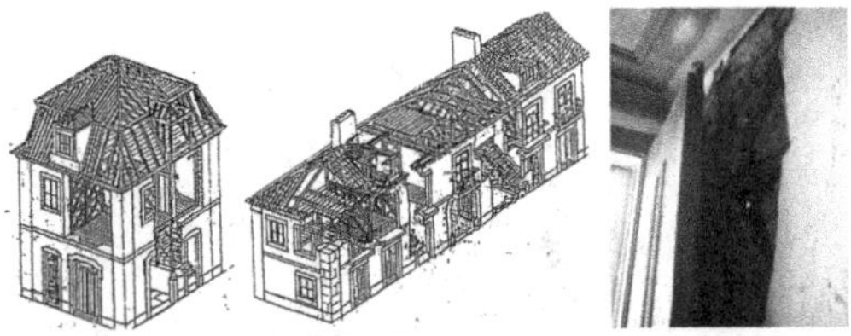

Figure 2. (left and middle) Use of *frontal* walls in two-story buildings in Vila Real de Santo António (Mascarenhas 1996); (right) *Frontais* preserved until today (Barros, 2011).

In terms of construction and materials, it should be firstly noted that the standardisation and prefabrication process, already applied for the Lisbon reconstruction plan, was used in the construction of *Vila Real de Santo António*. For example, the ashlars used for the quoins, plinths, opening lintels, steps or pilasters arrived already cut and worked, ready to be placed, as well as the wooden components, such as doors, windows, beams and floorboards (Correia 1997). Stone was transported by boat from neighbouring quarries, and it was used for the load bearing exterior and party masonry walls that are the main structural resisting element in all the buildings. Timber was used for the roof rafters and trusses, the floor beams and the timber frame partition walls of the upper floors. The timber roof structure was simply covered by wooden boards, on which ceramic tiles were laid. Sometimes reeds were also applied, intended to insulate the houses. Some ground floor rooms had vaulted ceilings, supporting the first floor, as a fire prevention measure, as it was done in Lisbon (Mascarenhas 1996). Ground floor partition walls were usually built in solid brick masonry. The buildings were finally plastered with lime and sand, and whitewashed. The construction of solid stone masonry building foundations using good quality lime mortar is an illustrative example of the generalised good quality and strength of the original buildings of *Vila Real de Santo António* (Oliveira 2009).

With respect to the seismic-resistant measures integrated in the buildings at the time of their construction, the most significant element was the incorporation of timber frame walls in the upper floors of the two-storey buildings, similar to those used in the *Pombaline* quarter in Lisbon. Given the low height of the buildings, the use of the whole seismic-resistant construction system developed for the reconstruction of Lisbon, known as *gaiola*, was not necessary. Nevertheless, the timber roof structure, and the timber floor structure were connected through these timber frame partition walls, known also as *frontal* walls (Correia 1997). These walls consisted of vertical, horizontal and diagonal timber members, filled with light masonry and they can still be observed in some of the buildings at the present (Fig. 2). The structure of the single storey dwellings simply consisted of load bearing stone masonry walls, and brick masonry partition walls. The roof was composed by timber rafters tied by collars, resting directly on the walls, and there

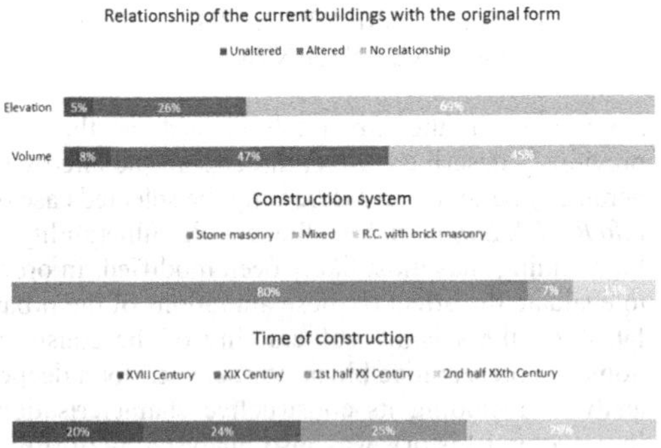

Figure 3. Characterization of the buildings in the *Pombaline* core of Vila Real de Santo António in relation with their original design (credits: J. Ortega adapted from Gonçalves, 2005).

was no ceiling. This simplicity and their regularity in plan and volume, together with their symmetry and small dimensions, only 2.7 m wall height, is enough for the buildings to have a significant resistance to earthquakes. Of course that the performance of these buildings to seismic action depends also on the quality of the connection between walls, as this is an important factor influencing the out-of-plane resistance.

3.3 *The current state of the Pombaline core of Vila Real de Santo António*

There has been a transformation process of the city, and the works carried out in the *Pombalino* core have led to an important depravation of their original characteristics. A previous important work was conducted aiming at analysing the transformation of the 1773 *Pombalino* core, within the *Plano de Pormenor de Salvaguarda do Núcleo Pombalino de Vila Real de Santo António* (SGU 2008). The work consisted of an analysis carried out on a building, intended to identify the remaining original *Pombalino* buildings and their morphological relationship with respect to the original design. Results indicated that only 155–164 buildings still possess *Pombalino* characteristics, illustrating the important alterations done in the historical built-up fabric, and the big degradation of the ideally original designed plan. Moreover, even the best preserved buildings still show important alterations in relation to their original form (Fig. 3). Generally, the transformation process has been characterized by a massive occupation of the blocks, leading to an extreme densification of the urban fabric, since patios were continuously occupied by additional constructions (Fig. 4).

Two-storey buildings located in the river front and the main square, together with the most significant buildings, present few alterations and preserve better their original design. On the contrary, the ordinary built-up fabric has considerably been more vulnerable to alterations. Thus, single storey buildings, both dwellings and salting factories, and warehouses, have been systematically exposed to demolitions, substitutions and large modifications, such as the addition

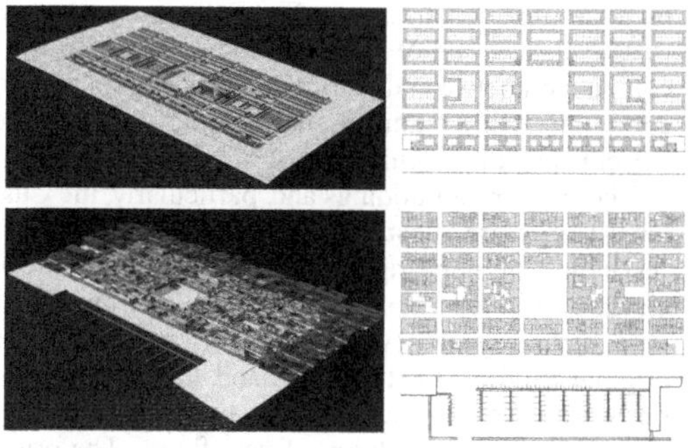

Figure 4. Comparison between the original plan and the current state of Vila Real de Santo António: (left) 3D model view (Gonçalves, 2009); (right) plan (credits: J. Ortega).

of new floors. This has also resulted in a significant increase in the average height of the buildings of the city.

The main alterations that can be found in the original constructions of the *Pombalino* core of *Vila Real de Santo António* involve the volume and the elevation. With respect to the volume, buildings were usually altered by the addition of new floors, or by the change of the eaves line. Regarding the elevation, modifications, these commonly consisted of the opening of new windows and doors, and the enlargement of the original openings. Other important common alteration reported, by the previously commented study has, been the substitution of the original roof typology, usually consisting of the replacement of the original hip roofs with flat ones (SGU 2008). Modifications of the construction system and structural elements, such as floors, can also be expected in most of the constructions. Most of these alterations are a normal consequence, resulting from the changes of the use of the buildings and the new needs of the users. Figure 5 shows some examples of the deep changes on the built-up environment, which have taken place during the last century in *Vila Real de Santo António*. With respect to the identification of traditional strengthening solutions, ties were usually observed in some buildings.

3.3.1 *Detailed study of the original Alfândega block*

The block that occupies the central position in the river front and contains the *Alfândega* or Customs House (Fig. 1) was selected for a more detailed study about the qualitative evaluation of the influence and of the alteration seen in the seismic vulnerability of the block building.

The Customs House is one of the most important buildings of the city, as the symbol of the *Pombalino* political power, and because of displaying an important economical role in the region. One of the main reasons for choosing this block is the fact that some of the buildings within it still preserve the original qualities in terms of dimensions, architectural layout, and constructions solutions and materials, namely the

Figure 5. Examples of alterations of the original constructions occurred mainly during the XXth century (Gonçalves, 2009).

Customs House, and the two-storey buildings facing the river.

Besides the Customs House, which is a unique building within the city, the selected block included two of the previously described architectural typologies: the salting factories, and the river front buildings. Figure 6 shows the original block plans, the elevations and general dimensions. The composition of the block was very regular. Buildings facing the river front had two stories plus an attic floor, while the single story salting factories were situated behind them, forming a patio with a U-shape plan configuration. The Customs House was composed by two independent two-story buildings. The main one was located at the riverside, aligned with the river front buildings, and had a mansard roof that rises up from the roof ridgeline of the neighbouring buildings. The other building was placed at the back of the block, and was also higher than the adjacent single story salting factories. There was a third patio in-between these two buildings.

The roof structure of the river front buildings consisted of simple roof timber trusses, covered by wooden boards (Mascarenhas 1996). The mansard roof of the Customs House was also built using a timber frame, but the structure is more complex (Fig. 2). The roof structure of the single story factories simply consisted of a couple of rafters resting directly on the masonry walls and timber joists, on top of which, wooden boards were placed, composing the ceiling (Fig. 7). Sometimes, instead of eaves, parapets were built along the edge of the roof, as an extension of the wall, for the finishing of the roof. Over the wooden boards of every building, regional ceramic tiles were laid.

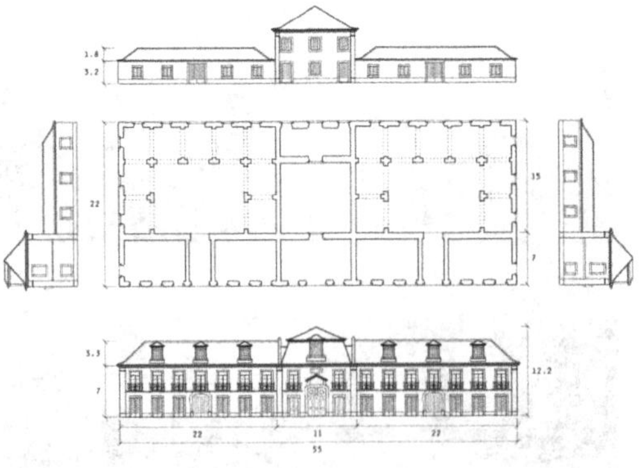

Figure 6. Original plans and elevation of the selected block (dimensions in meters) (credits: J. Ortega).

Figure 7. (left) system of arcades in the single story salting factories (Correia, 1997); (right) simple timber roof structure in the salting factories (Mascarenhas, 1996).

The floor structure of the river front buildings consisted also of timber beams and floorboards, being both elements pre-fabricated and imported from Lisbon, as previously commented. Assuming that the construction details were similar to those used in Lisbon, floor joists extended from one external wall to the other and their ends rested on transverse timber plates inserted within the walls, allowing a better distribution of stresses along the wall. The floorboards composing the flooring were placed perpendicular to the joists.

The timber frame partition walls of the upper floors in the river front buildings (Fig. 2) had also a structural function because, by connecting the different structural elements, i.e. roof, floor and exterior masonry walls, they contributed to the bracing of the building, improving its overall resistance. The partition walls at the ground floor level were heavier, built of solid brick masonry. The use of a lighter system in the upper level also helped to ensure a better structural behaviour, by lowering the centre of gravity of the building, and thus provide greater stability.

3.4 *Detailed study of the current Alfândega block*

The other main reason of choosing the *Anfândega* block is the fact that it is also a representative example of the transformation of the city. Some of the constructions within the block have suffered important alterations, some has been demolished and substituted and, at the same time, the block has been occupied by additional constructions. Figure 8 shows the constructive evolution of the case study, illustrated by a comparison between the original block and its current condition, while figure 9 shows some views of the current state of the building.

The river front buildings and, particularly, the Customs House are substantially less altered. They preserve most of their original characteristics in terms of elevation, but present some alterations in terms of volume, having some new constructions attached to the original building. On the other hand, the salting factories were either replaced with new buildings, or highly altered, mainly by addition of new floors. The original patios were also filled with new constructions and extensions of the original buildings.

All these alterations have increased the primary seismic vulnerability, because they affect directly most of the previously mentioned parameters that can be used to assess the seismic vulnerability of building aggregates. First of all, the original compact regular and rectangular plan configurations that typically behave better under seismic loading are completely modified. The accumulation of buildings within the patio has introduced setbacks and irregularities that may introduce torsional effects in the seismic response of the block.

This massive occupation of the patios, together with the addition of new floors to the original construction, has led to a significant increase in the height of the buildings within the block. The number of floors of a construction is already an important parameter that affects its vulnerability but the new buildings and additions do not respect the original eaves line, and introduce a big source of irregularity in elevation, leading to the presence of adjacent buildings with different heights, and thus increasing the seismic vulnerability. This type of alterations can also lead to the presence of staggered floors among the buildings within the block, which can be detrimental for the seismic response by introducing risk of hammering, particularly if the floors have been replaced with rigid reinforced concrete floors.

New buildings and alterations were built during the 19th and 20th centuries, but they were still built in stone masonry, which avoids the harmful effect that the typological heterogeneity between adjacent constructions can introduce. The original structure of the masonry arcades is, indeed, preserved in some of the old salting factories, and it can still be observed (Fig. 10), as well as some original vaults in the ground floor of the riverfront buildings. Roofs were mostly replaced or modified in every building, changing the original gable roofs for flat roofs (Fig. 10). The original eaves were also substituted by parapets, which are a non-structural element, more susceptible to topple.

Finally, there were many alterations in the original regular wall façade openings, both in their position, and the total area. The change in use of the buildings, and the commercial use of most of the ground floors of the buildings, led to the enlargement of the openings; and the opening of new ones, highly increasing

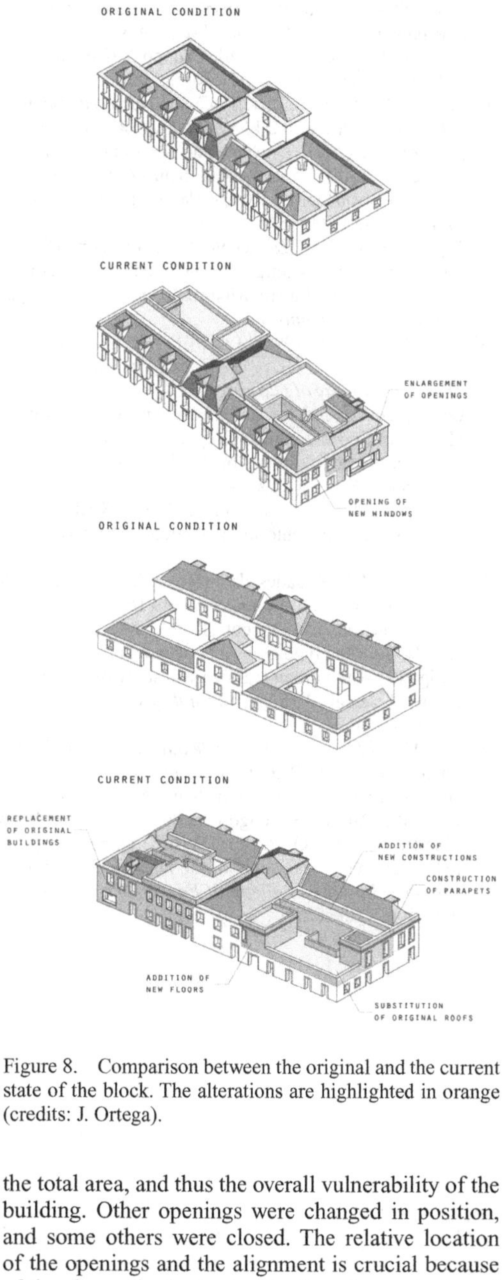

Figure 8. Comparison between the original and the current state of the block. The alterations are highlighted in orange (credits: J. Ortega).

Figure 9. Views of the block (SGU 2008): (left) Customs House façade; (middle) river front buildings; (right) current state of the old salting factories (credits: J. Ortega).

Figure 10. (left and middle) Remaining arches from the original salting factories (SGU, 2008); (right) commonly new flat roofs built replacing the originals (SGU, 2008).

the total area, and thus the overall vulnerability of the building. Other openings were changed in position, and some others were closed. The relative location of the openings and the alignment is crucial because of the change in the load path. Therefore, misaligned openings also increase the building's vulnerability. The fact that now some buildings within the block present a bigger number of openings than the adjacent ones, also cause a relative variation in the stiffness of neighbouring buildings, which is another parameter that may affect their seismic response.

4 CONCLUSIONS

The mischaracterisation of historical city centres is common everywhere. The particular case of *Vila Real de Santo António* has been addressed, since it exemplifies how an originally strongly designed seismic-resistant construction plan has been modified by successive transformations of the building form, resulting from new needs of the occupants. These alterations are not only detrimental in terms of loss of authenticity of an important architectural and urban heritage, but they also jeopardize the safety of the buildings, by affecting directly several parameters that are crucial in the seismic response of the building, and also introducing new sources of vulnerability.

The importance of assessing the seismic vulnerability of building aggregates in historical city centres lies precisely on this fact. Acknowledging that a deep characterisation of the constructions, including a comprehensive and detailing research on the geometry, structure, construction materials, not addressing just few critical parameters that can be recognised from outside, can provide an useful overview of how the seismic vulnerability of the studied block has increased notably, since its original condition, as it has been the objective of this study.

Vila Real de Santo António also exemplifies how the loss of the seismic awareness leads to the abandonment of Local Seismic Cultures. Here, the original seismic concern, arisen after the 1755's earthquake, resulted in an effective urban plan, and a carefully constructed architecture, which was eventually forgotten and compromised. Several reinforcement measures, at an urban level, such as reinforcing arches, buttresses or ties, can be commonly observed in other cities, where a Local Seismic Culture has been more active, because of the frequency of earthquakes, but these are missing in the case of *Vila Real*. The re-adoption of some of these traditional techniques can help reducing the seismic vulnerability of these constructions.

ACKOWLEDGEMENTS

The work presented in this paper was supported financially by the research project Seismic V – Vernacular Seismic Culture in Portugal (PTDC/ATP-AQI/3934/2012) funded by the National Foundation for Science and Technology.

REFERENCES

Barros, J.C. (2011). Núcleo Pombalino de Vila Real de Santo António: Salvaguarda, Desenvolvimento. *VIII Congresso Internacional das Cidades e Entidades do Iluminismo*, Vila Real de Santo António, Portugal.

Benedetti, D. & Petrini, V. (1984). Sulla vulnerabilità di edifici in muratura: Proposta di un método di valutazione. *L'Industria delle Costruzioni* 149(1), 66–74.

Calvi, G.M., Pinho, R., Magenes, G., Bommer, J.J., Restrepo-Vélez, L.F. & Crowley, H. (2006). Development of seismic vulnerability assessment methodologies over the past 30 years. *ISET Journal of Earthquake Technology* 34(472), 75–104.

Cardani, G., Giaimi, P., Bellucco & P., Binda, L. (2015). A simplified method of analysis to define the seismic vulnerability of complex buildings in historic centres. In *12th North American Masonry Conference*. Denver, Colorado, US.

Chester, D. & Chester, O. (2010). The impact of eighteenth century earthquakes on the Algarve region, southern Portugal. *Geographical Journal* 176(4), 350–370.

Correia, J. (1997). *Vila Real de Santo António. Urbanismo e Poder na Política Pombalina (PhD Thesis, Faculdade de Arquitectura da Universidade do Porto, Portugal).*

Ferreira, T.M., Vicente, R., Mendes da Silva, J.R., Varum, H. & Costa, A. (2013). Seismic vulnerability assessment of historical urban centres: case study of the old city centre in Seixal, Portugal. *Bulletin of Earthquake Engineering* 11(5), 1753–1773.

Formisano, A., Florio, G., Landolfo, R. & Mazzolani, F.M. (2011). Numerical calibration of a simplified procedure for the seismic behaviour assessment of masonry buildings aggregates. In *Proceedings of 13th Int. Conf. Civil Structural Environ. Eng. Comput.* Stirlingshire, Scotland, UK.

GNDT (1994). *Scheda di esposizione e vulnerabilità e di rilevamento danni di primo livello e secondo livello (muratura e cemento armato)*. Gruppo Nazionale per la Difesa dai Terremoti (GNDT). Rome, Italy.

Gonçalves, A. (2005). Caracterização do Núcleo Pombalino. *InPlanos – Salvaguarda Vila Real de Sto António: núcleo pombalino, ECDJ.9*. Universidade de Coimbra, Portugal.

Gonçalves, A. (2009). Vila Real de Santo António. Planeamento de pormenor e salvaguarda em desenvolvimento. *Monumentos* 30, 40–53.

Maio, R., Ferreira, T.M. & Vicente, R. (2015). A morfologia dos núcleos urbanos antigos: levantamento arquitectónico e construtivo do Bairro Ribeirinho de Faro, Portugal. *Conservar Património*.

Mascarenhas, J.M.D. (1996). *A study of the design and construction of buildings in the Pombaline quarter (PhD Thesis, University of Glamorgan, UK.*

Neves, F., Costa, A., Vicente, R., Oliveira, C.S. & Varum, H. (2012). Seismic vulnerability assessment and characterisation of the buildings on Faial Island, Azores. *Bull Earthquake Eng* 10(1), 27–44.

Oliveira, A. (2009). Casa da Câmara de Vila Real de Santo António. Levantamento arqueológico. *Monumentos* 30: 54–61.

Rossa, W. (2009). Cidades da razão: Vila Real de Santo António e arredores. *Monumentos* 30, 16–31.

Santos, C., Ferreira, T.M., Vicente, R. & Mendes da Silva, J.R. (2013). Building typologies identification to support risk mitigation at the urban scale – Case study of the old city centre of Seixal, Portugal. *Journal of Cultural Heritage* 14(6), 449–463.

SGU (2008). *Plano de Pormenor de Salvaguarda do Núcleo Pombalino de Vila Real de Santo António*. Sociedade de Gestão Urbana de Vila Real de Santo António (SGU).

Vicente, R., Parodi, S., Lagomarsino, S., Varum, H. & Mendes da Silva, J.R. (2011). Seismic vulnerability and risk assessment: case study of the historic city centre of Coimbra, Portugal. *Bulletin of Earthquake Engineering* 9(4), 1067–1096.

Part 6: Conclusions of the research

Seismic Retrofitting: Learning from Vernacular Architecture – Correia, Lourenço & Varum (Eds)
 ISBN 978-1-138-02892-0

Systematisation of seismic mitigation planning at urban scale

D.L. Viana, A. Lima, G.D. Carlos, F. Gomes & M.R. Correia
CI-ESG, Escola Superior Gallaecia, Vila Nova de Cerveira, Portugal

P.B. Lourenço
ISISE, Faculty of Engineering, University of Minho, Guimarães, Portugal

H. Varum
CONSTRUCT-LESE, Faculty of Engineering, University of Porto, Porto, Portugal

ABSTRACT: This chapter explores the relation between the seismic mitigation planning that was developed to contribute to minimise effects at urban scale, which arise within the seismic impact. In this sense, it deepens the notion that interdependence, solidarity and cohesion of the different morphological constituents promote a better prepared built fabric, when facing catastrophic phenomena of seismic origin. Preventive approaches and reactive measures arising from the testimony of earthquake-resistant culture in Portugal are also mentioned. From the recognition of the importance of local know-how, infers a set of logical that evidence conditions to transcend the respective vernacular context, making it possible to generically systematize them.

1 INITIAL REMARKS

1.1 *Morphology and seismic-resistant culture*

The socio-physical organisation of groups and communities obeys to multiple factors that determine the place of settlements and the corresponding collective structure. Among them, issues such as access to resources, the availability of food (collected or grown/produced), storage and distribution – and consequent transport processes and exchange -, in addition to aspects related to the overall security and the convenience of natural conditions of the territories (its topography and climate, for example.), are commonly regarded as decisive for the continued stay of population groups. Nevertheless, in many settlements, these logics face somewhat uncertain predictable phenomena, more difficult to frame, but with consequences and direct and indirect effects of great impact, implying risks – sometimes – not calculated associated with natural disasters, e.g. seismic activity. When this occurs, it is usually exposed the vulnerability of the settlements, and the need to mitigate the resulting damage. In this regard, it is pertinent to ask and witness the ways in which the relationship between the morphology of settlements and seismology was being addressed in order to – in good time – take shape a greater awareness of the significance of the last over the first, not only in terms of its configuration, but also in regards of the possible preventive measures and reactive approaches.

1.2 *Seismic-activity: A certainty*

The settlements configuration may take a variety of forms, which, in many cases, show contrasts in their built up fabric, the corresponding circulation networks and activities/land uses associated with them. Other variables such as the size of the sets, the housing density, grain size and porosity of mesh, the cadastral organisation and the level and type of instalment, not forgetting the range of public and/or collective space, conform aggregation systems with less or more aptitude to deal with seismic activity. The resulting community living, consolidating practices of earthquake-resistant culture, focuses on the expansion of collective spaces, the standardisation of roads, the provision of reserve areas for covering individuals, the foreshadowing of repeated and arranged architectural typologies of a linearly compact way . These references exemplify preventive approaches and reactive measures systematised in the research developed on the structure of the settlements – and its corresponding aggregation – of the case studies discussed throughout the research project 'SEISMIC-V' – Earthquake-Resistant Vernacular Culture in Portugal. It is also important to reinforce the notion that, in multiple contexts of settlements, such as those in different geographies of the country, it is urgent to rescue this "collective consciousness" for such a significant issue, whose doubt is not whether it will happen, but when it will happen.

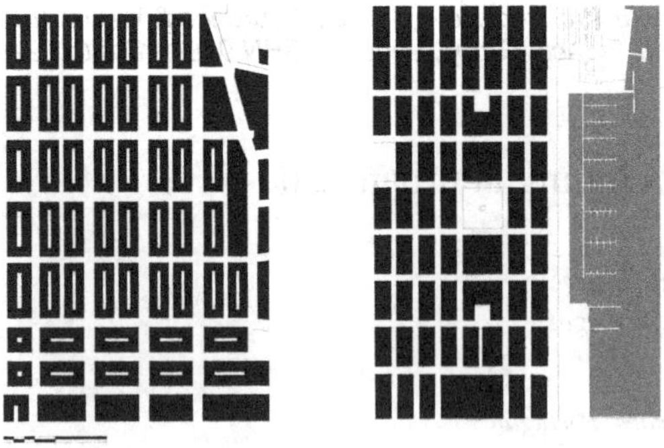

Figure 1. Orthogonal layout of the urban mesh in Lisbon (1) and Vila Real de Santo António (2) (credits: CIESG).

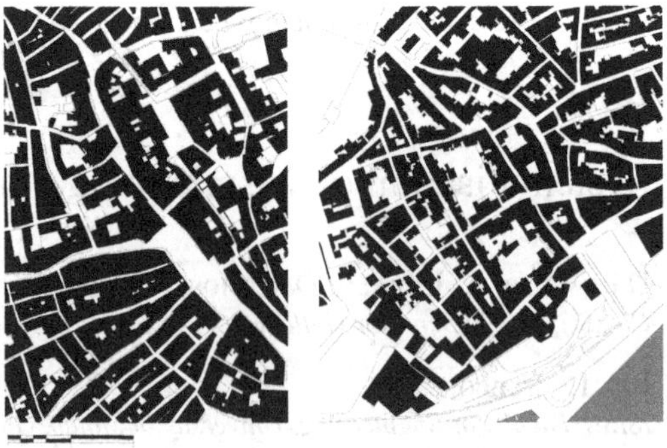

Figure 2. Historical Centre of *Évora* (1) and *Lagos* (2), consolidated centres of organic matrix (credits: CIESG).

2 SETTLEMENTS AND THEIR CONFIGURATIONS

2.1 *Compact settlements*

It is known that from the human conglomeration results organised settlements as compact settlements, due to the concentration of people, goods, resources, and built (infra)structures, in view to host community life, cultural ties, social cohesion, economic development, scientific knowledge and technical progress. These are sites and places with potential for the substantiation of solidarity and mutual aid logics, supported by the gathering of knowledge and experience feeding strategies aimed at predictable scenarios. Along the research developed under the project 'SEISMIC-V', the legacy of this type of configuration proved to be significant for the results obtained. Thus, with reference to cases like Lisbon and *Vila Real de Santo António*, it was revealed a framework of concerted actions initiated to prevent the impact of new extreme events, such as earthquakes. Either in Lisbon or in *Vila Real de Santo António* (Fig. 1), it was detected – following the recent most striking seismic events – a concern for planning spaces based on a strict orthogonal grid, the enlargement of arteries (aiming at broader streets and not so clogged) providing escaping routes and reserve areas. These preventive measures were not felt only in terms of morphology of the space system; it also had effects on the typological composition of the building fabric, conditioning volumetric, homogenising proportions, aligning heights and establishing a closer type-morphological relationship (e.g. forcing a greater interdependence between the conformation of buildings and their articulation with the surrounding streets). From this range of practices, progress was made in testing a theoretical *corpus* that requires both, attention and further developments.

2.2 *Consolidated centres*

The carried out survey also sought to focus on more empirical approaches to the problem of earthquakes. For this purpose, situations with "spontaneous" growth of settlements were taken into account, of which

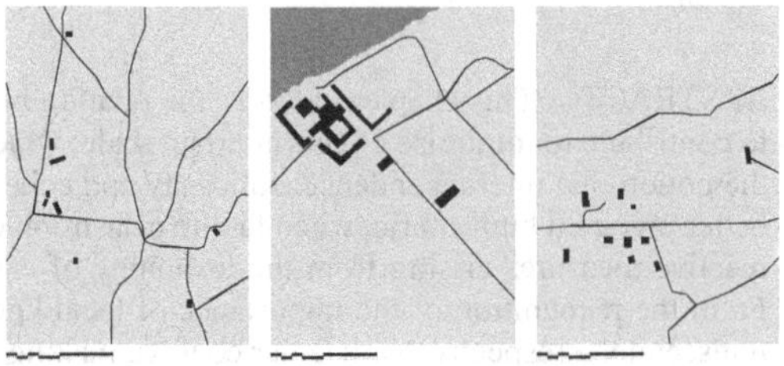

Figure 3. Settlement structure and built fabric. *Trindade* (1), *Coruche* (2) and *Melides* (3). (credits: CI-ESG.

consolidated centres of organic matrix resulted, presenting morphological more intricate and deployed settings. The process setting of this kind of space system occurs in time and over time, wherein the construct leads to the accumulation density of the building in confined spaces and without previous planning. The scarcity of this planning increases the impact of seismic activity, also raising the potential for possible damage at the level of the structure and building fabric. Thus, *Évora*, *Tavira* and *Lagos* are selected for analysis (Fig. 2).

Amongst these three case studies reactive measures to mitigate the effects produced by earthquakes are primarily verified. In opposition to preventive measures aforementioned – attempt to project the impact of future seismic events – the population implements (afterwards) a set of empirical solutions which aim to ensure the continuity and enhancement of existing constructions and the (infra)structures. These are logics that can be taken alone, or as a result of interventions that establish continuity between them, contributing to the overall solidarity of the settlements. Examples of this are actions on blocks, which are based on the physical cohesion of the same, promoting the distribution of counter-arches, relieving arches, and vaults and braced links connections. These initiatives are relatively combined and (sometimes) of small scale become relevant when considered all together, implying individual or collective efforts of private and/or public character. This empirical and local know-how can play a key role as complement the general planning strategies.

Figure 4. *Benavente* Plan, 1908 (1) and Plan and with the expansion area after the earthquake, 1909 (2). (credits: CI-ESG).

2.3 *Scattered structures*

The settlements with less continuous housing, morphological configuration were analysed. In the consistency of preventive approaches, even when generally seen, it was opposed a role of reactive measures. Considering it, in scattered structures, the empirical bias of the principles adopted anchored in the knowledge of the geo-morphological characteristics of the territory in which the settlements are distributed, an example being the use of geophysical features – such as rocky upwelling – for bases and *Poiais* (stone benches). When considering *Melides*, *Coruche* and *Trindade* (Fig. 3), it is possible to specify actions, such as reconstructive reinforcement with earthquake-resistant elements based on various devices, such as buttresses and tie rods.

In the setting of scattered structures, there is yet an articulation between typological/tectonic logics and the specificity of the territory, its topography and location. The relationship between morphology and earthquake-resistant culture is in the deepening interdependence between natural and artificial systems that make up the settlements. The aloofness complicates the planning of preventive strategies scale and/or widely concerted, however, it does not decrease the density of the collective and shared knowledge as a result of trial and error processes, which should be systematised.

2.4 *Mixed contexts*

In short, the settlements and types of forms indicated so far – based on what was said for the listed cases of study – exemplify aspects identified in the research that contribute to the generic definition of principles framed by a possible Earthquake-Resistant Vernacular Culture in Portugal. Noting the country, it is noticeable that it is composed of many free settlements, in which coexist the planned and the occupied, the rationality of the layouts and the organic nature of appropriate, the development of the (infra) structured and the so-called "spontaneous" growth. These are mixed contexts, whose morphology translates contrasts between collective policies and private initiatives, predictability of the land uses and actual activities, time and space, global agents and local actors of change, vigorous periods and recessionary times. The forms hybridisation mark these mixed contexts. In this framework, the research clarified that – in what case studies with this configuration matrix are concerned – the preventive and reactive measures to seismic activity (if any) share not only an empirical basis, but also theoretical one. As in everything else, the convergence of administratively planned strategies and the actions of people are ever-present.

Taking *Benavente* as an example, it may be said that these are combined in the different parts of the settlement: as regards with sectors said of "spontaneous" growth, of a more organic inclination (originating from earliest times), and in that, which relates to the areas of planned expansion for the resettlement of the population affected by the earthquake of 1909 (Fig. 4). In these is the adoption of measures, such as the design of linear orthogonal meshes, the implementation of the enlargement streets and the establishment of blocks of symmetrical plant. The confrontation matrixes and the regular and irregular morphological elements amplify the situations and aspects to be resolved in earthquake-resistant terms, making it necessary a panoramically perspective appropriate indicators of a hypothetical Portuguese earthquake-resistant culture.

3 BUILDING AND CONTINUOUS ROW SYSTEMS

3.1 *Continuous or row buildings*

More than in morphological terms, the research has consolidated the perception that if there is – effectively – an Earthquake-Resistant Vernacular Culture in Portugal, it will have being much more structured around typological/tectonic aspects and aggregation systems building, working towards the strengthening, cohesion and constructive solidarity of the fabric of the settlement and its adjacent space systems. It was evident that for the continuous and/or row building it is frequent the combination of aggregation systems, combining symmetric and regular plan constructions (with two orthogonal axes of symmetry) with earthquake-resistant elements, such as buttresses and the tie rods – not forgetting the aforementioned structural bearing walls (Fig. 5). The presence of these elements are noticeable in places like *Coruche*, where are also detected other types of capacity improvement measures, *poial* situations and the blind arches. In this way it is sought to guarantee a greater stability and mechanical solidarity of the construction, aiming to improve the performance of the building for seismic demands, and reducing their impact.

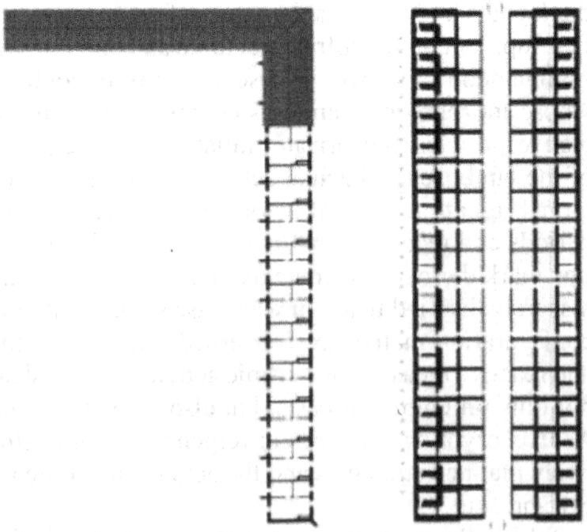

Figure 5. Row edification, *Couche* (1) Municipal district plan designed after the 1909 earthquake in *Benavente* (2) (credits: CI-ESG).

3.2 *Isolated building*

The isolated construction is characterised – within the seismic context – by the homogeneous and compact configuration of the respective plan (often rectangular matrix), with a reduced (and small) number of openings, not forgetting the gabled roof and the chimney on the facade. The isolated building emerges as the most common type in the areas under study.

3.3 *Blocks*

The structure and the logics of aggregation provided by type-morphological devices, such as blocks, contribute to the diversity of the forms of the settlements. Nevertheless, the research conducted within the multiple case studies converge towards substantiate the notion that – for seismic demands context – there are parallel alignments that form blocks of standardized plan, geometrically regular, and with an homogeneous architectural "language". In addition, aggregation systems building in symmetrical and longitudinal row are protagonists. Narrow streets or courtyards mediate the built fabric, providing small reserve spaces and/or relief.

The walls of separation between dwellings are also features, solidly built and with no span, looking for to ensure its maximum compactness and capacity. Further to this aspect, it is also to highlight the closed structures of the blocks and a consequent overall structural stability of the whole. Particularly in the case of *Évora*, it is to emphasize the occurrence of linkages between existing buildings (reactive measure, guaranteed by the use of counter-arcs) (Fig. 6) and to promote a greater stability of the whole of the buildings based in mechanical principles continuity and lateral displacement control. Prevails the notion that the built fabric more than understood as the sum of buildings, should behave like an interdependent system, reacting as a whole to the impact of seismic demands, so that – from scale logic – it is established a "solidarity" and "cohesion" between the buildings, which allow them to be better prepared to withstand the effects of earthquake demands.

Figure 6. Historical centre of *Évora*, placing of counter-arcs between the different blocks (credits: CI-ESG).

3.4 *Residential blocks*

The result of the impact of seismic activity may have multiple consequences, since the collapse of buildings, construction debris scattering, declining of objects, obstruction of traffic routes, disabling (infra) structures, the collapse of bridges and viaducts, floods, and fires spread between buildings, etc. Many of the measures presented so far attempt to project/mitigate these and other effects of earthquakes. The residential blocks are the typological plans most likely to suffer from these and other damages (given the respective scale, number of floors, and number of dwellings). Seen it, the attention to the residential buildings should be protagonist in seismic activity contexts. Given the relevance of this problem, along the research carried out, was considered a history of significant experiences that took place in Lisbon. Among them there are the walls separating property/plots of land (which rise beyond roof quota in order to prevent the spread of fire) and limiting the number of floors (in Lisbon, four below the cornice). But more important than was above mentioned, it's the conceptualisation of the built fabric based not in the individual logical entities, but divided into blocks (almost always of rectangular base), trimmed by a continuous front of streets. It is a solution with a better use of land and an urban facade that repeats the compositional effect of the urban design. It regularises views, clarifies the overall accessibility of the whole, contributing to the guidance of their users, also enhancing the spatial configuration of the unit system. These are aspects to be valued, in a post-seismic 'chaos'.

4 PREVENTIVE AND REACTIVE MEASURES

As already seen in seismic context, two types of actions taken by the people are to distinguish. A direct action, which aims to repair the damage already produced in building structures – a reactive approach – and another,

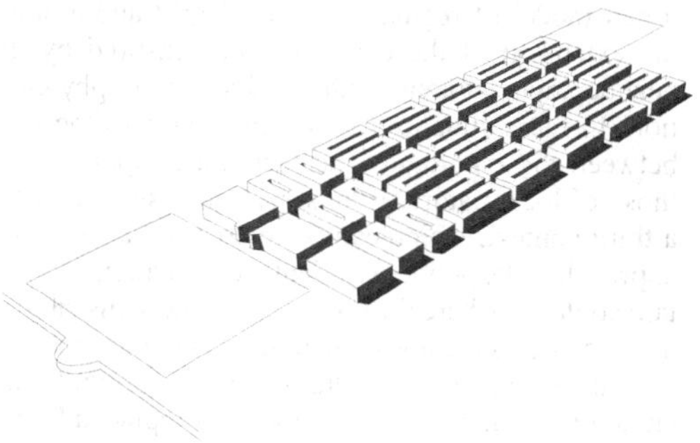

Figure 7. Orthogonal grid, N-S axis (direction of concussions), and spaces reservation, outlined in the proposal of Eugénio dos Santos and Carlos Mardel in Lisbon 1773. (credits: CI-ESG).

which is indirect, aiming at to reduce damages in future occurrences – the preventive approach.

In terms of urban planning as a process that anticipates the structuring of a given territory, the adopted measures are preventive in nature, such as Lisbon, *Vila Real de Santo António* and *Benavente*.

The reconstruction plan for Lisbon, drawn up in 1756 by the engineer Manuel da Maia, established a series of "new urban concepts" that became a reference. The first largely measure, adopted in the full organisation of the urban fabric, was the orientation of the blocks. The base structure of the building fabric would adopt a north-south direction, the main orientation of concussions, so as to, considering the conservation of energy rule, only the smaller facades would be exposed to the projection of seismic waves (Mascarenhas, 2009).

The geometrically structured and linearly regular urban grid designed to Lisbon, defined by rigorous and orthogonal alignment of facades, allows a quick and orderly evacuation system in case of earthquake (or fire). At the extremities of the block were implemented two large squares, as reserve spaces, the Commerce Square (south) and the *Rossio* Square (north) (Fig. 7).

The buildings were grouped in closed blocks so as to give greater structural stability to the whole. This would reduce the number of gables subject to shocks, reducing and restricting the rotation capacity of each urban element/building (since each would prevent the rotation of the other). Another of the great measures of reference, established by the Plan of Lisbon, was the widening of the streets: the main would have 60 spans (with a lane for cars with 50 spans and another one of 10 spans on each side for pedestrians); secondary streets would have about 40 spans.

The rule of proportion arises also as a reference element; i.e. based on the ratio between the width of the streets and the height of buildings, all blocks have the same symmetry in doors, windows and arch centrings. This space limitation would diminish the risk of collapse over the streets.

Any compositional elements outside facade plan were also banned, such as steps, windows or door protections, rings, etc., as well as the openings near the wedges – as these weaken the structural continuity of the built blocks and other equipment. The urban guidelines established by the Engineer Manuel Maia for the reconstruction of Lisbon were the basis for the planning of new urban structures as it is the case of *Vila Real de Santo António* The plan outlined in 1773 by the architect Reinaldo Manuel organises that city according to an orthogonal grid (north-south, parallel to the direction of concussions) and with the same width of streets. The urban fabric would be defined by blocks of 240x100 spans, one and two stories high (Lopes et al, 2004).

And such as Lisbon, *Vila Real de Santo António* would enjoy also a "reserve space": the *Marquês de Pombal* Square, located at the midpoint of the space system.

Benavente, presents itself as mixed context, marked by the existence of two distinct urban plots: one is part of the *Old Town*; the other part is composed of the extension area (post-earthquake, 1909). After the 1909's earthquake, it was designed a General Plan for Roads, specifically a Cadastral Plan of the Village, indicating the new measures to apply to the existing layout (*Old Town*), and the structuring of the new extension area. As for the *Old Town* it was proposed an enlargement and a process of correction of the communication routes. The *New Town*, an area of expansion, was structured according to a rationalist matrix, where large wide streets prevailed, based on an orthogonal logic. The matrix was composed of uniform and aggregated blocks in symmetrical row houses, in two parallel alignments, separated by an inner street (constituting the set of preventive measures taken). These were also implemented in *Benavente*, a series of reactive measures on the building fabric, incorporating earthquake-resistant elements applied to the built area, highlighting given to the tie rods and tethers (Vieira, 2009).

Consolidated urban centres, such as Évora and Lagos, are associated with reactive approaches. Évora features a matrix of medieval origins, which marks the consolidated historical centre, consisting of blocks of compact 'spontaneous' growth. The city is characterised by the one-off introduction of seismic reinforcement elements as counter-arches, buttress reinforcement at the tops of the arcades. These elements are intended to provide greater structural stability to all blocks. In the case of *Lagos*, the introduced approach ensures the same standards, that is, a consolidated building fabric, where blocks marked by introduction of protruding elements such as continuous basis (whose function is to reinforce the volume) emerge.

In dispersed settlements, generally of rural character, reinforcement reactive measures of greater implementation are: the perimeter lashing strap (linking several connected built structures) and the buttress, which appears, sometimes, in the angles of the extremities of continuous buildings.

5 FINAL REMARKS

The shape and the organisation of the analysed settlements undoubtedly refer to the state of general destruction, verified after the events of greater impact. It is in the context of higher occupational density that this assumption becomes clearer. The triggered action is closely related to the analysis of the regenerative capacity of the affected structures, and to solve their technological constraints. In cases where a positive assessment is obtained, as *Tavira* and *Évora*, collective processes of spontaneous nature are triggered up, assuming the reconstruction almost immediately after those events. In these cases there is no apparent rupture between local constructive cultures and the reconstruction techniques applied. The built typologies and the techniques employed, which showed reasonable resistance, thus justify its perpetuation. Naturally, there is not a redefinition of the pre-existing urban fabric logic. The intervention occurs only at the one-off articulation level of existing structures, usually in what the reinforcement or on mooring adjacent elements is concerned, aiming at to maximize the structural solidarity of the whole. In these situations, the arcades, the counter-arches and the masonry vaulted structures are the most frequently used elements.

Considering more severe situations, and whenever it becomes mandatory to maintain the geographical position, it is witnessed a concerted action, coordinated by the administrative services. The planning of the urban form arises logically associated with a classical and formal architecture. Its great principles relate to the regularity of the layout and the systematization of shapes and construction techniques. The reconstruction plan of Lisbon downtown arises naturally as the main reference in these circumstances. The layout, the aggregation system, the architectural typology and construction processes are imposed, by force of circumstances, to the target population. This is the case of *Vila Real de Santo António* and the relocation neighbourhoods *Benavente*. Both adopt the urban model of regular orthogonal grid and assume the extension of the existing roads, ensured by the new deployment areas with suitable topography election. In these situations it is clear a larger aloofness between the typologies and techniques applied, and those of local nature. It must be emphasised though, a third context, which can be interpreted as a hybrid approach to the previously identified situations. In certain contexts, where the population is more distributed and the effects are not so focused, it is witnessed what appears to be a gradual cultural contamination. In this situation the application of models is replaced by the one-off incorporation of strategies or elements under the filter of regional culture. The melting of techniques and procedures thus contributes to the development of combined characteristics, which will gradually consolidate, as shown, for example, in the cases examined in *Álcácer do Sal* and *Coruche*.

ACKNOWLEDGMENT

The authors gratefully acknowledge the support by the Portuguese Science and Technology Foundation (FCT) to the research project 'SEISMIC-V – Vernacular Seismic Culture in Portugal' (PTDC/ATP-AQI/3934/2012).

REFERENCES

Mascarenhas, J. (2009). *Sistemas de Construção – V: O Edifício de rendimento da baixa pombalina de Lisboa*. Técnicas de Construção. 3ª Ed. Lisboa, Portugal: Livros Horizonte.

Lopes, M. S., Bento, R. & Monteiro, M. (2004). *Análise Sísmica de um Quarteirão Pombalino*. Lisboa, Portugal: IST.

Vieira, R. (2009). *Do Terramoto de 23 de Abril de 1909 à Reconstrução da vila de Benavente – um processo de reformulação e expansão urbana*. Bevanente: CMB.

Seismic Retrofitting: Learning from Vernacular Architecture – Correia, Lourenço & Varum (Eds)
© 2015 Taylor & Francis Group, London, ISBN 978-1-138-02892-0

Systematisation of seismic retrofitting in vernacular architecture

A. Lima, M.R. Correia, F. Gomes, G.D. Carlos & D. Viana
CI-ESG, Escola Superior Gallaecia, Vila Nova de Cerveira, Portugal

P.B. Lourenço
ISISE, Faculty of Engineering, University of Minho, Guimarães, Portugal

H. Varum
CONSTRUCT-LESE, Faculty of Engineering, University of Porto, Porto, Portugal

ABSTRACT: This paper aims to identify and to systematize the data obtained during the working process of the research project 'SEISMIC-V – Vernacular Seismic Culture in Portugal' in terms of seismic-retrofitting elements earthquake-resistant reinforcement implemented in vernacular buildings. The data will be distinguished based on the structural seismic strengthening and according to the following criteria: strengthening elements, perimeter seismic-resistant elements, arches reinforcing elements, and combined reinforcing elements.

1 INTRODUCTION

Although Portugal is recognised as a country of moderate seismic risk, the fact is that future occurrences of earthquakes are likely and damage will be important. The most significant earthquakes of the last century occurred between 1531 and 1998, causing considerable damage in various regions. The most affected regions have developed, in specific moments, seismic cultures based on the introduction of reinforcement elements and techniques. The constructive culture of each region generated different approaches, both in reaction to the damages arising from seismic events, and as prevention for future damage repetitions. These consisted simply in the integration of structural elements in the building methods and in the traditional materials, consequently enhancing the constructive resistance.

Through the empirical knowledge of local populations different approaches were put into action: the morphological and the tectonic one, which contributed to the improvement of traditional building systems, due to the effects of seismic events over the past centuries. On the basis of the morphological approach is the specification of the different typologies of simplified dwellings construction, and the consequential introduction of a typology of compact, uniform and homogeneous features. This typological simplification allowed a greater control of the response of the volume in a seismic event. On the other hand, the tectonic approach develops the earthquake-resistant reinforcement elements hosted throughout the different occurrences, both in new housing constructions, and in already existing constructions that required a greater structural stability.

The elements identified throughout the different work assignments were divided into four distinct groups, considering their structural usefulness:

a) Strengthening elements of traditional building techniques, including rammed-earth (most characteristic construction technique of the study area).
b) Perimeter seismic-resistant elements. The perimeter elements appear mostly linked to the rural typology and aim at define a joint action of the walls and floors (i.e. the so-called box behaviour), ensuring the perimeter stability of the volume.
c) Arches reinforcing elements. The strengthening system usually arises associated with arches located on the ground floor, of traditional buildings, in urban or periurban areas, with typologies of more than one floor.
d) Combined reinforcing elements. This group comprises the structural elements applied between contiguous or continuous buildings.

2 MORPHOLOGICAL APPROACH

2.1 *Simple and compact*

One the most relevant earthquake-resistant strategies to the structural stability of the dwelling is its simplicity. The more compact, uniform and homogeneous a house plan and elevation are, the more stable it will be. Thus, a uniform distribution of structural elements prevents the torsional effects of the volume. The type of isolated earthquake-resistant housing, identified throughout the study area, has a ground floor, in general presenting a longitudinal layout, with few internal

Figure 1. House of compact design with reinforcing elements, in *Melides,* Portugal (credits: CI-ESG).

partitions. The featured gable roof allows the evenly sharing of actions on the rammed-earth (monolithic) or stone masonry bearing walls. A special concern with the relationship between the thickness and the height of the walls is notorious, so as to obtain adequate distribution of loading, to capacity of buttressing and to ensure its stability, in general. The lack of openings, typically a door and one or two windows, works towards the same end, as the greater the number of openings are, the larger is the area of structural vulnerability. The arrangement of the interior vertical walls demands for regular spacing in both directions, so as to provide bracing and a global structural behaviour.

Considering continuous housing (or compound arrangement, or semi-detached housing) this even distribution of forces was sought by symmetrical continuity of the volumes. The symmetry of the plan provides uniform distribution of the masses and stiffness, so as to avoid larger torsional effects. The longitudinal outer walls are braced by little spaced transverse walls, which function as partition walls. This arrangement allows the assembly to behave favourably and uniformly. The structure is simple, so that its response in case of a seismic event is able to distribute the forces evenly. When the row-housing plan adopts a T, L, or U shape, it is common that the inner walls are extended in both directions to prevent the different volumes to perform as isolated items. Similarly, protruding intersections may present buttresses in the corner, in order to prevent torsion and to avoid separation, points of special weakness for the stresses during an earthquake.

3 TECTONIC APPROACH

In regards to the tectonics of the housing, two types of actions taken by people can be distinguished: a reactive action, which attempts to mitigate the damage already produced by an earthquake on the building structure; and a preventive action, which aims to anticipate future events and limit damage. The analysed case studies indicate that it is not unusual to combine both actions in certain techniques and solutions. Moreover, after a seismic occurrence, people have adopted reactive and preventive techniques, simultaneously. This tectonic approach will be systematised in groups or "categories" of elements identified as earthquake-resistant. The first elements pointed out will be those used as reinforcing in traditional construction techniques; a second group will include those elements intended to tie together and provide safety to the system (box) of outer walls, ensuring the stability of the whole perimeter. Later on, a third group is composed of the buttressing elements that are responsible for strengthening the structural stability of the building; and finally, a fourth group will be set, which adopts structural elements of collaboration between contiguous buildings.

3.1 *Systems (elements) of structural reinforcement using traditional constructive techniques*

Rammed-earth is the most characteristic building system in the rural areas of our study area. One of the main reinforcements found at the level of the building system was the incorporation of horizontal reinforcement layers. Courses of tile, ceramic brick, stone or timber were observed among the monoliths blocks of rammed-earth, in order to, on the one hand, facilitate the transmission of bending action and inertia forces between the transverse walls (Fig. 2a). For this purpose the horizontal ties should intersect each other, connecting perpendicular walls, and preventing the separation of the walls in the corner. On the other hand, these elements also prevent the spread of vertical cracks, which would inhibit the collapse of large wall sections or even entire walls. This same reinforcement was also identified in clay brick walls that had horizontal wooden elements irregularly spaced in the wall.

Another element to strengthen the structural stability of the building volume, also identified in the study area, is the relieving arch. Integrated into the vertical faces, on the lintels of doors and windows, these arches are responsible for diverting the vertical loads carried by the wall or roof on these fragile areas. They are usually constructed of solid brick or stone (Fig. 2B).

Finally, it is relevant to mention the ***'pombalino'* system**, as a structural reinforcing system, built in. This system was used in the downtown of Lisbon and the downtown of Vila Real de Santo António, following the earthquake of 1755.

3.2 *Perimeter seismic-resistant elements.*

The **buttress** (*contraforte, gigante*) is the most common seismic reinforcement element identified among the case studies. It is used, both as a reactive and as a preventive action. Also known as giants or fence posts, the buttress provide stiffness and additional support to the construction, and its main purpose is to consolidate the exterior walls. Usually in the form of a triangle, it is perpendicularly arranged to the outer facades. Within the surveyed cases, the buttress disposal, if not arbitrary, can vary substantially. In the edge walls it often appears centered, until the height

Figure 2. a) Horizontal rammed-earth reinforced with fired brick, *Benavente*. b) Relieving arches in *Alcácer do Sal*, Portugal (credits: CI-ESG).

of the eaves, while another, usually of smaller dimension, can be laid in the continuation of the longitudinal facade.

The walls of greater length, in which the openings generally locate, may also have a buttress, which is almost always located in the middle of the facade or leveled with the interior walls. Buttresses at the corner were also identified, in the connection between two perpendicular walls (areas of greater structural risk). These were found in sets of contiguous buildings of a significant dimension. The identified buttresses are usually built in regular masonry stone and solid brick. When the buttress is adopted as a reactive measure, and does not constitute an original element, it is common to observe the lack of cohesion to the backed wall. Another element that adds to the structural stability of the building is the **strengthening of the corners**. The corners of the buildings, and likewise of the houses built in continuous row, are extremely weak points, being especially sensitive to the action of the two directions of the earthquakes. The corners should ensure the connection between the transverse walls, being therefore exposed to large forces. Identified in areas of urban and semi-urban setting, the strengthening of the corner arises on the enclosure walls, in a trapezoidal shape, up to of the first floor. It is defined as a proliferation of the corner section. It also works as a buttress at the corner, gathering the volume lengthwise and crosswise.

In Alentejo, it is common to have fired brick or stone masonry plinths, above the ground or floor level, due to the high capillarity arising from the soil. In some cases, it was detected an addition of masonry at the building's plinth, extending the wall width. This was for instance the case of some buildings in Evora historical center. The **reinforced plinth** helps spread the loads from the above walls and bring additional strength and stability to the building.

In *Alentejo* and *Ribatejo* housing typology it is common to observe the *'poial'*, an element, usually built in stone, near the outer wall, which in addition to helping its structural consolidation, it is also used as a bench. Throughout the analysed case studies, the masonry plinth acquires some relevance, being identified in

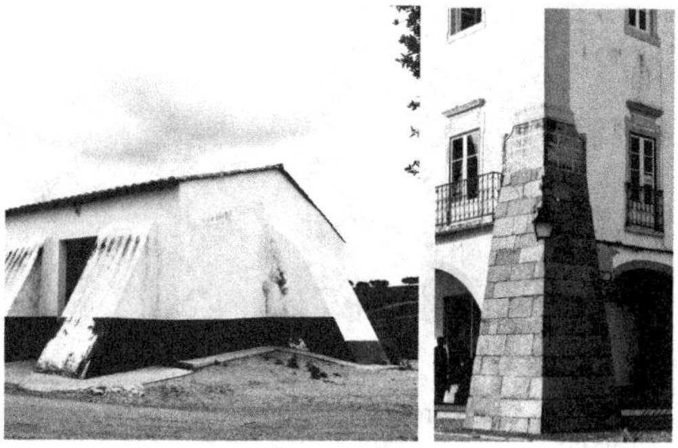

Figure 3. a) Reinforcement Buttresses, *Baleizão*. b) Corner reinforcement, *Évora*. (credits: CI-ESG).

Figure 4. a) House in *Benavente*, with tie rods and plinth as reinforcements. b) Metallic anchoring device (cat), *Melides*. (Credits: CI-ESG).

Melides, *Benavente*, *Évora*, *S. Brissos*, *Lagos* and the Azores.

In the case of *Melides*, the *poial* has an orthogonal section, and occupies the space between the top of the facade buttresses, as a structural consolidation complement.

In S. Brissos both models were identified: plinths of trapezoidal form, consolidating the gable at the edges of the built volume; and the *poial* reinforcing more the outer walls.

In *Benavente*, the plinth arises as a reinforcement element of the building set, closing the block in U-shaped, and consolidating the side gables. The plinth is also one of the most important elements in the Azores, where numerous case studies were identified. It is usual for the plinth to take full perimeter of the housing, thus resisting the longitudinal and transverse forces, and acting as an outer reinforcement ring.

The **tie-rods** are steel rods intersecting two opposite outer walls, to the height of the eave. These rods are anchored in the outer faces of the walls by metallic anchoring devices, confining the structure in the two directions. Cases were identified where this solution was applied in both the longitudinal and in transverse direction, thus connecting the four façades of the building.

Along with the buttress, the tie-rod is the structural reinforcing member with more presence within the study area. Tie-rods were identified in *Melides*, *Trindade*, *Alcacer do Sal*, *Coruche*, *Benavente*,

Figure 5. a) Arcades and vaults system in the historical centre of *Évora*. b) Flat arch or Jack Arch, *Castro Marim* (Credits: CI-ESG).

Figure 6. Different reinforcing counter-arches in the historical centre of *Évora* (credits: CI-ESG).

Baleizão, *Tavira* and *Lagos*. The tie rod appears as a reinforcement element, both in the preventive, and in the reactive approach. In cases, such as those identified in *Melides*, it is clearly adopted as a reactive strengthening solution, subsequent to a seismic occurrence, to reinforce a pre-existing structure.

In *Benavente*, the tie-rods placed next to the cover, or between the floors structures, when they are between floors, have been adopted as a preventive measure in the reconstruction and expansion of the urban structure after the 1909's earthquake.

Considering *Melides*, at the height of the tie rods were also identified continuous cornices, aiming at the increasing its resistance. These elements incorporate steel rods joined with a connector placed after the removal of the plaster along the contour of the housing, fixing the existing structural damage.

3.3 *Arches reinforcing elements.*

Considering urban or suburban typologies with more than one floor, the vertical forces of the building and of the entire set are transmitted through main (or primary) walls and pillars, which are connected either by **arches, or arches and vaults**. These solutions confer more resistance to the base of the building in a seismic occurrence, as forces are distributed more evenly.

The vaults that cover the ground floors are, generally, quadripartite vaults, consisting of four bent surfaces intersected by diagonal protruding ribs. The perimeters of the vaults encompass transverse arches that, in addition to the underpinning of the vaults, they also help to convey their loads to the walls and pillars (Fig 5.a).

Another arch-reinforced element identified in the study area is the **flat arch or jack arch**. These are elements that replace lintels, in order to achieve greater flexible resistance to the span set. The arches transfer vertical loads, distributing them laterally from the upper walls. Typically, they are constructed in fired brick, and take the entire thickness of the masonry (see Fig. 5.b).

3.4 *Combined reinforcement elements.*

The **counter-arch** is a structural element intended to block the breakdown mechanisms between adjoining buildings, typically of different heights. Its main function is to redistribute the horizontal movements at the floor level to the vertical walls of the nearby buildings. It is a kind of structural cooperation between adjacent volume structures, so that the collapse of one wall, does not foster a similar action in the adjacent buildings. For years it was assumed as a resource, widely used to consolidate structures that had been damaged, and where the collapse mechanism had already begun. Within this study area this element has been identified mainly in the Central Alentejo region, particularly in the area of *Evora*, as a structural reinforcement of cooperation between adjacent buildings, in order to ensure its entire blockade and a uniform behaviour towards an earthquake, in order to avoid the consecutive collapse of the built structure. (Fig. 6).

The **continuous cornice** (*cimalha*) appears as tying element between different units. This element allows the upper trim of the facade, functioning as a cross lintel to different walls in the same alignment. In addition to the vertical strengthening building system, it also allows the reception of the roof structure, either in the transverse or supporting the eaves framework. These elements generally comprise solid brick masonry or stone masonry, forming a sharp protrusion in contact with the roof eaves. In certain examples it may also be constituted by metal straps. When in stonework, they are usually left at sight, and when built with brick they are usually plastered and painted like the façade.

As well as ensuring a better horizontal leveling of the distribution of the load of the covering, and thus ensuring greater dispersion of forces in the volume, it is usually associated with a systematisation of arch centering and alignments of roof coverings. Thus, it also contributes to the dismissal of dissonant elements in the general volume. The continuous cornice is intended to neutralise the breakdown of contiguous walls, avoiding cracking at their contact edges, as well as the counter arch. It also prevents the consequent action of *'hammering'* (pounding), capable of occurring in adjacent structures (Fig. 7).

4 FINAL REMARKS

Considering the observation and analysis of the collected data, a series of structural reinforcing elements

Figure 7. Continuous cornice in *Bairro das Noticias, Benavente*. (Credits: CI-ESG).

used in the past as a response to a seismic event are identified. These are elements and techniques that are applied to reach increasing safety. Its implementation is directly linked to the intensity and the frequency of the occurrence, and it is not systematic, but occurs recurrently. The elements and techniques identified are, generally, common to the different studied areas. They can be distinguished as systems implemented, almost exclusively, in single-family housing, of isolated character; and also elements of a more urban character (application). Likewise, all systems, urban and peri-urban, and rural, can be classified according to its structural function.

Through the analysis carried out it is possible to conclude that, at an urban context implementation level, the elements and techniques applied point out more to the reinforcement of the set, than to the individual housing unit. The only elements identified with an isolated function were the tie rods. Most of the data collected point to structural cooperative systems between building blocks or contiguous dwellings: plinth at the base transversal to a set of buildings, the counter arch as a liaison (structural cohesion) between adjacent blocks, and homogeneous structural systems, such as arcades and vaults, or continuous cornice. In case of the isolated housing, the results point to the collaboration between different systems and strengthening techniques into the unit, as the buttresses, the tie rods, the *poial*, the masonry vaulted structure at the base, and the strengthening of the traditional building techniques.

The analysis can also synthesize a classification of the elements and techniques according to its structural function: perimeter reinforcing elements, cooperative reinforcing elements, reinforcing elements of from the constructive technique and arch-reinforcing elements. The results also reveal a greater integration of the perimeter reinforcing elements in the rural context than in the urban context. Moreover, when these elements are identified in any urban context, these always work as a structural element of cooperation between dwellings or blocks, and rarely in isolation. The cooperative reinforcing elements were so linked with the urban environment systems of a set.

The arch-reinforcing elements were found in both contexts, in rural areas as spans reinforcement system, and in urban context as structural continuous systems. The strengthening traditional building systems are also identified both in rural and urban areas.

ACKNOWLEDGMENT

The authors gratefully acknowledge the support by the Portuguese Science and Technology Foundation (FCT) to the research project 'SEISMIC-V – Vernacular Seismic Culture in Portugal' (PTDC/ATP-AQI/3934/2012).

REFERENCE

Correia M., Gomes, F. & G., Duarte Carlos, G. (2014). Projecto de Investigação Seismic-V: Reconhecimento da Cultura Sísmica Local em Portugal in Arquitectura de Tierra: Patrimonio y Sustentabilidade en regiones sísmicas, eds. M. Correia, C. Neves, R. Núñez, El Salvador, Imprimais.

Seismic Retrofitting: Learning from Vernacular Architecture – Correia, Lourenço & Varum (Eds)
© 2015 Taylor & Francis Group, London, ISBN 978-1-138-02892-0

Common damages and recommendations for the seismic retrofitting of vernacular dwellings

M.R. Correia
CI-ESG, Escola Superior Gallaecia, Vila Nova de Cerveira, Portugal

H. Varum
CONSTRUCT-LESE, Faculty of Engineering, University of Porto, Porto, Portugal

P.B. Lourenço
ISISE, Faculty of Engineering, University of Minho, Guimarães, Portugal

ABSTRACT: This closure chapter acknowledges the most common retrofitting elements and measures identified in vernacular architecture. It enhances common systematic structural damage and failure mechanisms during earthquakes, relating them with probable causes, and establishing general recommendations to consider in preventive and reactive actions. Also, recommendations for the design and detailing of new construction in prone earthquake areas are advanced.

1 INTRODUCTION

The earthquake impact in dwelings has been studied since long. The capacity of the structural elements and of the building are crucial for the survival of the dwellings under earthquake demands. The length of the earthquake, its epicentre location in relation to the dwelings, the local soil conditions, and the foundations system have certainly a relevant meaning in the survival of the building stock in earthquake prone regions.

This chapter presents systematic encompassed failures, identified in several of the vernacular dwellings located in seismic areas. It also presents preventive recommendations for new housing, as well as for the strengthening of existent building structures.

2 STRENGTHENING AND RETROFITTING ELEMENTS

Based on the research addressed during the Taversism project (Correia, 2002), and throughout the research project 'SEISMIC-V: Vernacular Seismic Culture in Portugal (Correia et al., 2014), the research project team identified several different elements and systems for seismic-retrofitting in vernacular architecture. Following are listed the most common elements and systems, observed during the entailed missions.

2.1 *Strengthening elements*

Retrofitting elements are commonly applied in vernacular construction, to improve the connection among units in masonry, and/or to improve the capacity of the walls: e.g. horizontal reinforcement in-between rammed earth layers/adobe masonry/stone masonry; wood connectors used on the interior of walls, *Pombalino* structural system.

2.2 *Perimeter seismic-resistant elements*

The perimeter elements aim at locking the built structure, defining a joint behaviour and response of the structural elements, such as walls, floors and roofs (i.e. to develop the so-called "box behaviour"). This ensures the connection and stability of the structural elements in the perimeter of the construction; e.g. buttresses, masonry quality improvement, and other strengthening solutions of the corners, reinforced plinth, and tie-rods.

2.3 *Arches reinforcing elements*

A combination of arches and vaults arises usually associated with traditional typologies. These are located at the ground floor, of two or more floor buildings, both in urban or peri-urban areas: e.g. arches and vaults on the ground floor; and flat arches (or jack arches) on the top of the openings.

2.4 *Combined reinforcing elements and solutions*

It comprises, for example, solutions, where structural elements are applied between contiguous buildings or row buildings: e.g. continuous cornice, and counter-arches.

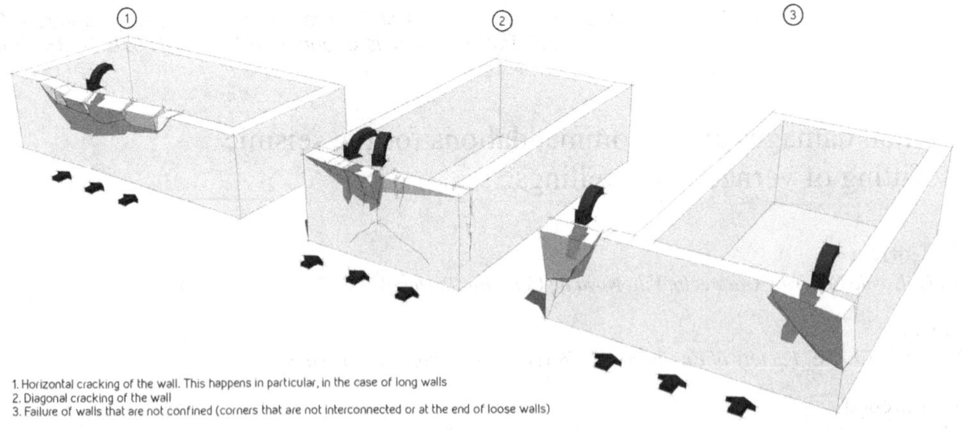

1
2
3
1. Horizontal cracking of the wall. This happens in particular, in the case of long walls
2. Diagonal cracking of the wall
3. Failure of walls that are not confined (corners that are not interconnected or at the end of loose walls)

Figure 1. Common out-of-plane failure mechanisms (credits: CI-ESG).

3 SYSTEMATIC FAILURE AND CAUSES

To favour the development of seismically safer constructions, recommendations should be addressed on improving and reinforcing protective measures. Attention should also be given to actions that may reduce the seismic impact. Several authors already address these relevant matters, such as Yamín Lacouture et al. (2007), López Trujillo (2014), Prieto Sánchez & Vargas Neumann (2014), among others.

Following are some of the recommendations that should be considered, when improving the seismic capacity of vernacular architecture and in new construction.

3.1 *Types of failure*

Following, are listed some common types of failure of structural components. The most probable causes, observed in vernacular constructions, when subjected to seismic actions, are also briefly defined:

a) *Out-of-plane collapse of exterior walls* due to poor connection to the perpendicular walls and to the horizontal structural systems (floors and roof); or due to the masonry poor quality (see Figure 1).

b) *Severe in-plane cracking of the walls* due to poor shear capacity to resist the horizontal in-plane demands, as a consequence of poor masonry quality, degradation, or large openings.

c) *Roof collapse* due to the collapse of it supports, due to poor connection with the exterior walls, or due to its heavy mass.

d) *Floors collapse*. Upper floors may collapse due to the collapse of their support system, or due to a weak connection with the exterior walls.

3.2 *Common errors and causes of failure*

Following are listed the common errors and causes of severe damage or failure, verified on one to two floors vernacular buildings, when seismically loaded (see also Figure 2):

a) *Heavy roofing systems*: Mass concentration on the top of the building may induce high concentrated heavy loads and damages.

b) *Poor connection between walls*: A weak interlocking in the connection of perpendicular walls may facilitate the out-of-plane of the walls, and decrease the global capacity of the building.

c) *Instable gables*: Absence or poor connection of the gables to the roof and floor structures.

d) *Insufficient strength capacity of the exterior walls*: This can be due to thin walls, or to walls with insufficient strength.

e) *Low quality of the construction materials or masonry system*: Weak units (adobe, stone) of the masonry system; poor mortar, lack of strength of the stone masonry system; masonry without mortar on the vertical joints; uneven compression when building the rammed earth; wicked rammed earth mixture, etc.

f) *Improper connection on the building corners*: When rammed earth, adobe masonry or stone masonry walls are not tied, or have a poor interlocking, at the building's corners.

g) *Irregularity on the openings distribution*: This may induce a concentration of stresses and demands on the walls, and on the existent openings.

h) *Poor or short support of the opening lintels*: It may collapse because of the stress concentration on the support, or due to damage of the earthquake demands.

i) *Excessive opening dimensions*: Too wide door or window openings may lead to stress concentrations and irregular behaviour of the building.

j) *Proximity of the openings to building corners*: Openings too close to the building corners may lead to a stress concentration in the wall.

k) *Windows or doors on the building corners*: The presence of openings in the building corner induces a discontinuity between perpendicular walls.

l) *Long walls*: Long walls, not properly connected to transversal walls and floor/roof structures tend to exhibit larger out-of-plane deformations.

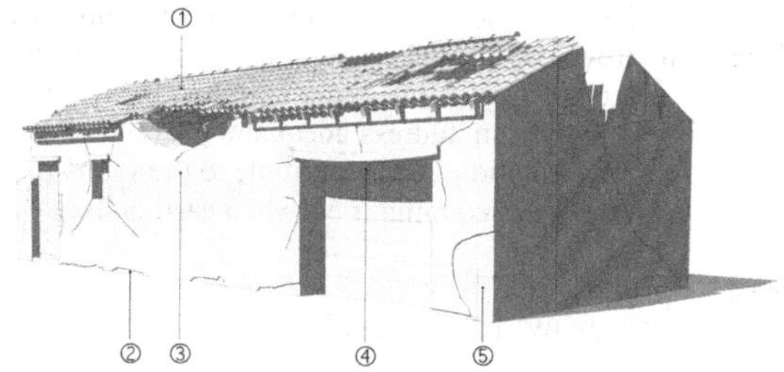

Figure 2. Common damages in masonry dwellings due to earthquake demands (credits: CI-ESG).

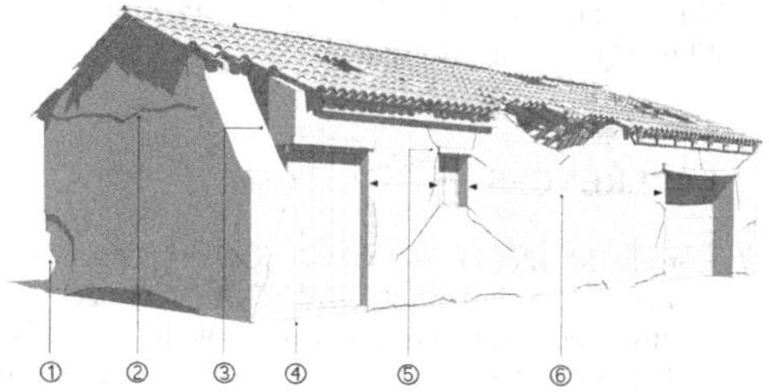

Figure 3. Common damages in masonry dwellings due to earthquake demands (credits: CI-ESG).

m) Absence or week foundation system: This may induce excessive deformations on the walls, and in the overall structure.

4 GENERAL RECOMMENDATIONS

A standard of general recommendations is addressed in terms of preventive actions and reactive actions, after the construction is erected (even before earthquakes), and/or criteria for design of new construction.

4.1 *Preventive actions*

First, it is fundamental to address a good construction in terms of materials and building systems. This relies on a suitable connection between elements, but also on materials with adequate properties: good compaction of the rammed earth, good mixture of the adobe earth, stone without salt problems, adequate mortars, etc.

a) Foundations: It is fundamental for dwellings to have strong and stable foundations, to ensure a good distribution of the vertical dead load, but also to be able to respond to earthquake lateral demands.

b) Wall plinth: It is relevant to assure a strong and adequate construction of the wall plinth, which guarantees a proper connection of the structure at the ground level.

c) Wall reinforcement: This should be considered between rammed earth horizontal layers or adobe/stone masonry horizontal joints. These retrofitting measures can be addressed, for example, through the use of internal horizontal wood elements, fired bricklayers, or stone layers, on the horizontal joints of the rammed earth.

d) Walls connection: During the construction, it is important to provide a good connection between interior walls and exterior walls, as well as between exterior walls at the corners. These connections can be improved through the integration of interconnected wood, stone slabs, or brick elements at the corners.

e) Ring beam: Usually applied on the top of the walls and/or at the floor levels. It promotes an adequate connection of the walls, and improves their out-of-plane capacity.

f) Lintels: Lintels with short support on the door and window openings should be avoided. Lintels should be property connected to the walls.

4.2 *Reactive actions*

a) Foundations: Consolidation or enlargement of foundations, but also careful integration of foundations (Correia & Merten, 2000), when they do not exist. All these measures will guarantee the adequate transmission of the vertical loads into the soil.

b) Reinforced wall plinth: To be applied externally to the walls, at their base.

c) Wall reinforcement: As it is difficult to insert horizontal connectors, in-between rammed earth layers or stone masonry joints, masonry reinforce ment can be considered, when external reinforcing solutions are introduced. For example, in Peru, Colombia and Portugal, several experiences have been made with geo-grids, raffia, etc. (Figueiredo et al., 2007).

d) Buttresses: To be applied in strategic weak points of the building or where out-of-plane may occurs.

e) *Tie-rods*: To connect opposite walls, at floor or roof levels, to assure their stability to out-of-plane demands.

4.3 *Design criteria*

Some recommendations regarding design criteria and detailing of buildings for prone-earthquake areas, follow herein:

a) *Plans should be regular*: Regular geometry in plan, as for example symmetrical plans, promotes a regular behaviour and distribution of demanding loads among the structural walls.
b) *Regularity in height*: Abrupt changes of strength, stiffness or mass in height should be avoided, in order to promote a regular response of the building to seismic demands.
c) *Roof structural system and roof coverings should be light*: Heavy roofing systems tend to develop high concentrated demands.
d) *Openings should not be too wide*: Dimensions of the openings should avoid high stress concentration, due to horizontal demands.
e) *Openings to close to corners should be avoided*: Openings should be located far enough from the corner, to avoid stress concentrations.
f) *Proper connections between the different elements should be promoted*: An adequate connection between walls, roof and walls, and floors and walls, tends to promote lower and better distribution of the demands of the earthquake among the structural elements.
g) *Reinforcement elements should be integrated in weak points of the walls*: Points, where high demands may be concentrated, should be reinforced with wooden, stone, or brick elements.
h) *Accurate constructive details should be provided*: Rigorous plans and drawings, including details of the structural elements, connections, etc. should be prepared, to avoid errors in the building construction that may affect its safety.
i) *Selection of adequate materials, rigorous construction and monitoring*: They all should be considered, so as to guarantee a safe dwelling.

5 FINAL REMARKS

From the analysis of vernacular architecture construction details in prone-earthquake areas derived recommendations for the behaviour and performance improvement of vernacular buildings. Simple, though effective measures can be taken at the design phase, such as structural regularity, continuity, etc. Additionally, addressing a consistent reinforcement of the housing can also promote an improved behaviour of the buildings under an earthquake impact.

The seismic safety of existing structures can be improved with these recommendations. Therefore, understanding why systematic failure occurs and how population can address adequate, preventive, or even reactive solutions, can contribute to the safety and survival of local communities, when earthquakes occur.

ACKNOWLEDGMENTS

The authors gratefully acknowledge the support by the Portuguese Science and Technology Foundation (FCT) to the research project 'SEISMIC-V – Vernacular Seismic Culture in Portugal' (PTDC/ATP-AQI/ 3934/2012).

REFERENCES

Correia, M. (2002). Preliminary Report of the Local Seismic Culture in Portugal. TAVERSISM project: La tutela attiva dell'edificato vernacolare nelle zone sismiche: Un'azione polivalente di valorizzazione del patrimonio, riduzione della vulnerabilità, di sviluppo locale. Ravello: UNIVEUR centre.

Correia, M., Carlos, G., Rocha, S., Lourenço, P.B., Vasconçelos, G., & Varum, H. (2014). Vernacular Seismic Culture in Portugal. In M. Correia, G. Carlos, & S. Sousa, (Eds). Vernacular Heritage and Earthen Architecture: Contribution to Sustainable Development. Proceedings of CIAV 2013 | 7°ATP | VerSus. London (UK): CRC Press/Balkema/Taylor & Francis Group.

Figueiredo, A., Varum, H., Costa, A., Silveira, D., Oliveira, C. (2013). Seismic retrofitting solution of an adobe masonry wall. In Materials and Structures, RILEM, doi 10.1617/s11527-012-9895-1, Vol. 46, N. 1–2, pp. 203–219.

Giuliani, H., Citrinovitz, A., Aladro, S. & Benavidez, H. (1987). Arquitectura sismo-resistente. Un nuevo enfoque para solución integral del problema sísmico. Informes de la Construcción, 38 (387), 64–69.

López Trujillo, O. A. (2014). La importancia de la investigación y reglamentación técnica para la construcción de vivienda con tierra. Projecto de Investigação Seismic-V: Reconhecimento da Cultura Sísmica Local em Portugal in Arquitectura de Tierra: Patrimonio y Sustentabilidade en regiones sísmicas, eds. M. Correia, C. Neves, R. Núñez, El Salvador: Imprimais.

Prieto Sánchez, R. & Vargas Neumann, J. (eds.) (2014). Fichas para la Reparación de Viviendas de Adobe. Perú: Ministerio de Vivienda, Construcción y Saneamiento.

Yamín Lacouture, L.E., Philips Bernal, C., Reyes Ortiz, J.C., Ruiz Valencia, D. (2007). Estudios de vulnerabilidad sísmica, rehabilitación y refuerzo de casa en adobe y tapia pisada [Seismic vulnerability studies, renovation and reinforcement of houses built with adobe brick and rammed earth]. In APUNTES Vol. 20, N°2, July–December 2007 (ISBN 1657-9763). Bogotá (Colombia): Pontificia Universidad Javeriana, pp. 286–303.

Seismic Retrofitting: Learning from Vernacular Architecture – Correia, Lourenço & Varum (Eds)
© 2015 Taylor & Francis Group, London, ISBN 978-1-138-02892-0

Author index

9781138028920